Western Technoscience

Western Technoscience

Not the Usual Story

Nathan Kowalsky

broadview press

BROADVIEW PRESS
Peterborough, Ontario, Canada

Founded in 1985, Broadview Press is a fully independent academic publishing house owned by approximately twenty-five shareholders—almost all of whom are either Broadview employees or Broadview authors. Broadview is supported by a collaboration with Trent University, a liberal arts university located in Peterborough, Ontario—the city where Broadview was founded and continues to operate. Broadview is committed to environmentally responsible publishing and fair business practices.

Library and Archives Canada Cataloguing in Publication

Title: Western technoscience : not the usual story / Nathan Kowalsky.
Names: Kowalsky, Nathan, author
Description: Includes bibliographical references and index.
Identifiers: Canadiana (print) 20250277794 | Canadiana (ebook) 20250277824 | ISBN 9781554816712 (softcover) |
ISBN 9781460408735 (EPUB) | ISBN 9781770489493 (PDF)
Subjects: LCSH: Science. | LCSH: Technology. | LCSH: Objectivity.

Classification: LCC Q158.5 .K69 2025 | DDC 500—dc23
Broadview Press handles its own distribution in Canada and the United States:
PO Box 1243, Peterborough, Ontario K9J 7H5, Canada
555 Riverwalk Parkway, Tonawanda, NY 14150, USA
Tel: (705) 482-5915
customerservice@broadviewpress.com

Broadview Press books are imported and distributed in the United Kingdom and European Union by:
Gazelle Book Services Ltd.
White Cross Mills, Hightown, Lancaster, Lancashire, LA1 4XS
sales@gazellebookservices.co.uk

European Union – Responsible Person (for official use only):
eucomply OÜ
Pärnu mnt 139b14
11317 Tallinn, Estonia
hello@eucompliancepartner.com
+33757690241

Canada

Broadview Press acknowledges the financial support of the Government of Canada for our publishing activities.

Edited by Michel Pharand
Book design by Michel Vrana

PRINTED IN CANADA

1 2 3 4 5 6 7 8 9 10 25 26 27 28 29 30

In Memory of Isaak Kornelsen
1991–2012

Can anything more than profound confusion be indicated by this admixture of diverse fields and concerns?
—Thomas Kuhn

Contents

Introduction

The Usual Story

THIS IS A BOOK ABOUT SCIENCE AND TECHNOLOGY, NOT A BOOK OF science and technology, although those exist. Reading one of those books might teach you knowledge or methods that you could use in becoming a scientist or engineer, but this book won't. This book will encourage you to *question* science and technology, rather than have you *do* science and technology. For example, what is science and technology? Or what are they good for? Questions like that are weird, which is probably why a lot of people don't read books like this. Everybody is supposed to 'already know' what science and technology are and what they're good for, because everybody has learned *the usual story* about science and technology pretty much since they were born.

But who is this 'everybody'? It's you and I, not least of all because you're reading this in English. This book is in English because of the imperial power of nineteenth-century Britain and twentieth-century America. The very fact that you and I share this language means that we're both part of a globalized Western modernity, even if we've got issues with that. Western modernity doesn't want us to have issues with it, though, which is why it's been telling people like us the usual story about science and technology ever since the Enlightenment. 'The Enlightenment' is what the usual story calls that period of time in Western Europe when thinkers championed political freedom and the power of reason to overcome what they saw as prior centuries of superstition and ignorance. Moreover, this emphasis on liberty and reason arose on the heels of the Scientific Revolution, so that scientific and technological progress became associated with this aforementioned freedom and reason. One scholar calls this "the dominant ideological perspective of modernism."[1] This is why we might already think we know everything we need to know about science and technology: we've been told for several hundred years that they provide us with reason, truth, freedom, and prosperity, and so there's no need to think about them anymore. All we need to do is get even better at *doing* science and technology.

Here are some examples of the usual story about science and technology:

In the city where I live, there was an event called "Logicon" which was supposed to bring "members of the scientific community together with the general public to celebrate logic, critical thinking, and the scientific method."[2] Although this event's self-description associates science with logic and critical thinking, it turns out that you usually don't actually study logic or critical thinking in the science faculty of a university (although you can use the scientific method there, of course). The odd thing about logic and critical thinking is that they are at home in the arts faculties of universities, specifically philosophy departments, where logic (sentential, predicate, syllogistic, etc.) and critical thinking (especially fallacy theory) are taught and researched by professional logicians—but not by scientists. So why would the Logicon event want to imply that it's the scientific community that primarily celebrates logic and critical thinking, and not philosophers?

Or, moreover, why would the so-called new atheists (e.g., Richard Dawkins, Christopher Hitchens, Sam Harris, Lawrence Krauss, or Neil deGrasse Tyson, most of whom are scientists) also call themselves the "rationalist" movement, when any historian of ideas would know that rationalism was a hugely influential *philosophical* school of thought dating back to the seventeenth century and having virtually nothing to do with atheism—and having a very complicated relationship with what we would today consider modern, empirical science? Only one thing is clear: we're not supposed to care about historical and philosophical questions like these because we 'already know' what the usual story has been telling us. Science is (supposedly) eminently *rational*, while things that are unscientific (like philosophy) are not only nonrational, but often completely irrational. At least, that's what we're usually told.

In the country where I live, our national leader was very interested in getting our landlocked oil deposits to overseas markets, and as such was enthusiastic about transcontinental pipelines. Many people protested these pipelines, and in response to the ensuing controversy, the country was told that "science, not politics, will ultimately determine whether the ... pipeline proceeds."[3] Presumably, this is because everybody knows that there are no right answers in politics, whereas in science, the answers are not only rational (as above), but objectively right. Politics (supposedly) gives us endless opinionated debate, but science (supposedly) gives us 'facts,' and they are (supposedly) true regardless of debate.

This is why the neurologist and podcaster Steven Novella could subtitle his book on scientific thinking as follows: "how to know what's really real in a world increasingly full of fake."[4] Science (supposedly) tells us the way things *actually* are, as they are in themselves, and with emphasis. Science is not supposed to be subjective; it's not fuzzy around the edges or open to dispute, like politics or religion or philosophy. No, science is *objective* knowledge of the truth and *neutral* with respect to opinions or feelings, and that's why we should want to do it instead of write poetry, read fiction, or worship fake gods. At least, that's what we're usually told.

But even if science is objective and neutral, that doesn't mean we are supposed to think that it lacks value or importance (even though value or importance are usually assumed to be subjective and biased things). Science is (supposedly) both objective and unbiased *and* valuable and important. That's why an astrophysicist could say that it's *better* to accept science's version of the Universe (with a capital U) than some unscientific version.[5] However, there's nothing neutral about science being *better*, or the way to know what's *really* real, because both imply that science is (objectively, truly) *awesome* (even though saying so sure sounds like a subjective opinion).

Nevertheless, the usual story sells science as *so* important that it not only (supposedly) provides humans with valuable objective knowledge about what ultimately matters, but also answers their deepest questions about the meaning of life. Thus *TIME* magazine has no trouble saying that "the discovery of the Higgs boson helps explain nothing less than why our existence is possible."[6] Technically, they are only talking about how the Higgs boson satisfies the last requirement of the standard model of particle physics, but the excitement and awe in the magazine's exclamation suggests implications of profound existential importance—as if the particle's discovery answered the question "why am I here?" rather than "by what process have I arrived here?"

Beyond satisfying existential questions of human significance, science is also supposed to both spring from and be the fulfilment of the deepest yearnings of the human spirit. In a somewhat nationalistic tone, Barack Obama announced that "[t]onight, on the planet Mars, the United States of America made history. It proves that even the longest odds are no match for our unique blend of ingenuity and determination."[7] America's national spirit, anyway, is (supposedly) fulfilled in a meaningful way by the Curiosity Lander's arrival on Martian soil. Science—and in this case, technology as well—provide (ironically neutral) *value*, which satisfies both the deepest desires of individual human beings and the aspirations of nations. At least, that's what we're usually told.

Speaking of technology, we already know (supposedly) what it is too: technology is applied science. That doesn't make science and technology the same thing, necessarily, but they certainly work together as a team. While science gives us valuable and neutral truths, technology takes that knowledge and puts it to use. That is, technology produces tools that we wouldn't have if we didn't have science to show us how to make them. Indeed, science that doesn't have a practical application is technically use-less, and as such not really valuable. The English word 'applied' shares the same etymological root as the French word 'plier,' which means *to fold*. So we could say that science should be folded or 'bent' towards technological use, and if it isn't applied, then it has no point.

Yet this is ironic, for the technologies we have as a result of this scientific bending are themselves supposed to be neutral: neither good nor bad. That is why we say they're "just a tool," especially if someone is concerned about them. The goodness or badness of technologies is simply a function of how they are used by someone, nothing more. They aren't objectively valuable one way or another, right? Except that they *are* supposed to be objectively useful, otherwise they wouldn't be made. We are supposed to make objectively useful technologies, and *better* ones, because otherwise technology (and science) would be useless and not valuable! So the usual story says that technological advancement or development is *intrinsically good*—'progress' is always better—even though technologies themselves are *value-neutral*. In either case, there's (supposedly) nothing to think about. At least, that's what we're usually told.

It's no surprise, then, that every time a new smart phone comes out, it's supposed to change your world.[8] But what is your world, or your life, that is supposed to get changed? Everybody knows (supposedly) that the world is physical or material, and our tools and technologies unsurprisingly move matter around. From bulldozers to artificial intelligence, technological progress boils down to material particles being placed in particular configurations—*better* configurations, that is. Technologies don't move *spirit* around, even though (ironically) technological progress (like science itself) is supposed to be the result of the human spirit! Even if we don't take 'spirit' literally (nobody should, anyway, in a technological culture where everything's particles), everybody knows (supposedly) that all humans everywhere have always wanted better technologies—that is, tools that move matter around more efficiently. Progress means more tool power to alter the material world, and surely all human societies since the dawn of time have wanted this power that we now have, because having it is good.

Of course not all of our technologies turn out to be as good as we want them to be at moving matter around. So how will we fix this problem? The solution-based approach is clear: we will fix technologies technologically, which is by moving matter around some more. Technologies are made of matter, after all, and so the only way to improve them is to apply material power to them. In this way, technology fixes itself. There isn't (supposedly) any other way to fix them. Much like how science was (supposedly) the standard of rationality, technology is the standard of rational problem-solving. In this way, technology tells us how to think: there is only one kind of reality (the physical) and there is only one way to encounter reality—by moving it around *more* (and more is better). So technology—or techno-logic—is the only way to think rationally about problems. At least, that's what we're usually told.

The Usual Story: Basic Points

In sum, what are the basic points of the usual story? First of all, the usual story is an intellectual framework. It is a kind of theory about how science and technology work and what they're good for, although it is not usually recognized as a theory: it's supposed to be taken for granted as fact, so that nobody notices that it's a theory. Frameworks or theories that are not recognized as such are called *ideologies*. The usual story is, therefore, an ideology about science and technology.

Second, the usual story is very old. It's older than modern science and technology, in fact. The dominant narrative of civilization has been claiming to naturally advance human progress in opposition to subhuman barbarians and savages since agriculture was invented about 10,000 years ago.[9] If you live in a so-called civilization, the antiquity of the usual story is probably why it sounds like second nature to you. It's been told to people like me and you for a really, really long time. The usual story about science and technology, meanwhile, just inserts the modern versions of science and technology into this larger progressive narrative.

Third, this means that the usual story is not static, but develops over time. After various twists and turns, the usual story about science and technology becomes pre-eminent in Europe about 400 years ago, right around when Europeans were colonizing the Americas, Africa, and Asia. Moreover, the usual story isn't just dynamic, but also filled with internal tensions and even contradictions. You might even say that it is a set of inconsistent stories, but that doesn't matter as long as nobody thinks *about* it. As long as you think *with* it, the usual story will explain science and technology to you so you don't have to pay attention to them at all, let alone read a book like this.

Fourth, the usual story positions itself against what it sees as savage, inhuman, wild, or untamed: namely *nature*, especially in its 'raw' state. Science and technology are supposed to overcome nature by knowing how it works, so that we can learn how to control and manipulate it. The people who might live in wild nature (if there's any left) are likewise considered to be inhuman or savage, at least in proportion to how far they have 'advanced' out of and away from nature towards 'civilization.' The beliefs of such people, moreover, are assumed to be mistaken. The usual story sees 'uncivilized' belief systems as forms of superstition and ignorance, usually religious or magical. Science and religion are supposed to replace these false or foolish beliefs with cold, hard truths that actually work.

Fifth, then, the usual story sees the upward trajectory of 'human culture' as being both continuous and discontinuous with what it replaces over time. Just as science and technology increasingly replace wild nature with civilization, so does scientific truth and technological power (supposedly) replace religious superstition and ineffective magic over time. The usual story sees science and technology as indicative of superior intelligence, enlightenment, and advancement. Science tells us what things are really like, and as such is 'objectively' true, as opposed to 'subjective' opinions or feelings about things. This understanding of objective truth can be so absolute that it is comparable to a 'God's eye perspective,' which is one reason why the usual story thinks science can replace religion.

Sixth, the objectivity of scientific truth is supposed to be neutral with respect to feelings or biased opinions, but it's not neutral with respect to value. It's *good* to know what the truth is, and as science progresses over time, science accumulates more and more of that goodness. This is 'progress.' The usual story sees science and technology as developing or advancing according to an inevitable, natural, instinctual, unstoppable, and ultimately infinite process of improvement. Progress is good, and what it leaves behind is bad. That's why the usual story is uninterested in the history of science and technology; the past is irrelevant compared to the glories of the present and the future. As we'll see in Chapter Fourteen, this is the 'textbook tradition' of science: we're taught in school that techno-scientific progress always moves forward to grasp the positive value of truth, while how it got there (i.e., its history) isn't worth mentioning at all.

Seventh, the usual story thinks technology progresses alongside science, because scientific truth is supposedly why technologies actually work. Technological power is assumed to be proof of scientific objectivity. Moreover, that objectivity means there's only one way for technology to develop: the way it has in fact developed in Western civilization. Western technology is how 'humans' conquer nature, how they conquer less advanced humans, and will eventually conquer other planets and stars. The technological progress of Western civilization is supposedly 'human nature,' just as tech bros and billionaires with spaceships assume it is. The usual story believes that everybody has wanted this technological power over everything in the universe, ever since there were people at all. Just as Western science is universally true, so is Western technology universally human—at least according to the usual story.

Finally, like science, technology in the usual story is both naturally objective and objectively good. On the one hand, everybody knows that a technology is 'just a tool,' not something with a mind of its own that we might have to think critically about. By contrast, anybody who worries about the social impact of a technology is just afraid of change or new things, and is probably mysticizing a neutral, dead object. Tools aren't good or bad; only people are. On the other hand, new technologies are supposed to be awesome! When we get more powerful tools, we can use them to do all manner of amazing things: solve our problems, create a better world, write essays for us in college so that we don't have to, etc. Technological power is *value*, and the more of it we get, the more advanced, developed, and *better* we are. We get more civilized with better technology, more decent than our primitive ancestors, and eventually technology will create for us a utopia! Of course there will be hiccups along the way, but these bad outcomes are just incidental side effects and nothing to worry about. We should not lose our faith in technological progress, because technological progress is good when almost nothing else is.

Or at least so goes the usual story.

Trouble in Paradise

The usual story is under a bit of stress, however. Its grand claims about scientific truth especially are viewed with increasing skepticism. In North America, for example, there is significant doubt about scientific truth claims, particularly about climate change and vaccinations, among people who identify with the right wing of the political spectrum. At the other end of the political spectrum, this skepticism is met with a further entrenching of the usual story, accusing right-wing skeptics of being 'anti-science' or uninformed about objective facts.

On the flip side, however, the grand claims about Western technology's universality are doubted by many on the left-wing of the political spectrum, seeing technology as a non-neutral tool for Western racism or colonialism. At the other end of the political spectrum (again), this skepticism is met with a further entrenching of the usual story, accusing left-wing skeptics of being relativists about truth and contributing to the downfall of civilization.

And yet both sides seem to accept the grand narrative of techno-scientific progress: advanced technology will be our salvation, be it carbon capture and storage, nuclear power, electric vehicles, renewable energy, or whatever. In spite of all this distrust and debate, then, the usual story is still the dominant framework that most of us work with when we think about science and technology in our cultures. Moreover, this distrust gets *deeper* because the usual story accuses anyone who disagrees with it of being irrational, ignorant, or even inhuman, rather than being able to see how the usual story can be subject to reasonable criticism without losing all that is good or promising about science and technology.

The usual story about science and technology requires a second look, therefore. You may have already suspected this before you picked up this book. Maybe science and technology aren't the only way to think rationally. Maybe science isn't necessarily objective. Maybe science is embedded in social projects that unavoidably include subjective dimensions like non-neutral value judgments. Maybe this isn't necessarily a problem. Maybe some of those social projects are nevertheless unjust: authoritarian, racist, sexist, colonialist, transphobic, or unecological. Maybe some of our deepest questions about the meaning of life are not answerable with explanations about particles. Maybe human beings haven't always thought that particle-based explanations were the best way to engage with the real world. Maybe science can be pursued for its own sake. Maybe human tools weren't always fabricated on the basis of scientific explanations of things. Maybe tools aren't value-neutral. Maybe technologies have value-implications beyond how people decide to use them. Maybe there's more to reality than material particles, and maybe there's more to goodness than the power to move material particles around. Maybe you don't only have to think like a tool.

Maybe this book isn't so weird after all. Maybe there is more to science and technology than what we're usually told. This book will argue that we do need to *think more about* science and technology instead of simply doing more science and technology, even though the usual story tells us that we don't need to bother. Therefore, this book will introduce you to other stories about science and technology, unusual or unfamiliar ones, ones that are alternatives to the usual story. With these unusual stories in place, you will be in a better position to come to your own informed and reasoned position on the value and meaning of science and technology in the society you happen to live in, without simply taking the usual story for granted. This will position you for engaging in something the usual story has never told you about: the intellectual study known as science, technology, and society.

Notes

1 Mick Smith, "The State of Nature: The Political Philosophy of Primitivism and the Culture of Contamination," *Environmental Values* 11 (2002): 409.

2 http://logicon.ca/about, accessed 7 September 2016.

3 Jason Fekete, "Federal Government Says Science Will Decide Fate of Northern Gateway Pipeline," *Ottawa Citizen*, 31 May 2013, http://o.canada.com/news/national/federal-government-says-science-not-provincial-wishes-will-decide-fate-of-gateway-pipeline.

4 Steven Novella, *The Skeptics' Guide to the Universe: How to Know What's Really Real in a World Increasingly Full of Fake* (Grand Central, 2018).

5 Carl Sagan, *Demon-Haunted World: Science as a Candle in the Dark* (Random House, 1995), 12: "It is far better to grasp the Universe as it really is than to persist in delusion, however satisfying and reassuring."

6 Jeffrey Kluger, "The Cathedral of Science: The Elusive Higgs Boson Is at Last Found and the Universe Gets a Little Less Mysterious," *TIME*, 23 July 2012, 28–31.

7 http://www.nasa.gov/mission_pages/msl/news/obama_statement_curiosity.html, accessed 7 September 2016.

8 "The Wireless Issue: 10 Ways Your Phone Is Changing the World," *TIME*, 27 August 2012, cover.

9 James C. Scott, *Against the Grain: A Deep History of the Earliest States* (Yale University Press, 2017), 1.

I

Definitions

Chapter One

What Is STS?

THIS BOOK IS AN INTRODUCTION TO THE SCHOLARLY STUDY OF SCIENCE and technology in society, otherwise known as "STS." Assuming that you've never heard of STS before, let me take this opportunity to explain what the discipline is. One reason you may have not heard of STS is that it is a relative newcomer to the academy: the first programs in STS date only as far back as 1969, which is not a very long time for an academic discipline to have been around. (History and philosophy, by comparison, have been around for roughly two-and-a-half *thousand* years.)

Although STS hasn't been around for a long time, the disciplinary specializations which comprise it have been. STS is an *interdisciplinary* field, which means that it brings together a number of other disciplines under the same topical umbrella. It is dominated by the *history*, the *philosophy*, and the *sociology* of science and technology, although not limited to these three. STS is also informed by perspectives from literary criticism, anthropology, fine art and design, etc. Interdisciplinarity is when your scholarship is not limited to a single discipline, but involves many disciplines interested in a single project. It is an attempt to overcome the fragmentation or isolation of human knowledge into discrete and sealed compartments, and in so doing regain the promise of the university as a "uni," a *unified* institution of understanding. Thus, in STS, philosophers collaborate with historians who collaborate with sociologists (etc.), and they collaborate around a central, unifying theme: the interrelation between science, technology, human societies, and the natural world they all inhabit.

As mentioned in the introduction, this central, unifying theme means STS is not in the business of *doing* science or technology. STS students and scholars are not necessarily scientists or engineers or physicians themselves. Rather, STS involves *thinking about* science and technology. Science and technology are *themselves the objects of study* in STS and, somewhat ironically, this allows for something of a 'scientific' attitude towards those activities. Classically understood, scientific study meant to stand at a distance from one's object of interest to allow for a degree of independence from that object. This, in turn, was supposed to help see things more 'objectively' and hopefully produce unique and impartial insights. For example, when you're riding a bike you 'get a feel for it' long before you know anything about the physics of gravity or balance. It would be distracting and unhelpful, in fact, to have to calculate your center of gravity (etc.) before or while riding a bike. So if you are going to do a physics calculation, it is better to *not* ride a bike when doing so, but rather to take a step back and watch a bike being ridden or look at an abstract diagram of bike riding. Another example of this sort of distance is when a police officer is being investigated for, say, the violation of civil rights; it seems more fair for the investigators to not be police officers themselves, or at least not from the same police force. Otherwise, the investigation would be an 'inside job' and its findings would be less credible. For similar reasons, then, STS stands in a *transcendent* relation towards science and technology—'over and above' them, as it were—so that its inquiries about science and technology may be more likely to ascertain facts objectively, and so that its evaluative judgments of science and technology may be arrived at more impartially. Of course, objectivity and impartiality are ideals not fully attained by any discipline; but like all other disciplines, STS strives towards these ideals to the best of its practitioners' abilities.

The Impetus behind STS

But *why bother* creating a new interdisciplinary academic field about science and technology? The mere fragmentation of university research into departmental cocoons isn't a sufficient reason for creating a field centering on science and technology in particular, when any other topic (such as religion or the environment) could serve a unifying purpose just as well. What was it about science and technology especially that gave scholars the impetus to unite their disparate studies around that topic in the late 1960s and 1970s? To answer this question, I need to tell a story. The story is—briefly—why not everybody thought everything was fine and dandy with science and technology, and especially not in the first half of the twentieth century—even though everyone was *supposed* to think that everything was fine and dandy with science and technology. After all, the usual story (as I called it in the introduction) suggested that the human race had recently entered into a new golden age of progress, prosperity, and enlightenment. The overcoming of superstition by scientific logic—and the increase of technological power through the application of scientific reason—was supposed to have led to several centuries of unstoppable and inevitable improvement in human society. Our manners and happiness could only get better as our tools got more powerful and our knowledge of the universe got more correct.

But to the shock and horror of many around the world, in 1914 Europe dragged the world (thanks to its colonization of almost the whole planet) into a violent and bloody war with casualties on an unprecedented scale. After roughly 200 years of supposedly inevitable improvement, it appeared that barbarism and savagery had returned in the face of *millions* of people dead in the space of four short years, entire economies obliterated, and monarchies tumbling like dominoes. But how was such wholesale slaughter even possible? Well, progress was certainly involved: trinitrotoluene (i.e., TNT, invented in 1863) and dynamite (invented by Alfred Nobel in 1867, who also owned the Swedish armament manufacturer Bofors), poison gas (chlorine gas, mustard gas, hydrogen sulfide, etc.), flying machines (zeppelins and airplanes), machine guns, tanks—none of these things could have been made without great scientific and technological advancement, and those millions of people could not have been killed or maimed without such advancements. People understandably wondered if such items and such slaughter were what we were supposed to get from centuries of scientific and technological progress, and of course the answer was 'no.' Thus they named the 1914–18 war "the Great War" because it was the worst war that had ever been conceived, and surely it was so bad that no war so 'great' would ever happen again.

Except, of course, that it did. This is why we now call it "World War I," because in 1939 the world got another 'great' war, during which *even more* people died or were maimed, and in addition to using and improving all the new weapons that were introduced in World War I (except for poison gas, which was improved as 'nerve gas' but not used), World War II introduced innovations such as massed civilian firebombing (Dresden in 1945), ballistic missiles (German V-1 and V-2 rockets), jet planes (the Messerschmitt Me-262 and Gloster Meteor), industrialized death camps (Auschwitz and many others), and atomic weapons (dropped on Hiroshima and Nagasaki). In the face of just that last item on the list, Robert Oppenheimer, director of the Manhattan Project, which created the first atomic weapons, borrowed a term from Christian theology to describe what he and his colleagues had done: "the physicists have known sin."[1] Rather than being objective and value-neutral, he saw physics as having done something analogous to committing an evil against the (supposed) Lord and Creator of the universe. He also borrowed a phrase from the Hindu scriptures, comparing the power his group had unleashed to that of Krishna, an avatar of Vishnu: "Now, I am become Death, the destroyer of worlds."[2]

Both of Oppenheimer's religious references reflected a more general impression that something wasn't quite right with how science and technology had turned out, yet North American culture got over that anxiety rather quickly. The United States emerged from World War II as an undeniable superpower, no small thanks to its industrial might. America prospered in the 1950s, and this prosperity was seen to derive, at least in part, from science and technology. From Betty Crocker cake mix (cake in a box!) to "better living through chemistry,"[3] science and technology were seen as the source of success, especially against the other superpower, the Soviet Union. But by the 1960s, cracks in the façade started to appear. Both the debacle of the Vietnam War—including rampant heroin use by GIs and their exposure to the cancer-causing defoliant Agent Orange—and the birth of the environmental movement—with Rachel Carson's attack on the pesticide DDT in her book *Silent Spring*[4]—contributed to a period of social upheaval where science and technology were no longer assumed to be objectively neutral or, for that matter, objectively good (i.e., the font of American greatness and prosperity). It is no surprise that the 1960s saw the birth of 'postmodernism' and its loss of faith in the idea of progress.

Amid this turmoil of the 1960s, then, the interdiscipline of STS arose as a way to adapt our educational institutions to "an aspect of social reality that had become too obvious and important to continue to ignore."[5] And the same might be said about today. In our time we also face issues of critical importance within which science and technology are implicated: e.g., the Deepwater Horizon oil spill in the Gulf of Mexico (2011); the Fukushima nuclear disaster (2011); ongoing protests over oil pipeline projects (Keystone XL, Northern Gateway, Kinder Morgan, Energy East, Dakota Access); the antivaccination movement; anthropogenic climate change; racist algorithms in predictive policing; the pathologization of gender and sexual identity by professional psychiatry and psychology; and large language models for generative artificial intelligence (none of which was used in the creation of this book). In each of these examples, there is real concern about how science or technology interrelate or interact with a variety of social factors. Why is it, for example, that antivaxxers distrust medical science so much? Are they simply irrational, or are there sociological explanations that include the way science is communicated to the public or integrated with the pharmaceutical industry? Or why is it that the scientific consensus on climate change is so politically volatile and disputed by roughly half of the American public?[6] It is not clear that the solution to these problems is simply more science or, for that matter, more technology. We need to think more critically about how to respond to such issues.

Crisis and Critical Thought

STS is not simply, therefore, an "interesting and complex" field of study, but is also a response to a perception that science and technology are "pervasive, potent, and problematic."[7] They are, after all, the dominant powers that shape the lives of most humans living on our planet today, and despite the assurances of the usual story, this influence is not always experienced as an unambiguous good. Some people might even say that the twentieth and twenty-first centuries (so far) were and are in a state of *crisis* (real or merely perceived), and that science and technology have contributed significantly to this state of crisis. Just as you're more likely to pay attention to a device when it is malfunctioning, "it is only when *things go wrong*, fail to meet expectations, or run into unexpected opposition that a society's ideological presuppositions are brought into full consciousness."[8] Crisis, therefore, stimulates *critical thinking*. Critical thinking about social issues means not simply accepting the way things are at present, but using human intelligence to see if there might be better ways for our societies to proceed. STS, then, is an attempt at rigorous critical thinking about science and technology as social issues, including the question of how science and technology *ought* to be pursued.

The question of 'the ought' is very important, as it is the primary modality of human language that can offer an alternative to 'the is.' Questions of the ought, however, are also very difficult to resolve, because the ought—or 'the should' or 'the must'—are *value-based claims* and not simple statements of fact. Unfortunately, science and technology do not address the questions of values (or morals or ethics) and indeed cannot. These disciplines' expertise lies in descriptions of the way things are, or instructions for moving things around. Therefore, values are not themselves within the purview of scientific or technological training. Values are, however, within the purview of the humanities and social sciences. This is another reason why STS is itself a humanistic and socially scientific discipline, and why it cannot be the same as simply performing technoscience. Furthermore, STS's engagement with values and the ought mean that it can never hope to pretend to be 'neutral' about science and technology; it will always have to have a point, which is to take a *position* on things and defend it. STS will always be saying something about science and technology, good or bad, but it really cannot be neutral.

Because a sense of crisis so readily awakens critical thinking, such thinking has a natural tendency towards negativity. A crisis, after all, is not a sense that everything is okay. But STS is not necessarily committed to a negative assessment of science and technology either. At one end of the spectrum, STS's critical thinking can *defend* the usual story wholesale. This stance is often called 'boosterism,' where science and technology are viewed as redemptive, our best (if not only) solution to the crises that may beset us. Robert McGinn characterizes boosterist STS scholars as defenders of

> the existing practice of science and technology in the universities—including, at times, in their own besieged professions, departments, and research laboratories. Such scholars often spoke of the need for "balanced" consideration of the impact of science and technology on society. However, to them this influence was almost wholly benign, and they sought to impart to their students a keen appreciation of this fact.[9]

At the other end of the spectrum, there are what Sergio Sismondo calls the "activist" roots of STS.[10] At this extreme, one would expect to see science and technology *attacked* rather than defended. McGinn refers to an "antitechnology mood" which subjects "recent scientific and technological practice to intense, sometimes tendentious critical analysis."[11] Broadly, the activist side of the spectrum of STS critique holds that science and technology ought *not* to continue in the way that they have gone, which is quite the opposite of boosterism.

Because these two options are extreme ends of a spectrum, most STS scholars fall somewhere in the middle, some leaning more towards boosterism, others towards activism. In either case, the point of STS is to come to a reasonable and defensible position on the value and trajectory of contemporary scientific and technological practice. The first step towards this goal is for STS researchers to "make their commitments explicit."[12] Otherwise, it won't be clear what perspective on science and technology is being offered up for discussion, making our own critical thinking more difficult. So what will be the perspective of this book? If you, as the reader, are to form your own reasonable and defensible position on the value and trajectory of contemporary technoscience, you will need to at least weigh your options. On the one hand, you have the usual story which says that science and technology are wonderful and there's nothing for you or anybody else to worry about. We have already received this message from powerful and pervasive voices in our cultures. But we don't often hear the other side of the story, which are those unusual stories that call the usual story into question.

And so this book will focus on those unusual stories, simply so that by the end of it, you will have at least two (if not more) options to choose from when making your own reasoned determinations about science and technology. This book will thus lean towards the activist side of things, an often negative appraisal of the status quo, so that you get to hear both sides. I do, in fact, think that this perspective is closer to the truth, but it's not your job to agree with me on that point. Rather than simply believing everything you read, your job is to think critically about what you read—and that includes this book. But if this book didn't have a point, a perspective to defend, it wouldn't really give you anything to think critically about. With that in mind, let us look more closely at how STS examines science and technology and what particular questions it tends to ask about them.

The Subject Matter of STS

When we think critically about science and technology as social realities, there are at least three different ways in which that can be done. First, we can study the ways in which science or technology have impacts on other social realities. For example, we could examine how (if at all) the widespread use of digital technologies influences or changes how people relate to each other, or even if it produces psychological effects. Or we could investigate whether science is a necessary or sufficient condition for a rational society free of superstition and ignorance. We could see if modern technology has a tendency to dehumanize people and degrade ecosystems. We could even inquire into science's authorizing function in public discourse, and see if reasonable comparisons could be drawn between scientific authority and the function of historical religions in maintaining certain forms of political order.

Second, we could look at impacts in the opposite direction: how might the broader social order (or elements thereof) impact science or technology? For example, we could ask if racist or fascist designers would come up with different technological applications than democratic or capitalist designers. Or we could investigate the limitations of human language and symbolic thought to see if that would make scientific objectivity or truth impossible to attain. We could see if scientific facts or

> **Externalist STS**: studying how science or technology have impacts on other social realities, or studying how the broader social order impacts science and technology.

technological artifacts depend on certain social arrangements to actually exist, without which they wouldn't even exist, or else just be useless. We could even inquire into whether a scientist's social context—e.g., industrialism, sexism, or colonialism—would shape the scientific theories and facts supposedly discovered by those scientists.

In both of these cases, society is seen as distinct from science and technology, and as such, these ways of doing STS are called 'externalist'—either science and technology are influencing society as if outside it, or society is influencing science and technology as if outside them. But STS can also be pursued in an 'internalist' way: what social realities exist *within* the institutions or practices of science, engineering, or medicine? We could examine whether science is a progressive institution of knowledge or a conservative establishment that seeks to maintain the status quo. We could find out why so much scientific and technical writing resembles a secret code, only accessible to a favored few. We could investigate how scientists resolve controversies in their own disciplines. Or we could even inquire into how technological novelties are identified as valuable or worthwhile, either by the designers who produce them or the consumers who buy them. All in all, it is difficult to put a limit on the fascinating and important questions one can pursue in STS.

Internalist STS: studying the social realities within the institutions of science or technology.

About This Book

Aside from telling you what STS is in the abstract, we can also get a sense for it by actually *doing* STS. And there are, of course, a lot of different ways to do that. This book, however, will go about it in the following manner. It will, as already mentioned, offer counter-narratives to the 'usual story' about science and technology. I think the usual story is false. I will seek to establish that science and technology are cultural products. They exist because a certain kind of culture makes them exist, and without that culture, we'd have (at least) different forms of science and technology. This book will also seek to establish that science and technology are cultural processes. They are—internally—a particular kind of human arrangement producing a particular kind of theory and artifact, while externally, they are human arrangements which function to uphold, maintain, or reinforce a broader sort of culture or society. At the same time, they also function to change, threaten, or dismantle other sorts of cultures or societies. Therefore, this book will seek to identify the cultures which science and technology *contributes* to, is *produced* by, and also *is*.

For example, the culture that I am currently living in (and you probably are too) seems to think that it is perfectly reasonable to ask the question, "What are you going to do with your education?" An education—and especially a university degree—is supposed to be *useful*, which usually means getting the student a better job after graduation. Moreover, this culture assumes that the 'STEM' disciplines (science, technology, engineering, and medicine) are the 'best' forms of education because they get you the 'best' jobs. Everybody seems to know that an Arts degree in the humanities, social sciences, or—gasp!—fine arts is useless. But it's important to recognize that not only is this an implicit claim about values—i.e., that economic usefulness is what counts as good (which is itself not a scientifically or technologically defensible claim)—but also that such pragmatism is rather peculiar to the culture of modernity. If you were to have asked an ancient Greek or medieval Frenchman what they were going to *do* with their education, you would likely have been met with an uncomprehending stare (even if you asked the question in the correct language). In those cultures, education was seen as distinct from training. Training was for jobs, of course, but you only worked if you had to. Education was what you pursued because you didn't *have to* work. Wealthy or aristocratic Greeks studied rhetoric and geometry not because they were going to get jobs as politicians or architects, but because democratic citizens were expected to be politically literate and scientifically informed. Feudal Europeans saw their society as comprised of various "estates": the French system famously had those who fight (the nobility), those who pray (the church), and those who work (the peasantry). Of course you'd go to university to get trained as a clergyman (or a lawyer or a doctor, but they didn't fit into the three estates model any better than the merchant class did), but being a priest or a monk wasn't a practical skill like being a farmer or shepherd. In these examples, then, being scientifically educated in theoretical disciplines was not considered important for labor or commerce, whereas the skills that earned you a wage—like being a stonemason—were an *art*, the practical knowledge necessary for doing your job. In sum: *science was economically useless while the arts were for economic survival.* Our current culture appears to have completely inverted that idea, such that science is valued for practical job skills while the arts are assumed to be useless. It behooves us to find out why this inversion occurred, because we will then learn something important about how our culture values science differently than many other cultures, including the cultures which gave birth to our own.

Or consider how Western medicine in my culture (and possibly yours) approaches the treatment of attention deficit (and hyperactivity) disorder or 'AD(H)D.' Standard medical practice is to treat this disease with drugs, as we know that (among other things) people who suffer from it experience an abnormal lack of dopamine, one of our brain's reward chemicals. But to treat AD(H)D with drugs is a symptomatic approach that does not take into account the possible social reasons for why this deficiency may exist in the first place. At least, that's what Dr. Gabor Maté argues; he claims that the disease itself is at least partially the result of the breakdown of communal supports for parents, and notes how many schoolchildren with AD(H)D perform better in school when under the supervision of a caring adult. Therefore, he claims, to treat it simply with medication ignores the complexity of human psychology. However, we might be wondering whether Dr. Maté's approach makes him a pseudoscientific quack. This skeptical response—which is, in principle, quite reasonable—assumes that the correct scientific approach to mental health is to view our psyche as a complex machine (perhaps a "three pound computer made out of meat")[13] with chemical components that should only be repaired using other chemicals. Why might this be? Is there something peculiar about a culture that reduces everything to brain biology and genetics, and uses pharmaceutical 'grease' to keep the 'wheels of the machine turning' instead of adopting a perspective which incorporates environmental or social factors into both diagnosis and treatment? What role might the pharmaceutical industries play in reinforcing the use of inorganic metaphors for organic beings? Historically, not all cultures have used dead things like machines to shape their theories of living things. Yet even today when we talk about 'how living things work' or 'operate,' we're using machine-based—and pragmatic—metaphors on the assumption that such metaphors are more scientific because they are, oddly, technological.

To grapple with such questions, we need to understand science and technology as culture. To do that, we must understand the culture it arose within. For better or for worse, that culture is most readily identified as 'the West.' Moreover, to understand Western culture, we will need to understand its intellectual history. Without this, we will be underequipped to understand science and technology in our own time and place. To be sure, this book cannot (and you cannot expect it to) provide an exhaustive account of the intellectual history of Western technoscience. It can only offer a brief account, and as such, can only give you an introduction. Even as a brief introduction, however, it will attempt to give equal coverage to history, philosophy, sociology, and sometimes anthropology, political science, literary criticism, and even theology of science and technology. If one of those disciplinary approaches interests you more than the others, there are a good number of texts that approach STS from those directions especially. If any particular school of thought therein interests you (say postcolonial STS or actor-network theory), you'll at least know what topics you'll want to investigate beyond what this book can offer. But in attempting to offer you equal billing across these disciplines and fields, this text hopes to offer you a sampling of the breadth of STS theory so that you can be prepared for more specialized delving later on in your career as a critical thinker.

This book will proceed according to the following outline. First, we will deepen our introduction to STS by examining the abstract questions of what science is, and what technology is, and whether they are distinct from each other, or instead interdependent and possibly indistinguishable. Second, we will move into an intellectual survey of modern Western technoscience with a section on *prehistory*, or that period of our human past for which we have no written records. By examining non-Western cultures that made (and still make) no claim to being scientific or 'technological,' nor express any desire in becoming that way, we will prepare the ground for seeing the cultural and technological shifts that preceded the first ancient scientific culture on record. Third, the section on *premodernity* will examine ancient societies' views of science and technology, especially those of the classical and medieval West, to see how certain technological and intellectual changes led to the emergence of modern Western culture, and its eventual scientific and industrial revolutions. Fourth, the section on *early modernity* will examine several characteristic aspects of modern Western culture, so that the flavor of modern technoscience will be more clearly identifiable. Fifth, the section on *classical modernity* covers some of the cracks in the façade as structural problems with the usual story came to the fore. The sixth section of the book will introduce a number of problematic and ongoing issues facing technoscience in our *late modern* age. The final section will show how STS theory can be applied to specific issues, what I call *case studies*. These last two sections cover major controversies in contemporary STS scholarship, which are really only the tip of the iceberg. Thus it would be premature to call this last part of the book an intellectual 'history' because the history of our time is still being written. But who knows—perhaps you will end up someday as one of the people who will write it?

Notes

1 Robert Oppenheimer, "Physics in the Contemporary World," Arthur D. Little memorial lecture, MIT, Cambridge, MA, 25 November 1947.

2 Robert Oppenheimer, quoting the *Bhagavad-Gita*, interview by Fred Freed, *The Decision to Drop the Bomb*, NBC, 1965, www.atomicarchive.com/Movies/Movie8.shtml.

3 A variation on a slogan for DuPont used from 1935 to 1982.

4 Rachel Carson, *Silent Spring* (Houghton Mifflin, 1962).

5 Robert E. McGinn, *Science, Technology, and Society* (Prentice Hall, 1991), 10.

6 Gayathri Vaidyanathan, "Big Gap between What Scientists Say and Americans Think about Climate Change," Scientific American, 30 January 2015, http://www.scientificamerican.com/article/big-gap-between-what-scientists-say-and-americans-think-about-climate-change/.

7 McGinn, *Science, Technology, and Society*, 8.

8 Mick Smith, "The State of Nature: The Political Philosophy of Primitivism and the Culture of Contamination," *Environmental Values* 11 (2002): 410.

9 McGinn, *Science, Technology, and Society*, 10.

10 Sergio Sismondo, *An Introduction to Science and Technology Studies*, 2nd ed. (Wiley-Blackwell, 2010), 134.

11 McGinn, *Science, Technology, and Society*, 10.

12 Sismondo, *Introduction to Science*, 134.

13 Marvin Minsky, quoted in James Wright, "A Note from IESBS [International Encyclopedia of the Social and Behavioral Sciences] Author James Wright," *SciTech Connect*, last modified 4 May 2015, http://scitechconnect.elsevier.com/a-note-from-iesbs-author/.

Chapter Two

What Is Science?

BEFORE WE CAN SENSIBLY INVESTIGATE THE INTERRELATIONSHIPS OF science and society, we need to make sure we know how to accurately identify what science is. Defining science is, in fact, an active and ongoing project within STS itself—part of what we called 'internalist' STS in Chapter One. Within STS, the specific project of defining science and distinguishing it from pseudoscience and non-science is known as the "demarcation problem." It is an important problem, because calling something "scientific" is a value-laden term of approval, just as calling something "pseudoscience" is a clear expression of disapproval. What counts as science is not, therefore, simply an issue of (scientific!) fact, but a controversial conceptual and moral question. Of course, a brief introduction (which this book is) cannot hope to conclusively resolve such controversies. However, we can explore various angles of the discussion so that you, the reader, can be better informed of what is at issue when working towards a definition of science. In this chapter we will first look at the two major candidates for the definition of science. Then we will consider a number of supplementary notions which seek to refine the definition of science in an explicitly modern sense. This will involve a brief mention of a particular school of thought within STS, as well as a broader definition of science which can include non-modern and non-Western science. The chapter will conclude with a look at the sort of critical questions usually asked about science within STS.

The English word "science" is etymologically derived from the Latin word *scientia*, but that doesn't mean the two words actually mean the same thing. The meaning and usage of words can change over time, but these changes can help us to better understand our own society's vision of science in contradistinction to those of other societies. The

Romans used *scientia* to translate the Greek word *epistemē*, which simply translates into English as "knowledge." At its most basic, science just means "knowledge." However, *epistemē* wasn't the only Greek word for knowledge; their word *gnosis* also meant knowledge, but it implied secret knowledge or knowledge of the arcane (it is present in the English word "agnostic," for someone who doesn't know if the god[s] exist[s] or not). The assumption behind *gnosis* was that a person couldn't gain that kind of knowledge simply by investigating or thinking a lot. Rather, *gnosis* had to be revealed to a person who underwent certain procedures or initiations, often secret or mysterious ones. *Gnosis* was not in principle accessible to just anybody.

Epistemē, on the other hand, was knowledge that was in principle accessible to anybody. It could be acquired by study or mastery of learning. You didn't need to join a secret society to get it. You'd just have to work hard, develop your talents, direct your mind towards the cosmos, and then you'd be a scientist. The English word *epistemic*, meanwhile, broadly means "pertaining to knowledge," while *epistemology* refers to the study of how we can know anything at all. Science is an epistemology or way of knowing, therefore, and its knowledge is of a certain type. In principle, scientific knowing could be about anything at all. *Scientia* or *epistemē* "were applicable to any system of belief characterized by rigor and certainty, whether or not it had anything to do with nature."[1] So Aristotle called his study of the ultimate reasons for things "the divine science,"[2] while in the Christian middle ages, theology (the study of God) was honored as the Queen of the sciences. Even today, in Continental Europe, scientific research is not limited to what in English we call the natural or social sciences, but includes liberal arts such as literary criticism or philosophy. The German word for science is *Wissenschaft* (which is clearly not derived from the Latin *scientia*, but would literally translate into English as "knowing-ness"), and it refers broadly to any "systematic theoretical inquiry," not necessarily to the study of nature.[3]

Epistemic: pertaining to knowledge.
Epistemology: the study of how we can know anything at all.

Classical Definition of Science: any rigorous and certain body of knowledge acquired through study and mastery of learning.

The contemporary English use of the word "scientist" to exclusively denote a practitioner of the natural sciences (at least) is very recent; it was invented by William Whewell in 1833. This means that even Sir Isaac Newton, the famous Englishman who (among other things) formulated the theory of gravity, would not have considered himself a "scientist." Rather, he was a *natural philosopher*. Therefore, even if we were to now decide, perhaps, that "science" means systematic theoretical inquiry *into nature*, we would have to recognize not only that this conception of science is incredibly recent, but also that it is geographically unique to the English-speaking world. The definition of science which might be assumed to be common sense turns out to be not very common after all.

Science as Technology

If, however, we take the Anglo-American historical and geographical (etc.) social situation for granted—and, along with it, a working definition of science as *systematic theoretical inquiry into nature*—we still would need to identify the characteristics of scientific inquiry. There are a number of different possible components of (modern) "science" (in English), some of which are complementary, others which are in tension with yet others. Richard Lewontin, the renowned evolutionary geneticist, defines science in terms of two functions. The first function of science, he says, is that it

> provides us with new ways of manipulating the material world by producing a set of techniques, practices, and inventions by which new things are produced and by which the quality of our lives is changed. These are the aspects of science to which scientists appeal when they try to get money from governments or when they appear on the front pages of newspapers in their public relations efforts to maintain their prosperity.[4]

Furthermore, the quantum physicist Don Page says that "[t]he more questions we ask about the laws of nature, the better technology we can develop.... The long-term benefits of theoretical physics come from the technologies it makes possible."[5] Both Lewontin and Page are saying that science is fundamentally technological. Even though we might theoretically distinguish between science and technology, they aren't separable in practice: for Lewontin, science *is* the process by which new manipulative techniques are produced, while for Page, better tools are the justification for doing science at all. Lewontin goes on to claim that if something "does not make it possible to manipulate the world for our own benefit," then it is "untenable" as a science per se.[6]

In these examples, then, we see the idea that systematic inquiry into nature involves those "pattern[s] of behavior by which humans have gained control over their environment."[7] On this understanding, then, we could say that "medieval science" would have included the coats of mail armor worn by thirteenth-century knights, that "classical science" would have included the aqueducts which piped water into Rome, and that "prehistoric science" would have included the self bows and flint-tipped arrows invented after the end of the last Ice Age. Likewise, "modern science" would include sending people to the moon, or quantum computing. Science, by this definition, is *know how*.

> **Science Defined as Technology:** science is how any organism copes with the natural environment.

But what about other animals and plants, all of whom have their own abilities "to cope with the natural environment"?[8] If science is simply defined in terms of manipulating the material world, it is arbitrary to exclude nonhumans from the definition. Cats *know how* to catch mice, purr, and see in the dark, while cactuses are quite good at tolerating droughts and discouraging animals from biting them. Now if the idea of cat scientists or cactus scientists strikes you as absurd, then perhaps this definition of science is too broad. Perhaps we should refine this definition such that the scientific manipulation of the material world needs to be performed by *tools*, i.e. things which are (unlike cat eyes or cactus spines) not integral parts of an organism's body. And yet certain chimpanzee groups use sticks to extract delicious termites from their concrete-hard nest mounds, while otters can float on their backs holding oysters which they crack open by hitting them with rocks against their chests. Neither sticks nor rocks are integral to the bodies of chimpanzees or otters (respectively). They're tools, and if tool use makes you scientific, then we have chimpanzee and otter scientists. If that implication also strikes you as absurd, then either there is something categorically unique about human tool use which separates it from nonhuman tool use, or your definition of science is not broad enough. Maybe technology is not a sufficient condition for science, even though it might be a necessary condition. For the time being, let us table this discussion for our chapter on technology (Chapter Three), and examine the second major definition of science on offer: science *without* technology.

Science as Pure Theory

The potential shortcomings of defining science solely in terms of technology might be why Lewontin discusses a second function of science: explanation. Science is supposed to provide "a deep understanding of how the world works."[9] Science may not necessarily be about providing us with new tools or techniques, but rather with providing us with reasons, facts, or truths. This sort of knowledge is not practical know how but rather *theoretical*, in the sense that it provides abstract models of how the components of the universe function, regardless of whether we use this knowledge for practical purposes.

If science is fundamentally a form of explanation, then it doesn't have to be associated with technology. We can have tools that work even without explanations of how they work, just like riding a bike is possible without knowing how balancing works in physics. Science would therefore not be the use of tools to extract termites, but rather explaining why chimps need to use sticks to do that, or how hard termite mounds are in comparison to actual concrete. Lewontin points out that "it is remarkable how much important practical science [in his earlier sense of science-as-technology] has been quite independent of theory."[10] He gives the examples of agricultural hybrids and cancer therapies as two very successful technological advancements that do not, in fact, have much if anything to do with scientific modeling or abstract theory. Rather, they remain "essentially an empirical process in which one does what works."[11]

Furthermore, science can be understood as distinct from technology in that the former can be pursued without any concern for the latter. This separation between technology and science is seen in the phrase "pursui[ng] knowledge for its own sake."[12] Purely theoretical science is investigated just because scientists want to know the truth about something, not because they want to make something out of it. In this sense, science is theoretical because it doesn't have to have a use value or a technological application. The very word "theory" implies as much; *theoria* was the Greek word for being a spectator, somebody who stood at a distance or observed sporting events from the seats, giving them a degree of detachment or objectivity which they wouldn't have if they were physically involved in performing the activity under examination itself.

So by this definition, science in its *pure* form is "a body of theoretical knowledge" about the natural world, whereas *applied* science is the technological "application of theoretical knowledge to the solution of practical problems."[13] It's noteworthy that while we seem capable of articulating a notion of purely theoretical science without necessarily requiring it to have any pragmatic importance, it is still difficult to separate entirely our notions of science and technology. Both Lewontin and Page cannot speak of pure theory without also invoking technological implications in some form or other. It would indeed be difficult in the contemporary world to justify funding for theoretical scientific research if it didn't have some possibility for technical or (even better) economic application down the road. So while defining science exclusively in terms of technology seemed to be too broad, it also seems that defining science without any reference to technology is too narrow. Simply calling science "systematic theoretical inquiry into nature" misses something that our contemporary cultural context (at least in the mainstream Western world) assumes or expects from what people call "science" in English.

Science Defined as Theory: science is the abstract modeling of the objective nature of things, done for its own sake.

Other Possible Components of a Definition of Science

Even if we were to define science as systematic theoretical inquiry into nature without any commitment to technological application, we may want to hone it more precisely to reflect *modern* approaches to science rather than allowing it to include, say, classical Greek or ancient Chinese theories of nature. STS scholars therefore have a number of different possible refinements to consider. One possibility is that science should provide *lawlike* explanations of natural processes. True science would identify patterns or processes in nature that have universal validity across time and space: e.g., laws of nature such as gravitation or the conservation of energy. Moreover, such universal statements would be "preferably expressed in the language of mathematics. Thus Boyle's law (formulated by Robert Boyle in the seventeenth century) states that the pressure in a gas is inversely proportional to its volume if everything else remains constant."[14] If an attempt at systematic theoretical inquiry about nature did not reveal such lawlike statements, then it would be considered pseudoscience instead.

Another possibility is that science should follow a particular *method*, namely "the scientific method." In the words of David Lindberg, an historian of science, this definition associates science "with a set of procedures, usually experimental, for exploring nature's secrets and confirming or disconfirming theories about her behavior."[15] So non-experimental sciences—say, for example, the Aristotelian approach which viewed experimentation as unnatural and so unsuited for learning about nature[16]—would be pseudoscience, regardless of how universal or mathematical its statements might be. Other STS researchers argue that "there is no such thing as *the* scientific method," but that "the myth of a common methodology" is what makes doing science as a unified discipline possible.[17] So even if there are a variety of ways to identify scientific methods, this definition can still delineate what (especially modern) science is supposed to be.

Related to the notions of lawlike and experimental science is a third possibility: that scientific claims must possess a certain *epistemic status*. Now that you know what *epistemē* means (i.e., pertaining to knowledge), you can see that "epistemic status" refers to the kind or character of knowledge. Here, the issue is that knowledge claims made by scientists must have a certain quality, both in how they are supported and how they are believed. In the first instance, scientific knowledge must be justified by empirical evidence: evidence that is ultimately reducible to what humans can detect through their five senses. True science, then, cannot include beliefs that are grounded in intuition, logical deduction, tradition (etc.). In the second instance, scientific knowledge must also not be dogmatic. Because it is always possible that new evidence may come to light, scientists must always be ready to revise or reject their previously held theories. All scientific statements have to be believed tentatively and provisionally, not absolutely, because evidential warrant cannot be absolute. Thus, Bertrand Russell claims that "it is not *what* the man of science believes that distinguishes him, but *how* and *why* he believes it."[18]

Possible Stipulations for a Definition of "Modern" Science:
- lawlike explanation
- experimental method
- empirical justification
- tentative or provisional belief

A further element to the epistemic status of scientific claims forms an entire school of thought within in STS: *Mertonian structural-functionalism*. This is the notion, propounded by the sociologist Robert K. Merton, that science is a social structure (or an "institution") which should function to provide society as a whole (of which science is a part) with *certified* knowledge. This certified knowledge is supposed to help improve the functioning of society as a whole. So not only is certification itself an epistemic status, it also has a clearly social role to play by virtue of this epistemic status. Science is supposed to supply society with a valued sort of knowledge by conforming to certain regulatory norms: it should produce universally true statements (à la lawlike explanations earlier); it should be "communist," in the sense that its publications are truly public—publicly accessible

Mertonian Structural-Functionalism: science is a social institution whose function is to provide knowledge that maintains the larger social structures of which science is a part.

throughout the community; it should be disinterested or politically neutral (as in the original Greek sense of *theoria*, the detached spectator); and it should engage in organized skepticism, so that knowledge claims are rigorously tested for truth (compare with experimental procedure and provisional acceptance of evidentially warranted truths). Following such norms does not guarantee the absolute truth of science, but it does (according to structural-functionalism) provide a stamp of approval that larger social structures (such as governments) can find useful. All this assumes, however, that one's larger social context is something deemed worthy of supporting with usefully certified truths. But if a society is oppressive or unjust, the Mertonian scientist's support for it—in providing authoritative knowledge for that society to use—can certainly be problematic (which is itself a potential problem with the norm of political disinterestedness). It may be that science is not above the social fray after all.[19]

Non-Modern Natural Philosophy

Be all that as it may, these various refinements of the definition of science have the implication (often intentional) that there was no such thing as science before the so-called Scientific Revolution. So while the definition of science as technology had the problematic implication of "chimpanzee scientists," these latter definitions of science have the problem that there could be no such thing as, say, ancient Greek scientists. Whatever the Pharaonic Egyptians or Ming Chinese were doing with respect to nature, it wouldn't be science, and might even be pseudoscience. But it's not clear that this is a particularly helpful way to describe what they were doing. If our definitions require that science be seen as *true* before it even counts as science—such that anything which turns out to be false is thereby defined as pseudoscience—then, if we are to hold our scientific beliefs non-dogmatically or provisionally, what we now believe to be "science" may actually prove to be pseudoscience from the perspective of some future science. At any given moment in the present we can never be sure that what we think is science is actually "science," because we have failed to disassociate our definition of science from the concept of truth.

Therefore, if we are to have a *stable* definition of science which does not include a dogmatic assumption about science always having to be true, we should broaden our definition to include non-modern forms of science (and reserve most of the foregoing definitions for, at most, modern science). Science in the broad sense can include any culture's *beliefs about nature*, even if we might not currently think those beliefs are true. This allows for the conceptual possibility of non-modern science (e.g., the ancient Near-Eastern belief that the sky was a hard shell over the Earth called the "firmament")[20] and the conceptual possibility of false science (e.g., "reflexology," the belief that massaging the nerve ends in your feet can have health benefits for the rest of your body). By distinguishing beliefs about nature from the truth about nature, we can study various forms of science as social realities, and we can do so in a systematic and rigorous (i.e., scientific) way.

Science Defined as Natural Philosophy: science in the broad sense is any culture or society's beliefs or theories about nature.

Working Definition (in English) of Modern Science: systematic theoretical inquiry into nature.

In terms of history, this means we "take the past for what it was" and "respect the way earlier generations approached nature" without chastising our ancestors for not measuring up to the standards we currently hold ourselves to.[21] Indeed if we want to say (as the usual story does) that modern science is simply a development of the human spirit and a result of the accumulation of many incremental improvements over much older forms of science, then what we now do scientifically *must* be in some sense continuous with past practice! For this reason, Lindberg identifies the object of his study as "natural philosophy" or a culture's "philosophy of nature." Such terminology unambiguously includes any beliefs a society may have about nature without confusing it with the term "science" and its many modernist connotations (although it risks conflating 'philosophy' and 'theory' with 'beliefs,' an issue we will confront in Chapter Four). On the other hand, it can be difficult to see any continuity between natural philosophy and science if we assume (as is commonplace) that "philosophy" is very nearly the opposite of science! The fact of the matter is that there is a very strong connection between what philosophy is and what science is, both historically and conceptually. Indeed, "systematic theoretical inquiry into nature" is precisely what natural philosophy means. Therefore, in this book I will happily use the term "natural philosophy" interchangeably with "science" in the broad sense, although for ease of exposition I will just as easily refer to "ancient science" or "modern science" (etc.), the latter of which clearly does not think of itself as a form of philosophy. In any case, the relationship between philosophy and science is itself an issue that is part of STS, and we will grapple with it throughout this book.

Critical STS Inquiry into Science

As you have seen, simply coming to an understanding of what the word "science" is supposed to mean is an exercise in critical thinking and one which comprises at least a part of the field of STS. But STS is also much more than the attempt to define science. It also proceeds—tentatively, on the basis of the preceding definitional work—to analyze and evaluate science as a social phenomenon. There are myriad ways to do this, of course, and our introductory look at STS in this chapter can only hope to cover a paradigmatic example of how STS submits science to critical scrutiny. The *epistemological status of science* is one such example, and it is the question of the nature of scientific knowledge itself. In the case of modern science, does lawlike explanation, experimental method, and evidential warrant offer a guarantee (or an authoritative certification) that scientific knowledge will be 'objectively' true? Lewontin explains this supposed objectivity as follows:

> Not only the methods and institutions of science are said to be above ordinary human relations but, of course, the product of science is claimed to be a kind of universal truth. The secrets of nature are unlocked. Once the truth about nature is revealed, one must accept the facts of life. When science speaks, let no dog bark.[22]

This is how the usual story typically views science: it is neutral, universal, absolute, and viewer-independent, standing outside the social fray.

Lewontin argues that this sort of objectivity can be claimed for science because it appears to possess several features. First, science can be thought to *transcend human bias*: "the institution as a whole must appear to derive from sources outside of ordinary human social struggle. It must not seem to be the creation of political, economic, or social forces, but to descend into society from a supra-human source." Second, science can be thought to *transcend human error*: "the ideas, pronouncements, rules, and results of the institution's activity must have a validity and a transcendent truth that goes beyond any possibility of human compromise or human error. Its explanations and pronouncements must seem to be true in an absolute sense and to derive somehow from an absolute source." Third and finally, science can be thought of as an *elite social group* wherein unbiased and error-free knowledge can be incubated: "the institution must have a certain mystical and veiled quality so that its innermost operation is not completely transparent to everyone. It must have an esoteric language, which needs to be explained to the ordinary person by those who are especially knowledgeable and who can intervene between everyday life and mysterious sources of understanding and knowledge."[23] If modern science has these characteristics, then it is not surprising that we can be tempted to define it in terms of objective truth and to define all other attempts at science as misguided or erroneous pseudoscience.

> **Objectivity:** how an object is, in actual fact, not as its perceiver would have it to be; understanding a truth without error or distortion; knowing something as if the knower of the thing has no effect on the object known; also known as "the view from nowhere" or "the God's-eye perspective."

Thus, on the one hand, science can be represented as a social institution with a function in the broader social context to provide objective truths, which is the gist of Mertonian structural-functionalism. On the other hand, STS scholarship can critically investigate this very picture of scientific objectivity. Lewontin points out that the vision of science outlined above correlates extremely closely to the social role of state-enforced religion, more like *gnosis* than *scientia*, even. An elite social group which transcends human bias and error could just as well be a description of the role Christianity played in medieval Europe, or any ideology whose job it is to keep society moving smoothly by giving its citizens a unified political narrative to live by. So the social function of Mertonian science—the production of certified knowledge—is anything but objective and neutral. Rather, it has the *value-laden*, human-relative purpose of providing "social legitimation" for the status quo.[24] In this respect, science may be in the business of keeping certain social interests in power by providing those interests with authoritative knowledge that solidifies their social position. Science may be, therefore, no more objective or neutral than "olde tyme religion" ever was.

STS scholars also call into question the objectivity of the products of science themselves, namely scientific *facts*. Some argue that "non-scientific" factors determine the form of scientific explanation and thus the sort of facts we get. Lorraine Daston, director of the Max Planck Institute for the history of science in Berlin, argues that

> both government and corporate funding have come with strings attached. These strings have not only encumbered scientific research with bureaucracy and distorted incentives (was the war against cancer really a reasonable scientific, as opposed to political, goal?); they have denatured the rhythms and internal ethos of research. The relentless three- to five-year tempo of grant proposals; the pressure to choose solid safe bets rather than more venturesome topics of investigation; the mentality of public accountability that has led to ever greater reliance on mechanical quantitative indicators of quality such as citation indices—all of this results in a system that can and has been gamed.[25]

If this is so, science as a culture takes on the shape of the bureaucratic culture which funds it, and as such produces equally bureaucratic results which surreptitiously make bureaucratic culture sound scientific! Daston continues:

> The dangers of research being driven by commercial agendas are even more obvious: bans on the open publication of data; outright manipulation of results; and, at least in the case of pharmaceutical companies, ghostwriting of articles in scientific journals. In both the biomedical sciences and social sciences, the number of published results that can be replicated has plummeted; the incidence of fraud has skyrocketed. No amount of external professional policing can replace an internalized ethos of inquiry—not the ethics of science's responsibility to society but that of science's responsibility to itself. Science needs [a] counterweight to the values of the market for its own sake.[26]

In this way, science can also take on the character of the economic system which funds it, and this involves new values which even go *against* Mertonian norms of disinterestedness, public access, and rigorous testing. If so, how could science be even remotely objective or universally true?

Daston does not think science must inevitably succumb to these financial pressures. Rather, even though it all too often does, science's entanglement here is a pathological distortion that can and should be fixed. Lewontin, however, argues that science can be ideological (rather than objective) in even more subtle ways, "in the form of basic assumptions of which scientists themselves are usually not aware yet which have profound effect on the forms of explanations and which, in turn, serve to reinforce the social attitudes that gave rise to those assumptions in the first place."[27] Modern science, for example, "sees the world, both living and dead, as a large and complicated system of gears and levers."[28] This mechanistic way of viewing the world assumes that everything can and will be better understood if it is reduced to its component parts, what Lewontin calls "the reductionist view" of all things.[29] In his own field of biology, this is exemplified by the idea of genetic determinism:

> Genes make individuals and individuals make society, and so genes make society. If one society is different from another, that is because the genes of the individuals in one society are different from those in another.... That is why molecular biologists urge us to spend as much money as necessary to discover the sequence of the DNA of a human being. They say that when we know the sequence of the molecule that makes up all our genes, we will know what it is to be human.[30]

But the reductionist view doesn't just magically appear to the mind of modern scientists as if it were another law of logic. Rather, Lewontin argues that it comes from the collapse of medieval communal political structures whereby people had understood themselves in terms of their social class. Only with the rise of economic individualism and factory capitalism could reductive explanations of wholes in terms of their parts gain any traction. So scientific reduction doesn't come down to society from on high; science rather reflects and reinforces "the dominant values and views of society at each epoch."[31] In this case, "all of modern science takes as its informing metaphor the clock mechanism described by René Descartes in Part V of his *Discourses*,"[32] and the mechanistic science that results grants scientific legitimacy to the notion of economic individualism and to the industrialized factory system that utilizes human beings as replaceable parts.

As a molecular biologist himself, Lewontin isn't suggesting that modern science is hopelessly ideological and should be completely abandoned in favor of alternative or older scientific theories. Just because scientists "view nature through a lens that has been molded by their social experience"[33] doesn't mean "we should give up science in favor of, say, astrology or thinking beautiful thoughts."[34] However, the epistemological status of scientific knowledge will always be subject to critical evaluation in STS. If the usual story presents science as objective and absolute, a critical appraisal will ask whether science really is objective and absolute. All Lewontin hopes to do is "acquaint the reader with the truth about science as a social activity and to promote a reasonable skepticism about the sweeping claims that modern science makes to an understanding of human existence."[35] Science will inevitably operate on social assumptions and, as such, can and should be only provisionally trusted in so far as those social assumptions themselves can be reasonably trusted. Science cannot be separated from the society it legitimates, and to not think critically about the social world we live in will default to letting science function as religious or political dogma. Lewontin does not want us "to leave science to the experts," such that we're "mystified by it," but wants us to "demand a sophisticated scientific understanding in which everyone can share."[36] That is, in a nutshell, what STS is, and the goal we will pursue in the following pages.

Notes

1 David C. Lindberg, *The Beginnings of Western Science: The European Scientific Tradition in Philosophical, Religious, and Institutional Context, 600 B.C. to A.D. 1450* (University of Chicago Press, 1992), 4.
2 Aristotle, *Metaphysics*, 983a5.
3 Robert E. McGinn, *Science, Technology, and Society* (Prentice Hall, 1991), 15.
4 R.C. Lewontin, *Biology as Ideology: The Doctrine of DNA* (Anansi, 1991), 4.
5 Quoted in Brian Murphy, "If Time Existed ... Theoretical Physics at the UofA Would Be 50," *Folio*, 24 September 2010, 2.
6 Lewontin, *Biology as Ideology*, 15.
7 Lindberg, *Beginnings of Western Science*, 1.
8 Lindberg, *Beginnings of Western Science*, 4.
9 Lewontin, *Biology as Ideology*, 5.
10 Lewontin, *Biology as Ideology*, 4.
11 Lewontin, *Biology as Ideology*, 5.
12 Murphy, "If Time Existed," 2.
13 Lindberg, *Beginnings of Western Science*, 1. Keep in mind that while science *may* be applied in a technological way, not all technologies are applications of scientific knowledge. Many technologies are not dependent on a scientific understanding of how they work and were certainly not invented on the basis of any scientific explanation.
14 Lindberg, *Beginnings of Western Science*, 1.
15 Lindberg, *Beginnings of Western Science*, 2.
16 Andrew Ede and Lesley B. Cormack, *A History of Science in Society: From Philosophy to Utility*, 4th ed. (University of Toronto Press, 2022), 22.
17 Wenda K. Bauchspies et al., *Science, Technology, and Society: A Sociological Approach* (Blackwell, 2006), 21; original emphasis.
18 Quoted in Lindberg, *Beginnings of Western Science*, 2; original emphasis.
19 For a chapter-length introduction to Mertonian structural functionalism, see the third chapter of either edition of Sergio Sismondo, *An Introduction to Science and Technology Studies* (Wiley-Blackwell, 2004 [1st ed.]; 2010 [2nd ed.]).
20 For a science fiction treatment of this cosmology, read the short story by Ted Chiang, "Tower of Babylon," in *Stories of Your Life and Others* (Vintage Books, 2002), 1–28.
21 Lindberg, *Beginnings of Western Science*, 3.
22 Lewontin, *Biology as Ideology*, 8.
23 Lewontin, *Biology as Ideology*, 7.
24 Lewontin, *Biology as Ideology*, 6.
25 Lorraine Daston, "Can Liberal Education Save the Sciences?" *The Point*, 25 May 2016, https://thepointmag.com/2016/examined-life/can-liberal-education-save-the-sciences.
26 Daston, "Can Liberal Education Save the Sciences?"
27 Lewontin, *Biology as Ideology*, 10.
28 Lewontin, *Biology as Ideology*, 12.
29 Lewontin, *Biology as Ideology*, 12.
30 Lewontin, *Biology as Ideology*, 14.
31 Lewontin, *Biology as Ideology*, 9.

32 Lewontin, *Biology as Ideology*, 12.
33 Lewontin, *Biology as Ideology*, 3.
34 Lewontin, *Biology as Ideology*, 16.
35 Lewontin, *Biology as Ideology*, 16.
36 Lewontin, *Biology as Ideology*, 16.

Chapter Three

What Is Technology?

IN THE PRECEDING CHAPTER, WE EXAMINED VARIOUS DEFINITIONS (AND critiques) of science so that we could be better prepared to investigate the interrelationship between science and society. In like manner, we will need to carefully identify what technology is if we are going to sensibly investigate its interrelationship with society, and with science. Indeed, we have already seen that science is sometimes defined *as* technology—as if the two were the same thing—and so in this case as well, defining technology is an active and ongoing project within STS itself. In the first part of this chapter we will look at several candidate definitions of technology, including the notion that technology and science are, if not the same, then inextricably linked. In the second part of the chapter, we will take a broad look at the sort of critical questions usually asked in STS about technology. With that, the introductory phase of this book will come to an end. After this chapter, we will begin exploring alternative perspectives on how science and technology have interacted with society in human history, and how those accounts challenge 'the usual story' about science and technology in society.

It's likely that you have more direct contact with technology than you do with science. Unless we happen to be scientists ourselves, laboratories and formulas are probably not part of our everyday experience. Rather, we have direct experience of all manner of technologies in our everyday lives, from the most sophisticated iterations of portable computing to the most rudimentary implements used for gardening. At virtually every moment of our lives, we interact with and are surrounded by tools of all sorts: "In each lived moment of our waking and sleeping, we are technological civilisation."[1] Technology is certainly more visibly ubiquitous than science is. This may be why, when people turn a critical eye towards the social world they live in, technology is more frequently implicated in their critiques than science. That is, we're more likely to complain about the effect of technology on our lives than we are to complain about science.

It may come as a surprise, then, to learn that while the "philosophy, history, and sociology of science are fairly long-standing academic specialties, ... the history and especially the philosophy and sociology of technology are still in the early stages of development."[2] This might be due to the fact that academic scholarship has (understandably) seen itself as an intellectual pursuit, and as such it would presumably focus on the study of other intellectual pursuits (like science) rather than practical pursuits (like technology). Moreover, the advent of modern science was a major intellectual shift in Western thought, a revolutionary change that warranted much serious attention if it was to be adequately understood. Revolutionary technological change, by contrast, is less clearly an intellectual effect, and industrialization (for example) occurred quite a bit after the scientific revolutions. (For example, Descartes and Bacon were forcefully advocating for a radically new approach to science over 400 years ago, whereas the first effective railway networks appeared less than 200 years ago.) In any case, the issue of technology is a particularly stimulating aspect of STS, both because of its comparatively immediate relevance to our daily lives, and because there's more room to do exciting research on technology in STS when not as much has been said about it so far. For example, what *is* technology anyway?

Technology as Leverage over the Material World

One philosopher of technology has suggested that we define technology as "all the ways in which intelligence implements practical purposes."[3] *Purposes* and *intelligence* both imply consciousness, such that technologies (according to this definition) could not be created by beings which lack consciousness. So, for example, lettuces cannot fabricate (let alone use) tools. On the other hand, the compound notion of *practical implementation* takes technology beyond the realm of consciousness or the intellect into the physical world. "Practical" usually means that which pertains to the performance of an *action* or getting something *done* in the physical world which (presumably) exists outside of our minds. That we use the word "implement" for this process of making something happen in 'the real world'[4] is particularly telling, because when we use the word as a noun instead of a verb, "implement" means exactly the same thing as a tool.

> **Technology as Leverage over the Material World:** technology is all the ways in which intelligence implements practical purposes.

Technology, then, cannot simply be an intellectual pursuit. It must have a practical application. Moreover, this application must give us some ability to effect some kind of change in the physical world. If, for example, we intend to chop down a tree, simply thinking about chopping that tree down will not make it fall down. We need a *thing* to actually cut the tree down, like an axe or a chainsaw. Our minds, all by themselves, do not have the power to exercise any control or power over matter (even though it would be cool to bend a spoon just with your mind). But with *tools*, we can exercise leverage in a wide variety of ways over the material world. That is why we use axes or chainsaws to cut down trees—without such items, we could not interact with our environment at all.

Or could we? Even though lettuces presumably lack consciousness, they certainly do interact with their environment. They appropriate nutrients from the soil, along with moisture and sunlight. With these resources at their disposal, lettuces grow—expanding in space—and eventually reproduce, creating more lettuces. They compete for space and resources with other plants and have, in their own unique lettuce way, certain measures to counteract predation. Yet they do all this without using tools. Perhaps we cannot say that lettuces 'do' anything, if we define action (doing) as something which requires consciousness, but the fact remains that lettuces have effects on the physical world even though we would not say that they use technologies.[5] Lettuces interact with their environment by virtue of their *bodies* and nothing else. We humans share at least this rudimentary fact with lettuces. We can also interact with our environment using nothing but our own bodies. We might find it very difficult to chop down a tree without an axe or a saw, but we can try to use our bodies for that purpose nevertheless. The point is not that biting or punching a tree is an ineffective way for humans to chop it down (even though that's true), but rather that—by the definition of technology as *mere leverage over the physical world*—our bodies are 'technologies' just as much as axes or saws are. If so, then a beaver—which is probably conscious—can implement its practical purposes in the world by chopping a tree down by using the 'technology' of its teeth.

The question, then, is whether it is reasonable to suggest that "technology" includes one's own body, and whether STS should include the social study of using our teeth to achieve certain purposes (such as chewing). Perhaps we should say that our bodies *are* technologies, which would allow us to more readily claim that other technologies (e.g., prosthetic implants) are *not* fundamentally different from teeth, fingernails, or hair. There would be no significant distinction to be drawn, then, between a cyborg and a completely naked human being. The trouble with this move is that the term "technology" almost becomes meaningless, for it gives us no ground for distinguishing anything from technology: everything (except our disembodied minds) is a technology for something. (Some STS scholars, for example, define technology as "physical objects," which is basically all the things.)[6] As such, there's nothing unique or interesting about it and there's no point in studying technology as opposed to studying picking your nose.

Another problem with this definition of technology is that it entangles STS in a very old philosophical quandary: the mind-body problem. Recall that we cannot bend spoons with our minds; this was one of the reasons why human beings needed tools to implement their practical purposes. But I can bend the spoon with my mind: my mind tells my hands to take the spoon and bend it! That implies that my body is the tool of my mind. But how is my mind supposed to use my *body* as a tool if my mind, by itself, cannot bend a *spoon* (which is itself another tool)? To define your material body as a tool is to correspondingly define your mind as a kind of immaterial ghost which uses the tool. The problem with ghosts is that, contrary to stories about poltergeists rattling toasters or turning off the lights, ghosts are immaterial beings (if they're anything at all) and as such would pass right through matter rather than being able to move it around. Your ghost would be just as incapable of moving your body around as you are incapable of moving a spoon around with just your mind.[7] So if your body is defined as a technology, your mind would be, by definition, incapable of using it.

Therefore, to *merely* define technology in terms of intelligence implementing practical purposes seems too broad. It requires us to accept that beavers' mouths are technologies, that our own bodies are technologies which are in principle incapable of being used by our consciousness, and that everything (other than our consciousness) is a technology, in which case the very notion of technology identifies nothing in particular. Besides, almost nobody uses the term 'technology' to refer to their own body, unless they're trying too hard to be clever. Therefore, I propose that we at least supplement or modify this definition of technology to get closer to a useable and plausible definition of the term.

Technology as Augmentation

The concept of *supplement* or *modification* can help supply a supplementary and modified definition of technology: technology is anything which intelligence uses to *augment* the body's powers to exert leverage over material things. Fingernails, for example, are an integral part of a human body, and so they aren't an augmentation. Brass knuckles, on the other hand, are an augmentation because they add something to the power of the human hand which it otherwise wouldn't have. Similarly, ears are not a body augmentation whereas hearing aids are. So by the definition of technology as body augmentation, fingernails and ears are not technologies, but brass knuckles and hearing aids are. Anything from a pointy stick to a stick of lipstick is a technology, just as much as a spaceship or clothing is, because they all allow us to interact with the world of physical things in ways that we could not if we were limited to what our unassisted bodies could provide.

> **Technology as Augmentation: technology is anything intelligence uses to amplify the body's natural abilities to implement practical purposes.**

This distinction between the unassisted body and technological augmentation presupposes a very ancient distinction between *the natural* and *the artificial.* We will say more about the artificial later in this chapter, but for now we should note that the English word "natural" derives etymologically from the Latin word *natura*, which is the equivalent of the Greek word *physis*, the etymological root of "physical" in English. Both *natura* and *physis* refer to that which gives birth to itself (which is why English words like "nativity" and "neonatal" refer to birth). The natural or the physical world, to the Greco-Roman mind, is that kind of reality which takes care of its own reproduction. We could

Classical Definition of Physical Nature: any type of thing which brings itself into being; this includes biological reproduction, geological formation, and cosmological evolution.

say that turtles are 'natural' or 'physical' because you or I don't have to do anything for there to be turtles in the world. Humans don't have to implement any practical purposes for turtles to exist! Rather, it is the turtles themselves that ensure there will be turtles in the world, all else being equal. Turtles take care of the business of turtling. The same generally holds for rocks, water, or human ears. Intelligence does not have to form the purpose of "let there be rocks" or "I want to have two ears" for there to actually be rocks, or for my body to have two ears. My two ears (all else being equal) are on my head because 'nature' put them there (speaking anthropomorphically, anyway). I was fortunate enough to have been born with two ears, and they have managed to stay attached to my head at least up until the time of my writing this chapter. We might be justified in saying, therefore, that it's 'natural' that I have two ears.[8]

By contrast, it would be unusual to say that my having hearing aids is 'natural,' because my body didn't automatically 'give birth' to the hearing aids in a sense similar to how my body (or my genetic code) furnished me with ears.[9] Hearing aids do not create themselves by virtue of their own nature, even if my body needs them. (It might actually be natural for me to lose my hearing over time.) Something *more* is required to create hearing aids, and that more-ness (if we can call it that) is why hearing aids are an augmentation, not simply because they *amplify* a body's natural capacity, but also because they require the purposeful application of consciousness to exist as such. Somebody had to *invent* hearing aids, and somebody had to *make* them, because hearing aids do not invent or make themselves. That is why they are a technology, according to this definition.

We can (and in this chapter, we already have) use(d) the word *tool* to mean exactly the same thing. Tools are the same as technologies, because they both *add* something to the natural capacities of physical bodies—they add power (amplification) to one's bodily ability to exert leverage over the physical world—and they require the contribution of a conscious intelligence to both conceive of and create them. In both these ways, tools or technologies are body augmentations.

It is very helpful to define technologies as tools—i.e., useful objects not intrinsic to bodies—because then we can make sense of the claim that human beings (at least) have always been a technological species. Tool use among hominids is first attested among *homo habilis*, a species which evolutionarily preceded our own by approximately 2.4 million years. However, by this definition we must also accept that other (nonhuman) species use technologies, including the chimpanzees and otters mentioned in Chapter Two, which use twigs or rocks to help them get at food which is otherwise difficult to access (termites or oyster meat, respectively). Chimps and otters are (presumably) conscious at some level, and it requires some intelligence to use the tool to achieve the desired end. (In the case of chimpanzees, only some troops know how to use sticks to access termites within their mounds, suggesting that tool use among that species is a cultural innovation transmitted within the group by communication, rather than being instinctual across the entire species.)

Nevertheless, STS is justifiably more interested in investigating human technology and its social aspects, even though nonhuman technology is certainly a fascinating field of inquiry on its own. Maybe chimp sticks or otter rocks are just found objects, repurposed without any significant alterations, whereas we'd expect technologies or tools worthy of the name to have been intentionally modified from their natural state. This notion of alteration, coupled with the aforementioned notion that tools require the intervention of an intelligence to bring them about, leads some STS scholars to define tools in a very precise way with a very precise term: *technics*. Technics are "material products of human making or fabrication," and "generic types or kinds" of technology rather than individual items.[10] A technic is thus a kind of thing (though not any one example of that kind of thing) which a human has purposefully produced. For example, 'the computer' is a technic (it didn't bring itself forth on its own accord, and it was made by humans, not chimpanzees), whereas the computer I'm using to type out this chapter is only a specific instance of the technical category (technic) we call 'the computer.'

> **Technics:** general categories of material products made, fabricated, or produced by human beings.

But STS scholars also point out that defining technology in terms of technics is only half of the picture. A technic is just an abstract category of thing, most items of which would simply sit still until somebody *used* them. A technic will always and only make sense in relationship to people; the social isolation of technics or tools makes no practical sense. In fact, a technology would be useless or functionless if it existed in isolation from beings who knew how to operate it. Therefore, there is always a *social dimension* to technologies without which technologies would be functionless. The other half of technology, then, is *technique*, the actions or know-how required to make or use a tool or a technic. (I once had a boss who didn't know how to use the mouse attached to his desktop computer; he lacked the technique to properly operate the technic! I have a similar problem with trackpads on laptop computers.) Technologies don't make sense without also speaking to the technical knowledge bound up with them.

A technique, precisely defined, is "the complex of knowledge, methods, materials, and if applicable, constituent parts (themselves technics) used in making [and operating] a certain kind of technic."[11] If we are speaking about 'computer technology,' therefore, we cannot simply mean 'the computer' as a technic, but also the information and components required to make the computer, the skills necessary to put those two aspects together and actually bring the computer into existence, and the skills required to actually use the computer.

> **Technique:** the information, components, and skills required to produce or operate a technic.

All of this is *technique* which, unlike technics, exists on the 'inside' of a conscious mind rather than in the 'outside' world of material things. As a combination of technics and techniques, technologies bridge the divide between mental and physical realities just as our own embodied human identities do. STS scholars will just as readily speak about techniques as they would about technics when examining technology (understood generally as body augmentation), and that is because STS does not examine technology (or science) in isolation from social networks.

Technology as Quintessentially Modern

Chances are, however, that what interests us most about technologies are not simple tools or techniques, but the highly advanced technologies that we engage with on a regular basis and towards which we direct so much of our current attention. You have probably heard someone exclaim in a moment of frustration that they "hate technology!" when their fancy digital device doesn't work as intended. Such exclamations wouldn't be particularly meaningful if, by technology, they meant technics and techniques in general, because then that person could mean that they hate basic tools like shoes or fire, and basic techniques like doing up buttons or wearing a hat. So is there any sense in identifying or defining 'technology' as the *modern or contemporary* tools and techniques characterized by high levels of advancement or complexity? Some STS scholars think so. After all, if what we mean by technology is simply technics and techniques, then why would the English language have yet another word—i.e., 'technology'—for it?

George Grant, in his classic essay "Thinking about Technology," argues that the "current use of the word 'technology' in North America" is a "novelty" for, as the Europeans point out, "our [North American] usage confuses us [North Americans] by distorting the literal meaning" of the word.[12] The word 'technology' was invented in 1829 by Jacob Bigelow (four years before the word 'scientist' was coined) to indicate *the scientific* study of the application of science to practical problems. This is what the word's component parts mean: *technē* and *logos* are the Greek words for 'arts and crafts' (*technē*) on the one hand, and 'the systematic study of something' (*logos*) on the other hand. Arts and crafts are completely reasonable concepts to associate with technology; they require skill (technique) to create and use, and they are themselves material products of human fabrication (technics). Indeed, the Greek notion of *technē* is the basic contrast to *physis*; *technē* does *not* bring itself into being, but rather requires an external agent to create it. So while we might consider turtles to be 'physical' or 'natural' because turtles are the agents by which other turtles come to be, turtle soup would be artificial (an adjective) or an artifact (a noun) because it requires artifice (a verb) to create it. That's

Classical Definition of Technē: that which does not bring itself into being, but requires the skilled craft of an artisan to produce; cf. artificial, artifice, artifact.

why making turtle soup is an 'art' or a craft. In this sense, turtle soup is artificial: turtle soup certainly doesn't make itself, and turtles don't make it either. *Technē* as skilled art or craft certainly makes sense in terms of body-augmenting technics and techniques, no doubt because those words are derived from *technē*.

But the concept of *logos* doesn't seem to have anything to do with arts and crafts at all. At a very rudimentary level, *logos* is the Greek word for 'word,' but more specifically it refers to *rational* and *ordered* words. That's why in English we use the word 'logic' for the structure and rules of correct thinking. To the Greek mind (broadly speaking), the *logos* was the heart of the cosmos, itself a rational and ordered place. The *logos* was that very principle of rationality and order. Therefore, if a science was directed towards the ultimate nature of things, its rigorous study (see Chapter Two) would eventually discover and know the *logos* itself. That is why so many of our branches of science end with the suffix '-logy.' Biology is the science of living things (bios-logos), zoology is the science of animals (zoē-logos), anthropology is the science of humanity (anthropos-logos), psychology is the science of the soul (psychē-logos), and so on.

But recall from Chapter Two that science was classically understood to be *theoretical*: theory is the abstract modeling of the objective nature of things done for its own sake. In this case, then, the literal or etymological meaning of 'techno-logy' would be the *theoretical science of practical arts and crafts*, as intended by Bigelow around 200 years ago. We also saw in Chapter Two that some people have defined science in terms of technology (i.e., how any organism copes with the natural environment), and here we see the inverse: technology defined as a science! But just as we found some flaws with the definition of science as technology, there is something rather odd with the definition of technology as the scientific study of crafts. Theoretical science is done for its own sake and not for any practical purpose, whereas arts and crafts (technics and techniques) are rarely done for any other reason except practical purposes. The startling nature of the term 'technology' is that it combines the practical with the impractical, theoretical knowledge with hands-on know-how, distance (transcendence) with contact (immanence). Even if science is not supposed to be the same thing as technology, the very word 'technology' literally means that they are inseparable from each other. In fact, we could just call it 'technoscience,' which many STS scholars actually do (myself included, as you can tell from the title of this book), especially when referring to characteristically modern science and technology.

Literal Definition of Technology: the theoretical science of practical arts and crafts.

Perhaps a solution to this terminological muddle would be to stipulate 'technology' as those technics and tools that we possess because science has made them possible. For example, it is difficult (if not impossible) to invent certain tools (like flash memory drives) without a knowledge of subatomic physics (like quantum tunneling). Likewise, it's difficult (if not impossible) to have a knowledge of subatomic physics without certain tools (like bubble chambers to track paths of particle decay). We might say that science, tools, and techniques go hand-in-hand (even if they're not interchangeable with each other), and that's why we'd have a word like 'techno-logy.' Moreover, if we're talking about *modern* or highly advanced science, then its intertwining with modern technics and techniques would explain the temptation to use the word 'technology' to refer to the highly complex technical devices which increasingly dominate our modern lives. That would give some justification for saying that 'technology' means *modern technology*, the kind of technology we get because of modern science.

In fact, Grant argues that "we westerners willed to develop a new and unique co-penetration of the arts and sciences, a co-penetration which has never before existed. What is given in the neologism [of 'technology']—consciously or not—is the idea that modern civilization is distinguished from all previous civilizations because our own activities of knowing and making have been brought together in a way which does not allow the once-clear distinguishing of them."[13] When technology is understood to include science, and when science is understood to include technology, we need to recognize that this understanding is unique and new in the history of known societies. One of the things that makes modernity modern, according to Grant, is that for the first time in recorded history, science is combined with technology and technology is combined with science. Whereas ancient sciences could and did exist independently of ancient technologies (and vice versa), modern science cannot exist without modern technology and modern technology cannot exist without modern science. This is why *techno-logos* is the perfect way to identify 'modern technology,' and why it can actually make sense to say that we've only had 'technology' for around 200 years. Therefore, it is coherent to complain about technology and still appreciate and use shoes, fire, buttons, or hats. However, because we do not normally intend the word 'technology' to mean only modern technics and techniques, I will use the weird word 'technoscience' (sometimes 'techno-logos' or even 'technologic') to remind us of the weird fact that technology is a uniquely modern word masquerading as something timelessly human. (I'll also use it to refer to 'science and technology' when I want to use one word instead of three!)

Technology Defined as Technologic or Technoscience: technology is the quintessentially modern interpenetration of science, technics, and techniques.

But if technology (defined as techno-logos or technoscience) is quintessentially modern, this means that there is a discontinuity between modern technology and the technics and tools of other, non-modern societies. Of course, inasmuch as modern techno-logos is a subset of technics and techniques in general, there will be continuity between it and any other culture's particular subset of technics and techniques. But inasmuch as other known societies do *not* combine their technical activities with their scientific activities (if any), then modern technology will be discontinuous with those societies' technics and tools. Modern technology is, in this sense, something new under the sun.

This discontinuity is a challenge to the usual story, however. The usual story just defines technology as technics and techniques. However, simply defining modern technologies as technics and techniques

> too easily leaves the impression that our understanding of what constitutes knowing [*logos*] and making [*technē*] is not radically different from that of previous civilisations. In fact, the modern 'technique' may seem at first to suggest the same kind of meaning as what is given in the Greek 'technē,' as if we have simply progressed in efficiency of making. We then attribute our greater efficiency to the modern scientists, who guaranteed the progress of knowledge by clarifying its sure methods, and through that objective knowledge achieved greater ability to make things happen.[14]

This usual story assumes "progressing continuity"[15] between modern technoscience and the technologies and sciences that preceded it. According to this way of thinking, what we currently do in our modern era is essentially the same as what all cultures across history and the globe have wanted to do, it's just that we finally achieved it after a long but inevitable process of cultural evolution (otherwise known as 'progress'; see Chapter Ten). The "time was not accidentally ripe" for other cultures, but it was for ours; "those [other] peoples were not evolved enough to discover the sure path of science," which would have allowed them to have what we have.[16] But of course (so we tell ourselves), they nevertheless would have wanted what we in fact have.

The discontinuity between modern techno-logos and other forms of technics and techniques challenges this usual story of inevitable progress towards modern technoscience. Progressive "'histories' of the race ... close down on the startling novelty of the modern enterprise, and hide the difficulty of thinking it. We close down on the fact that modern technology is not simply an extension of human making through the power of a perfected science, but is a new account of what it is to know and to make in which both activities are changed by their co-penetration."[17] If we think that "nothing much has changed" technologically since the Stone Age,[18] we are assuming that the only way to understand *other* cultures' science and tool-regimes is in terms of *our* own modern and Western technological frameworks. However, such an approach does not do justice to the historical or anthropological record, nor to the legitimacy of any culture other than the one which currently dominates the global mainstream. The usual story also fails to do justice to the intuition that there's something particularly interesting about modern technology. Therefore, one of the 'unusual stories' we will investigate in this book will be the quintessentially modern unification of science and technology, and the contrast between that technoscientific vision and other, alternative visions of technology (and science).

Critical STS Inquiry into Technology

As with science, simply coming to an understanding of what the concept 'technology' means is an exercise in critical thinking and one which comprises an internalist aspect of the field of STS. But STS can also proceed on a presumptively commonsense definition of technology in order to examine other critical questions about its social significance. As an introduction, neither this chapter nor this book can offer a comprehensive overview of all the possible questions one could ask about technology and society. However, I can hazard a summary of one basic and nearly universal issue in the critical theory of technology: *value-ladenness* vs. *value-neutrality*. Posed as a question, this is asking "what is the value of a technology?"

Typically, people are not expected to inquire after the value of technology. The usual story is supposed to give us clear indications that technology is either good or neutral, and in either case, there's no reason to question it further. If technology is either fine and dandy, or a neutral means to whatever ends someone might choose to pursue, then there is no *crisis* of technology that would require us to *critique* it. You may have heard the expression "it's just a tool." This is a way of saying there is nothing critical about a tool. There is nothing more to the tool than our use of it. In itself, the tool presupposes only use-value, or its 'instrumental value,' the value that *we* get out of it or impose upon it. There isn't anything of significance in that tool beyond what we find it useful for. That's why we use the term 'instrument' to describe this kind of value.

Technological Value-Neutrality: the value or importance of a tool is entirely dependent on how it is used; no value is inherent in the tool itself; the tool is only of instrumental value.

This, then, is the view that tools are *value-neutral*. While being 'useful' is itself valuable, it is an entirely open-ended value that can be directed in any way and towards any end that you may choose. Thus, as a computer scientist once said, "The computer does not impose upon us the ways it should be used."[19] The computer, like any other tool, does not tempt us towards any particular action or form of action; it has no mystical powers that control our minds. It is just an object and, in this respect, technological neutrality is an analogue of scientific objectivity (see Chapter Two). The presence of a tool is not an influence on our decisions any more than the absence of that tool is. Rather, a tool simply helps us carry out whatever course of action we have independently decided to pursue.

This neutral approach to tool use could be characterized as 'managerial,' because all the powers contained in any tool are simply potentialities that need to be managed well. Good managers are usually experts who have comprehensive knowledge of all the 'technicalities' involved in using tools to their full potential. If anything untoward or unpleasant appears to result from the use of a technology, these will be considered 'side effects,' yet another dimension to be managed by experts but certainly not something we should find intrinsically troubling (that's why it's called an 'effect' on the 'side').

Because there is nothing fundamentally problematic about the technology itself, experts and managers can fix any side effects that might pop up, and we can rest easy and be at peace. Crisis averted!

Recently, my local city council has been debating whether they should install a bulletproof barrier between themselves and the public gallery in the council chambers of city hall. Some councilors are worried it would send the wrong message to their constituents. Would the barrier tell citizens that their government views them as a potential threat, or that their representatives are no longer easily accessible? Other councilors argue that the barrier should not be 'politicized,' and that technical experts rather than politicians should decide if the barrier should be installed or not. An STS perspective would have us ask: what *are* the social and political implications of installing the barrier or not? Are there value implications intrinsic to the barrier itself, or do they only depend on how the barrier is used, implemented, or viewed? If the decision to install the barrier is a technical rather than a political decision, does that mean the barrier is "imposing on us" *that* (if not how) it should be used? And if the barrier *should* be installed, does this imperative mean that the barrier is not neutral at all, but intrinsically *good*?

Contrary to the value-neutral approach to technology, we might then approach technological value at least at the level of *design*, because designs are always purpose-driven, and purposes are not neutral. A purpose is an end or a goal for something; it is a value, 'the point' of doing anything at all. Nobody designs a pointless technology, because that would make it valueless and useless (remember our first candidate for a definition of technology: how intelligence implements practical purposes). At the most rudimentary level, usefulness is a value, and not all uses are morally equal. The use-value of bulletproof barriers is stopping bullets, whereas the use-value of bullets is not being stopped (bullets are designed to penetrate organic matter at usually supersonic speeds, transferring kinetic energy so as to cause massive tissue damage). Both of these tools are designed to achieve their respective purposes, and if they do in fact achieve those purposes, then we say that they were designed *well*. That is a value-assessment of a *good* design.

Moreover, this value is *embodied* in the tool itself. You don't have to use bulletproof barriers in a particular way in order for them to be bulletproof. Rather, they are bulletproof or they aren't, in and of themselves. They will be bulletproof if you stand in front of them or behind them (even though it's more effective to stand behind them, *if* you want them to be bulletproof *for you*), because bulletproofing is an objective quality. But this means that technologies can be objectively *value-laden*, loaded or burdened with a value dimension that isn't necessarily dependent on how it is used: "all technologies ... embody characteristic values that always go before and define practical aims. Every artifact is the incarnation of some value, positive or negative."[20] Merely at the level of design, then,

Technological Value-Ladenness: the values or importance of a tool are also embodied in the object; some value is inherent in the tool itself; the tool has an intrinsic value in addition to how it is used.

the value of technologies can be assessed—*evaluated*—or critiqued. If it were otherwise (which is what the usual story typically says), then technologies would be neutral to the point of utter functionlessness.

Insofar as STS offers alternatives to the perspective of the usual story, technologies can be critically assessed at the level of values. But this assessment is not necessarily limited to designs. Like a bulletproof barrier that might send the wrong political message, technologies can have value-laden social impacts beyond what they were designed to do. Bulletproof barriers, anyway, are certainly *not* designed to send any political messages! And yet if we study the rôle technologies play in social contexts (including municipal politics), we frequently find that they carry with them value-burdens that exceed their design purposes. For example, when automobiles were invented, nobody invented them for the purpose of facilitating suburban sprawl.[21] Yet this and a myriad of other social changes had to occur because of the ascendancy of the automobile:

> If we are to use automobiles, we need techniques to find oil and other raw materials required to make an automobile and to use it; highways must be built, requiring the manufacture and development of materials needed for the task; service stations must be developed; governments must set up licensing procedures, must take steps to insure safety, legislative and judicial determinations must be made regarding punishments to be meted out to drivers who injure or kill others. In short, the automobile as a means has its own weight, sending us out in search of a vast array of other means.[22]

In a sociological sense, the automobile *does* impose upon us the ways in which it should be used, and in this respect the meaning of the automobile is in excess of the purpose for which it was designed.

One concept that comes up in relation to both design values and unintended technological effects is *affordance*. An affordance is a use that presents itself to a human user, something that it is evidently 'good for.' Designers may want to build affordances into their product to make it obvious and easy for people to use, but objects can have affordances even if nobody designed them that way. Maybe you've picked up a piece of wood before and noticed that it's really good as a walking stick, even though nobody modified it to be one. This is an affordance in the stick; it lends itself to being used as a walking stick, regardless of whether that was the purpose of the stick's shape. Even designed objects can have affordances that the designer did not intend, like how social media lends itself to cyberbullying in ways that email doesn't. All affordances exist only in relation to users, but the affordances are 'there' in the object—in the technology—even if nobody put them there on purpose. In this way, the affordances of technologies, systems, and

Affordance: the use that something offers or lends itself towards, regardless of whether that use was part of the thing's design; cf. p. 265.

environments are one way to think of values embodied in the world that go beyond the mere instrumental use of neutral tools.

However, the usual story does not want us to think that there is more to technology than its use as a neutral tool. Value-embodiment in design and unintended technological effects or affordances are supposed to be relegated to the category of side effects which, as we've seen, are to be managed by experts. The usual story represents side effects as "unanticipated by the linear methods of knowing that were used in designing them."[23] But this approach also *reinforces* the linear method of knowing so often presumed in our technologies: side effects "require still more technological solutions, … [and] have also tended to be justified by the 'bottom line' of quantifiable, material considerations—often measured in money, sometimes in ever-higher speeds or in comparative megadeaths."[24] The usual story is thus a theory of knowledge: it's telling us how we should solve problems, what methods are useful, and what answers are acceptable. These epistemological perspectives can be embodied in technologies. For example, the mechanistic reductionism assumed in much of modern science (see Chapter Two) can be embodied in technologies like factory farms, alarm clocks, workday commutes, and eight-hour shifts. These are all ways of technologically organizing the world as if it's a machine (see Chapter Eleven). Technological systems can therefore embody worldviews in addition to intended and unintended value-implications.

This is why Grant was so concerned with showing how the definition of technology as the quintessentially modern unification of science with tools was historically unique. That combination, he argues, carries with it a modern ideology about knowledge as power (see Chapter Nine). Techno-logos is a theory of knowledge, for Grant, because technology assumes that *we know* that the solution to technological problems is "an even greater mobilization of technology.… More technology is needed to meet the emergencies which technology has produced."[25] We (supposedly) know that *all* solutions are technological solutions, that there is no way to solve any problem without using technology. Even social arrangements are viewed as technological problems, where human beings in society are treated as inert material particles that can be best controlled by employing expert techniques like 'human resource management.'

Grant argues that unlike the sciences of ancient Greece, China, or India, the modern and Western science of tools tries to get "knowledge of the kind that put[s] the energies of nature at [our] disposal, as does modern western physics."[26] Techno-logos is the "modern western will to be the masters of the earth."[27] But this is a value-laden system of ideas worthy of our critical scrutiny rather than an objectively rational and neutral 'solution-based approach' deserving of uncritical acceptance. For Grant, technologic is so pervasive that it is the framework within which we moderns think, and that this framework is embodied in the tools we surround ourselves with. "In each lived moment of our waking and sleeping," he says, "we are technological civilisation."[28] That is an historically unique sociological situation of considerable philosophical importance, and as such, a paradigmatic (though not the only) example of how technology could be examined and assessed as value-laden by STS scholars.

Technology could be value-neutral. It could instead be value-laden, loaded down with a variety of inherent values. It could be value-laden at the level of purposive design, the level of undesigned social effects, the level of worldview embodiment, or even the level of cultural enframement wherein the intellectual baggage of the Western world exerts an incognito dominance over our ways of thinking and acting. These possibilities for technology mirror those we considered for science, namely whether it is objective and absolute or ideological and subjective. These are, I think, the central and core issues at the heart of STS, and this textbook will strive to give you, the reader, a better footing for coming to your own conclusions on such matters. On the assumption that you already come to this text with a background largely informed by what I've been calling 'the usual story' (see the Introduction), I will present *unusual stories* to counterbalance those presuppositions: historical, philosophical, and sociological alternatives to the dominant picture of science and technology that we have received from our cultural surroundings.

The place to start is a social context as far removed from modern technoscience as one can reasonably imagine: human prehistory, wherein neither modern nor Western science or technology existed or would have remotely made sense. That is the topic of our next chapter.

Notes

1 George Grant, *Technology and Justice* (University of Notre Dame Press, 1986), 11.

2 Robert E. McGinn, *Science, Technology, and Society* (Prentice Hall, 1991), 8.

3 Frederick Ferré, "Technological Faith and Christian Doubt," *Faith and Philosophy* 8, no. 2 (April 1991): 220.

4 We will examine this notion of 'the real world' later, in Chapters Six and Seven.

5 The question of unconscious or nonhuman action in 'the outside world' is one of the issues that STS theorists explore. Actor-network theory (most often associated with Bruno Latour; see Chapter Fifteen) holds that nonhuman and inanimate objects are not inert, as the term 'object' would normally suggest, because they can in fact 'act' in a network alongside us humans. Both science and technology are constituted by such networks between human beings and other 'objects,' regardless of whether those objects have a conscious intellect or not. For more, see Bruno Latour, "A Collective of Humans and Nonhumans: Following Daedalus' Labyrinth," in *Pandora's Hope: Essays on the Reality of Science Studies* (Harvard University Press, 1999) or Sergio Sismondo, *An Introduction to Science and Technology Studies*, 2nd ed. (Wiley-Blackwell, 2010), chapter 8.

6 Wenda K. Bauchspies et al., *Science, Technology, and Society: A Sociological Approach* (Blackwell, 2006), 7; paraphrasing Wiebe E. Bijker et al., eds., *The Social Construction of Technological Systems: New Directions in the Sociology and History of Technology* (MIT Press, 1987).

7 You might try to avoid this ghostly problem by arguing that your mind is not immaterial, but completely made out of matter—exactly the same stuff that your body is made out of—and thus capable of interacting with it as a tool. But in that case, your mind would lack the 'inside' dimension that distinguishes it from the 'outside' world that you can exert

leverage over with your tools. That is, you'd be committing yourself to saying that you have effectively the same level of consciousness as a lettuce. The lettuce's body (its tool) would be the same as the lettuce's mind. In this case, technology wouldn't involve intelligence or purposes at all. This is in addition to the fact that you still can't bend a spoon with your mind, even if your mind is made entirely out of matter. Far be it from me to claim that I've solved the mind–body problem, but material (un)consciousness is not the way to do it.

8 I do not mean to suggest, however, that the concept of 'nature' is not controversial within STS. It rather is, both in being defined independently of human action (for more on that, see Chapter Fifteen) and in having certain (usually positive) value connotations (for more on that, see Chapter Ten). For the record, I do think it is good (or at least fortunate) that I have my two ears and bad that I have incurable tinnitus, but that evaluation is not relevant with respect to our present project of defining what technology is.

9 I certainly do not want to imply that technologies are "unnatural" in the value-negative sense. That is yet another issue of STS concern, but not one which concerns us at the level of definition. In fact, one could argue that hearing aids (like many, but not necessarily all technologies) are *required* of us *by nature*, at least in cases where hearing loss has occurred (and especially by no fault of one's own). In such cases, technologies step in to address a deficiency in one's own nature, even replicating or restoring a natural process. This view of technology as 'natural' in an extended sense is characteristically Greek, and we will consider it in more detail in Chapter Seven.

10 McGinn, *Science, Technology, and Society*, 14; emphasis removed.

11 McGinn, *Science, Technology, and Society*, 14.

12 Grant, *Technology and Justice*, 11.

13 Grant, *Technology and Justice*, 12.

14 Grant, *Technology and Justice*, 12.

15 Grant, *Technology and Justice*, 12.

16 Grant, *Technology and Justice*, 13.

17 Grant, *Technology and Justice*, 13.

18 Elizabeth Skakoon, "Nature and Human Identity," *Environmental Ethics* 30, no. 1 (Spring 2008): 49.

19 Quoted in Grant, *Technology and Justice*, 19.

20 Ferré, "Technological Faith," 221; emphasis removed.

21 James Howard Kunstler, *The Geography of Nowhere: The Rise and Decline of America's Man-Made Landscape* (Simon and Schuster, 1993).

22 Vincent C. Punzo, "Jacques Ellul on the Technical System and the Challenge of Christian Hope," *Proceedings of the American Catholic Philosophical Association* 70 (1996): 24.

23 Ferré, "Technological Faith," 222.

24 Ferré, "Technological Faith," 222.

25 Grant, *Technology and Justice*, 16.

26 Grant, *Technology and Justice*, 13.

27 Grant, *Technology and Justice*, 12–13.

28 Grant, *Technology and Justice*, 11.

II

Prehistory

Chapter Four

Myth and Ritual

NOW THAT WE HAVE OUR PRELIMINARY DEFINITIONS OF SCIENCE AND technology out of the way, we are ready to begin this book's short intellectual history of modern Western technoscience. We are going to start as far back as we can reasonably go: the prehistory of our human species, when neither modern nor Western science or technology existed or would have remotely made sense to people who were nevertheless fully human. This sharp contrast will highlight the uniqueness of technoscience as a social phenomenon. We will try to understand a cultural context within which *myth* is the primary vehicle of human intellectual endeavor, and cultures that did not aspire to Western scientific rationality. Rather than assume that the "desire for technological innovation [is] universal in human society," we will consider human "ways of life that simply do not place great value on technological change"[1]—where *rituals* have more importance than the latest digital gadgets, for example. We will try "to understand a different way of life" from the one of our own time,[2] so that we can understand technoscience as a social phenomenon. Otherwise, we will be 'unscientific' in our understanding of science, because we will be dismissing anthropological, historical, and sociological facts in order to hide "the startling novelty of the modern enterprise."[3]

I am proposing, then, a contrast between myth and ritual on the one hand, and the technoscience of our age on the other hand—or at least with the ideal of technoscience given to us by the usual story. If we can gain a sophisticated understanding of myth and ritual, we will begin to sophisticate our understanding of technoscience. Historically speaking, myth and ritual clearly predominate in 'prehistoric' human societies. The notion of 'prehistory' is problematic, however for at least three reasons. First, how can something be historically *pre*-historic? Was there a time before history? This irony is due to an ambiguity in our use of the word 'history.' Colloquially, we refer to all of the past as 'history,' no matter how far back we go. Professionally, however, historians define 'history' as the intellectual reconstruction of the past based on *written records*. Technically, therefore, 'prehistory' is that period of the past (or present) before the invention of writing. Reconstructing that period of time requires archaeological and anthropological methods, because there are no written records to rely on from back then.

The second reason why the notion of prehistory is problematic is because the prefix "pre" often brings with it a connotation of anticipation, as if prehistoric cultures were waiting or hoping for written history but just 'hadn't made it yet' to our level of cultural 'advancement.' This is an inappropriate and inaccurate way to understand different cultures, because it falsely projects our peculiar cultural assumptions about 'progress' or 'development' onto people who were not trying—and may not have wanted—to think and do things the way we might currently. For example, the "Polynesian islanders studied by anthropologist Raymond Firth in the 1920s ... acknowledged the superiority of the 'white man's' artifacts, [but] were not envious of foreign technological success nor did they seek to emulate it."[4] Not every human culture has wanted to be modern. Therefore, in considering prehistoric attitudes towards thinking and doing, we must resist the temptation to assume that such people were trying to be modern like we think we are.

Prehistory: the period of human history that lacks written records and requires archaeological and anthropological methods to study.

The third reason why the notion of prehistory is problematic is because it suggests that the people who came before written history weren't really people. 'History' sounds suitable for human beings like ourselves, while 'prehistory' sounds suitable for what the usual story calls 'savages,' 'cave men,' or 'the missing link' between ourselves and our subhuman, apelike ancestors. However, defining who was human and who wasn't was one of the intellectual tools used to justify colonization and slavery.[5] In this and other respects, Western archaeological science has contributed to "stories that have dehumanized Indigenous people" of the western hemisphere,[6] and also Africa and parts of Asia. The usual story sees prehistoric people as "a part of nature, not culture, [and] as simple hunter-gatherers lacking creativity, science, and intellectualism."[7] It is far too common to assume that Indigenous people were and still are unsophisticated, ignorant, and crude 'Others' compared to the scientifically enlightened and technologically superior Europeans who conquered or colonized virtually every part of the world. We will return to the theme of Indigeneity at the end of this chapter, and colonialism and scientific racism in Chapter Ten.

Myth and Technoscience

The usual story can be an obstacle, therefore, to understanding the differences in perspective across human cultures. This is because it proposes a story of "progressing continuity" between modern Western technoscience and the thought and action of other (and especially older) human cultures.[8] It sees itself as a natural outgrowth and culmination of what it assumes human beings have always been like and always wanted. For example, the science fiction writer Arthur C. Clarke famously said that "any sufficiently advanced technology is indistinguishable from magic."[9] Similarly, some T-shirts and memes declare that "science is like magic, but real." There is an irony to this continuity, however, for if it's true, then there is no fundamental difference between myth and science, or between ritual and technology. And yet the usual story *also* wants to say that myth is the exact opposite of scientific 'fact' and that ritual (especially if it's associated with magic) is the exact opposite of 'effective' technology. Technoscience truly works, whereas myth and ritual don't—right? After all, we devote entire television programs to "mythbusting" legends and superstitions.[10] We do not need a dictionary to tell us that myth means "a widely held but false notion" or a story featuring "supernatural or imaginary persons" who don't actually exist.[11] We also 'already know' that rituals or magic are empty gestures or repetitive and ineffective performances that fail to achieve any meaningful goals. Traditional cultures can be erased, not because they lacked a sophisticated intellectual life (they didn't), but because their intellectual life was (supposedly) *wrong*—a mistaken, mythological attempt at science—waiting to be corrected and replaced by the technoscience of colonists. This tension in the usual story—between progressing continuity and utter incommensurability between prehistoric cultures and modern technoscience—requires a second look.

Robert Segal, a scholar of religious studies, defines myth "as simply a story about something significant."[12] This emphasis on significance distinguishes myths from lighthearted legends or folk tales,[13] but for our purposes we only need note that myths are supposed to convey something important or meaningful to their intended audiences; myths are not supposed to be value-neutral. Segal also notes that while myths may "express a conviction" about what some people may believe to be true, factual truth or falsehood is not essential to the definition of myth. Rather, a myth need only be a story that is "held tenaciously by adherents,"[14] that tenacity presumably being connected to the perceived significance or profound value imparted by the story itself—another kind of truth, perhaps. Finally, the main figures in mythological stories are personalities, such that impersonal forces (like gravity) will not play central roles in myths. These personalities are not necessarily human, however; they may also be divinities or animals, plants or even landforms.

Myth: a story about something significant, held tenaciously by its adherents and featuring personalities as its main figures.

Rituals, unlike myths, are actions rather than stories, even though they may often (if not always) be connected to myths. Though they are actions, however, rituals are never straightforward attempts at exerting leverage over the material world; simply scratching an itch is not a ritual. Because rituals are physical behaviors that do not display any clear intention to manipulate physical matter, it would be basic and easy to view them as misdirected or mistaken techniques. But a less simplistic explanation would be to see rituals as *symbolic behaviors*. There is always a tantalizing gap between a symbol and what it symbolizes; the two are never the same thing. For example, the flag of a country is never the same thing as the country itself, although there is a rather mysterious (you might say mythological) connection between a flag and its country. Waving a flag is a symbolic action, and burning a flag is a rather different sort of symbolic action, but they both convey particular meanings. Obviously, however, they are not attempts at directly modifying the physical state of a country.

A ritual, therefore, is a kind of action that conveys a particular set of meanings (burning a flag is not necessarily a ritual, but there are rituals surrounding the display of flags), but as a symbolic action, there is always a distinction between what the ritual behavior looks like it's doing and whatever the ritual is trying to do. For example, the ritual washing of a Christian priest's hands before (the ritual of) Holy Communion merely gets the priest's fingers wet; as actual handwashing it's very ineffective at sanitization. But this is because ritual hand washing isn't actually supposed to be about hand sanitization, but rather about spiritual purification (whatever that might be). Another example is the ritual of adoption of a child by having the child crawl through the mother's skirts as if she were giving birth. In the forthright words of Ludwig Wittgenstein, "it is crazy to think there is an *error* in this and that she believes she has borne the child."[15] Clearly, the ritual is symbolic of the child's reception into the family, and not a confused attempt at actually giving birth to a child.

Ritual: procedural action that conveys meaning through symbolism.

The scholarly definition of myth, therefore, makes the factual truth or falsehood of myth a secondary question at best, whereas ritual's symbolic function has more to do with meaning than technological effectiveness at manipulating matter. This subsumes ritual under myth's primary function of conveying meaning. In both cases, neither appears to be about the same thing technoscience is about. Myth appears *not* to be an attempt at giving a factual description of the physical world any more than ritual is an attempt to effectively alter the physical world. It's only when we assume that myth and ritual are preliminary attempts at achieving technoscience that myth and ritual appear to be false and useless technoscientific failures: "Baptism as washing—this is a mistake only if magic is presented as science."[16] The tension in the usual story between

Broad Understanding of Myth: includes legends, folktales, oral traditions, fiction, ceremonies, rituals, magic, and religion.

progressing continuity and the denigration of myth and ritual is solved, therefore, by denying that myth and ritual are essentially mistaken forms of technoscience.

However, the conflict between science and myth has a fairly long and complex history:

> Nineteenth-century theories [of myth] tended to see the subject matter of myth as the physical world.... Myth was typically taken to be the 'primitive' counterpart to science, which was assumed to be wholly modern. Science rendered myth not merely redundant but outright incompatible, so that moderns, who are by definition scientific, had to reject myth.[17]

As Segal later notes, however, the Western tradition's challenge to myth goes as far back as the ancient Greek philosophers (see Chapter Seven). The usual story, which dismisses myth "for explaining the world unscientifically,"[18] has considerable sociological momentum, therefore, and won't likely go away simply because it doesn't make philosophical or anthropological sense.

But myth hasn't gone away either. Segal argues that one of the problems with complaining that myth is unscientific and doomed to be "a victim of the process of secularization that constitutes modernity"[19] is that this view "conspicuously fails to account for the retention of myth in the wake of science."[20] The onward march of science into the twenty-first century has not irresistibly and inevitably led to the defeat and obliteration of superstition or prejudice (these are often associated with myth and religion, perhaps unfairly), but has rather seen the rise of religiously motivated terrorist movements and populist nationalist political movements, both to the chagrin of self-styled scientific rationalists and liberal progressives. While the usual story might explain these aberrations away as merely temporary setbacks, more puzzling is the fact that myth and science exist *alongside* each other in our time, often without any apparent conflict. Hollywood films are in the business of telling stories, the overwhelming majority of which are *not true*. Fiction quite dependably produces the most successful summer blockbusters, and yet nobody seems to care that such stories are 'unscientific.' Even science fiction replicates classical mythological archetypes—such as the hero's quest—spawning fandoms whose devotees display a commitment that at least borders on the religious. And yet the most fervent 'science nerds' seem perfectly content with enjoying science fiction and fantasy realms, apparently unconcerned with these mythological and ritualistic features. Is fandom really a failure of technoscientific rationality? Is there something problematic about enjoying a *Star Wars* film or reading *The Lord of the Rings*? It is difficult to see why there would be! Rather, it appears that broadly mythological storytelling is still highly valued in our contemporary world, even to the point of being *more interesting* (measured simply by public participation and discretionary spending) than factual scientific literature or documentary cinema. All claims to strangeness aside, truth may not be as satisfying as fiction.

Not only is science less easy to sell than myth, science is also the new kid on the block in comparison to myth, which might account for some of myth's aforementioned staying power. When Segal mentioned earlier that "myth was typically taken to be the 'primitive' counterpart to science," the word 'primitive' reminds us that (according to our current scientific understanding) the human species is about 300,000 years old, whereas science (in the modern sense) is at best only 500 years old. While we only have written records of myths that go back roughly 6,000 years, it is highly likely that mythology goes as far back in time as human beings do. That is to say, while only a very small fraction of human history could be reasonably considered culturally scientific, the overwhelming majority of human history should reasonably be considered culturally mythological (even if we exclude our contemporary societies from that categorization, which we shouldn't). The cultural resilience of myth across the aeons is a fact in need of explanation—at least if we are inclined to seek 'scientific' explanations of the world as it appears to us. For 99.998 percent of our species' past, what was myth good for, if it wasn't trying to offer factually accurate descriptions of the physical world or produce increasingly effective technologies?

Social Explanatory Power

If we can account for the longevity of myth, we will be able to explain what it is actually trying to do rather than define it in terms of what it wasn't trying to do. We have seen that the myths and rituals of prehistoric human societies weren't trying to replicate technoscientific rationality or efficiency. Technologically, of course, prehistoric peoples knew how to survive and flourish in their comparatively wild environments far better than a contemporary city dweller (like myself) would. From making fire without matches to killing large herbivores with flint-tipped darts, prehistoric people were clearly intelligent and skilled. While this technical proficiency might have been situated within a broader ceremonial or even religious context, the technical operations themselves weren't symbolic rituals or understood as myths instead of facts. As David Lindberg points out, "this brings us back to the distinction between technology and theoretical science [see Chapter Two]. It is one thing to know how to do things, another to know why they behave as they do.... [P]ractical rules of thumb can be effectively employed even in the face of total ignorance of the theoretical principles that lie behind them."[21] Just as technology can be successfully used without a scientific or theoretical understanding of how it actually works, techniques and tools can be successfully used without believing that mythology provides factual explanations of how they work. Prehistoric people knew how to do things very well, but myth was not the science behind their technologies. If you have an intellectual desire to understand the scientific principles that lie behind successful technology, mythology is ill-suited for satisfying that desire. On the flip side, in as much as myth requires value-communication, conviction, and personification, it would be "crazy" to think that it offered a systematic, theoretical account of material properties the way science tries to.

Restating the above in a provocative way, Wittgenstein says that the surprising thing about prehistoric people "is that [they] do not act from *opinions* [they] hold about things."[22] What he means by "opinion" here is equivalent to what we defined as *theory* in Chapter Two: "a deep understanding of how the world works."[23] Wittgenstein complains that the nineteenth-century "account of the magical and religious notions of [people] is unsatisfactory: it makes these notions appear as *mistakes*.... But *none* of them was making a mistake except where [they were] putting forward a theory," a model or intellectual picture of what was supposed to be the actual workings of the physical universe.[24]

Myths are not, therefore, explanations of things which help us improve our control over material things. They're more closely connected to things like broken hearts. Broken hearts aren't literally malfunctioning blood pumps, but are symbols that refer to a deeper level of human experience. If you've had your heart broken, you'll need your friends to comfort you emotionally, not explain how hearts or hormones work: "Every explanation is an hypothesis. But for someone broken up by love an explanatory hypothesis won't help much.—It will not bring peace."[25] Myths operate at levels like emotional comfort, for example, "bringing peace" at least in a psychological (if not also in a spiritual or religious) sense, rather than being technoscientific descriptions of scientific realities. Myths are not theories which attempt to address the human need to know facts; rather, they address much more profound human needs for meaning and significance.

Consider the following inspirational text:

> A modern mythology played in real time. Life, unscripted. The greatest drama. Every night a war. Giants colliding. Unmasking life's great mystery. What it means to be changed. We are links in a chain. A universal code instilled upon generations. Loyalty and courage, patience, balance. Stories of change, alive and igniting our hearts and binding our souls. A game played with artistry enhanced by technology. Man, machine, the beast transforming. Tonight you bear witness to years of evolution driven by obsession, every instinct, every impulse. This is our game, our code.[26]

This set of words is not a theory. It is supposed to get 18,000 people pumped for hockey—it was the narration accompanying the video montage shown at the beginning of home games for my city's NHL hockey team. It doesn't matter if this narrative is 'true' or 'false'—it's not even clear if it *could* be true or false! But regardless of truth or falsehood, it may speak to you at a deep level, depending on how much you love the sport or the team. If you are a fan, these words might help motivate you to cheer along with the crowd in the arena. That's what they're supposed to do. They're supposed to 'speak' to your 'heart.'

Lindberg argues that this sort of inspiration or motivation is the function of myth: "explaining, and thereby justifying, the present state and structure of the community, supplying the community with a continuously evolving social charter."[27] The meaning-full stories of myth are shared by a community, and inspirationally hold that community together. The hockey narrative quoted above is geared towards motivating hockey fans in a particular city and surrounding geographical area to feel a social bond with thousands of otherwise complete strangers. Hockey in general functions as a catalyst for unifying the citizens of my city. We like to think of ourselves as a 'hockey city,' even though we all know that we don't literally have hockey in our DNA. Even if people can't attend the game live, people all over the city will be wearing their team colors, waving flags, and even the transit busses will proclaim their allegiance. Not everybody in my city is a hockey fan (plenty of people find hockey annoying here), but it's almost impossible to imagine Edmonton, Alberta without the Oilers (now you know where I live). Even the name of the team uses a key driver of the local economy—oil—as a symbol around which citizens can unite. Hockey slogans here proclaim that "this is Oil Country" even when everybody knows that oil has nothing to do with hockey, scientifically speaking. But oil is a socio-economic symbol just as much as hockey is, and the two of them supply a large part of the meaning which is supposed to hold my local culture together.

Myth, therefore, is the "principal repository" of the "values of the community." It may "legitimate current leadership roles, property rights, or the present distribution of privileges and obligations."[28] The meaning of mythological stories primarily explains a person's place in the current web of social relations, to provide everyone with a powerful sense of social identity, regardless of whether we are people of hockey, people of the oil, or people of something else. Rather than providing a theoretical picture of the physical world, myth is supposed to provide a compelling explanation of your *social context* to you. Rather than being technoscientific, myth is sociological.

What, then, about all those myths which talk about the gods or the creation of the world? Aren't those attempts at describing physical facts instead of explaining social contexts? The answer comes back to the issue of meaning and significance. Meaning is a function of connections; words only have meaning if they have connections to other things. For example, the word "dromology" was meaningless to me until very recently when someone pointed out that it is related to the Greek word *dromos* for racetrack (as in "velodrome," a bicycle race track), after which it made sense as a systematic investigation of speed. The word "whale" only means something to you and me because we unconsciously connect it to very large marine mammals (which are, in turn, words connected to other things). But simply stating the word "whale" isn't going to mean very much at all unless it is connected to other words in a grammatically correct sentence: e.g., "there is a whale in the back of your truck."

Similarly, myths connect you to your community members, those members to a cultural or political structure, and those structures to 'the world' or even 'the entire universe' and all that may be in it, including the spirits (who may or may not be a part of your universe). Myths attempt to convey the meaning of your life by connecting you to all the things imaginable (including the gods and the creation of all things). Myths are the vehicles for a 'worldview,' a comprehensive schematic through which all of our experiences are interpreted and explained in a way that makes sense to us (even if we aren't aware that we have a worldview). Regardless of what you're experiencing—be it a flooding river, the death of a loved one, sexual tension, food insecurity, the stress of parenthood, violence and war, or the meaning of life itself—the process of making these things fit into a grand scheme of things is the process of meaning-making and signification. This is the function of myth: "to offer satisfying explanations of the major features of the world as experienced by the community."[29] *Myth is supposed to make your world make sense to you.* Your 'world' is so much more than physical facts. It could include spiritual beings. It could even include technoscience! In fact, the usual story could itself be a myth, in as much as it presents technoscience as the meaning and significance of all things (but more on that later).

Experiential Explanatory Power

The philosopher John Gray claims that myths are not a symptom of humanity's immaturity, as if in our species' childhood we explained things using fictional stories but eventually grew out of that stage as we became more scientific. Yes, of course we could accurately describe hockey in terms of human kinesiology, the friction coefficient of frozen water, and the velocity of small rubber discs, but that would (at best) only be part of the picture. Natural scientific description would not be able to account for the meaning of hockey, even though natural scientific descriptions of hockey could be assessed as 'true' or 'false' depending on how well they correspond to physical facts about hockey.

But how would we assess the merits of myths and their meanings if we cannot assess them by their scientific correspondence with physical facts? Gray suggests the following:

> Myths can't be verified or falsified in the ways theories can be. But they can be more or less truthful to human experience, and I've no doubt that some of the ancient myths we inherit from religion are far more truthful than the stories the modern world tells about itself.[30]

What might it mean to be "truthful to human experience"? Lindberg suggests that myths are experientially true when they make sense or are psychologically satisfying within a human frame of reference. Theories do not operate within a human frame of reference; theories are abstractions that present reality as if you aren't there. Theories give a model or 'world picture' of how things are 'objectively,' as if they were ironically seen from the view from nowhere (see Chapter Two).[31] This is probably why you know more about science than you do about scientists. In science, we like to pretend that the scientists aren't there doing anything. Scientists are the invisible or hidden beings that produce scientific knowledge. They're not in the picture; only the truth is. That's (supposedly) the universal perspective, the perspective of the universe. An experiential perspective, by contrast, is not how reality would look from a god's-eye point of view, but how it actually looks from our human point of view (assuming we aren't gods).

The View from Nowhere: a perspective that imagines perception from all perspectives, not one; a universal, objective, impersonal viewpoint; not really a perspective at all.

Therefore, from an experiential perspective, the world isn't round unless it actually looks that way to you. And unless you are currently an astronaut or alien in orbit around the planet, it doesn't. This doesn't mean the world isn't actually round. It only means that the world doesn't look round and that any extrapolations we might want to make about the world's actual roundness are abstractions derived from observational experience (which is how classical Greco-Roman and medieval European scientists postulated the Earth's roundness without the aid of space flight). Experientially (and chronologically), the subjective or first-person human frame of reference *always comes first.*

Think of this like a first-person shooter video game; when you play it, you don't see your character's head or face, just their hands 'out front,' as if your character's hands were your own. This is how myths work and offer psychological satisfaction: rather than providing abstract facts and expecting those to be meaningful to people, myth uses the first-person perspective that every human being intuitively and inevitably sees the world through. So for example, a myth about the universe would probably describe the Earth as flat or the stars as tiny, not because that's how they are objectively, but because that's how they are subjectively, as you see them. But the framework of your own experience is not limited to how things simply appear to your five senses. It also includes all the elements of your everyday life, things that happen to almost everybody and the mundane objects that are ready to hand in your culture. "Thus the beginning of the universe is typically described in terms of birth," because childbirth is a fairly common human experience. Another example is an "African myth [which] describes the earth as a mat that has been unrolled but remains tilted, thereby explaining upstream and downstream—an illustration of the general tendency to describe the universe in terms of familiar objects and processes."[32] Taking another everyday activity as its point of departure, a "Babylonian myth attributes the origin of the world to the sexual activity of Enki, god of the waters [who] impregnated the goddess of the earth or soil, Ninhursag."[33] None of these stories have to be literally true in order for us to 'get it.' They each connect grand themes of the cosmos with things people are familiar with—like birth, sex, and floor coverings—and if they work in that way, then they're truthful in Gray's sense.

An Experiential Perspective: how things appear to a perceiving subject; first-person perspective; the personal relevance of everyday experiences and objects.

Myths don't have to be about the cosmos, either. The story of the Buffalo Woman among various Indigenous nations of the North American Great Plains tells of a hunter who married a woman who was actually a shapeshifting bison. He had a son with her, but his family did not accept them and drove her and her son out of the community. Her husband followed them to Buffalo Nation, where he was accepted into their community and transformed into a bison. If this story were understood as a scientific theory, it would be false; humans and bison cannot transform into each other or successfully interbreed. But positive and negative family relationships are real, more basic in terms of subjective human experience than reproductive incompatibility between many species, and thus make immediate experiential sense. Understood as a myth, the story means that hunter societies and the prey on which they depend are connected through a powerful kinship relation that conveys values like gratitude and reciprocity. In this sense, the myth is truthful regardless of so-called scientific descriptions of abstract biological facts.

The story of the Buffalo Woman, and the characters of Enki and Ninhursag mentioned earlier, also exemplify another familiar aspect of life that characterizes mythological accounts: the personal dimension. Lindberg claims that mythological accounts project "human or biological traits onto objects and events that seem to us devoid not only of humanity but also of life."[34] (This emphasis recapitulates the personality component of Segal's earlier definition of myth.) When the wind blows through the trees, we often say that the trees are "sighing," because the sound of the wind in the trees is reminiscent of the sound humans make when they sigh. But did you notice what happened in the sentence which initially referred to wind blowing in the trees? "Blowing" is also a personification of the wind; the wind doesn't actually have a mouth with which to blow air through the trees, but we commonly speak in this way using a compelling metaphor derived from human experience. Anthropomorphizing happens in modern speech even when we don't realize it.

Moreover, anthropomorphizing is another way in which myths make sense. Lindberg points out that there is "an inclination in preliterate cultures not only to personalize but also to individualize causes, to suppose that things happen as they do because they have been willed to do so."[35] Random occurrences usually don't have meaning within a mythological framework. Therefore, mythological "conceptions of space and time are not (like those of modern physics) abstract and mathematical, but are invested with meaning and value drawn from the experience of the community."[36] Modern technoscience is not only different from myth and ritual; it would be *meaningless* and *ineffective* as an explanation within a social context dominated by myth and ritual. Stated boldly, impersonal 'explanations' don't really explain anything to people because *people don't experience things impersonally*. This is why Wittgenstein said earlier that a scientific explanation will not bring peace to someone who's been heartbroken by a lost love. Myths work better because they meet people where they're at. That's why Gray could say that "some of the ancient myths which we inherit from religion are more truthful than the stories the modern world tells about itself." You may not necessarily agree about ancient myths being more true than modern ones, but at least you should be able to see how myth and science can be different without myth being automatically false in comparison to science.

Non-Material Needs

Sociologically, we have seen the function of myth as communicating the central values of a community to its members. Experientially, we have seen the success of myth in communicating those values in a non-abstract and personal way. In neither case is myth a clear threat to what technoscience sees as its own role; myth and technoscience appear to have quite separate goals. We might say that technoscience is directly concerned with the physical world, whereas myth is concerned with a broad sociological and psychological sense of 'world': the realm of values, social connections, and psychological needs, even though both of these aspects incorporate the physical world at a variety of levels. For example:

> Burning in effigy. Kissing the picture of a loved one. This is obviously *not* based on a belief that it will have a definite effect on the object which the picture represents. It aims at some satisfaction and it achieves it. Or rather, it does not *aim* at anything; we act in this way and then feel satisfied.[37]

Wittgenstein isn't saying that mythologies consciously set forth certain sociological and psychological goals and then construct myths or rituals to achieve those goals. It appears to be more spontaneous than that. When you're frustrated, you just hit a pillow; when you burn your finger, you curse. Neither these actions nor these words 'explain' anything; they rather *express* something, and expression is one of the things humans do in addition to explaining things. Expressions like these, be they words or actions, may (or may not) help us feel a bit better. But they don't necessarily alter the physical world. If anything they alter our mood or help us cope with what is happening in our life (physically or otherwise). Mythical speech and ritual behavior, therefore, address *non-material needs*, even when those needs are related to material things.

Wittgenstein makes this point most strikingly with respect to magic:

> In magical healing one *indicates* to an illness that it should leave the patient. After the description of any such magical cure we'd like to add: If the illness doesn't understand *that*, then I don't know *how* one ought to say it.[38]

Magic isn't an attempt to be as effective as using pharmaceuticals or surgery to combat a disease. It's rather an expression of a community's evaluation of that disease and the expression of a desire that the disease 'go away.' Magic is an articulate and complexly networked way of saying 'fuck off' to cancer (if this sounds excessively vulgar, keep in mind that it's also the name of a 501(c)3 nonprofit organization using "wit, edge, and humor [to change] the way people think and talk about cancer ...").[39] Magic is a form of wish expression.[40] On this account, therefore, sacrificing a chicken and sprinkling its blood on your fields to ensure fertile soil isn't necessarily a confused attempt at using fertilizer or a mistaken understanding of soil science. Rather, it can be an expression (to the universe, to the gods, to whomever the connections would be most meaningful) that you really would like the soil to be fertile—and if the gods don't understand *that*, then how else are you supposed to say it? According to Wittgenstein, human beings do this sort of thing because it feels right—because it's just the sort of thing that human beings do. Humans are a "ceremonious animal,"[41] and the "purpose of any ceremony is to build stronger relationships or bridge the distance between aspects of our cosmos and ourselves."[42]

Therefore, the fact that magic 'doesn't work' completely misses the point. When a child pretends to have a tea party with its teddy bears by serving up empty air in tiny plastic cups, the child knows that the point is not to actually drink tea. This doesn't mean that myth or ritual are like immature child's play, something that humanity grows out of as it 'matures' into Western technoscience. It means that even little kids understand the difference between symbolic and literal expression, and so adult Westerners really ought to be able to know the difference too. This is why, Wittgenstein says, it's stupid of us moderns to think that rituals like the rain dance or a sun dance are actually about making it rain or making the sun rise:

> I read, amongst many similar examples, of a rain-king in Africa to whom the people appeal for rain *when the rainy season comes*. But surely this means that they do not actually think he can make rain, otherwise they would do it in the dry periods in which the land is "a parched and arid desert." For if we do assume that it is stupidity that once led the people to institute the office of Rain King, still they obviously knew from experience that the rains begin in March, and it would have been the Rain King's duty to perform in other periods of the year. Or again: towards morning, when the sun is about to rise, people celebrate rites of the coming of the day, but not at night, for then they simply burn lamps.[43]

It is stupid to take myths and rituals literally, *and* it is stupid for us to think that prehistorical people did take them literally: "it never becomes plausible that people do all this out of sheer stupidity."[44] In other words, it's "not that ancient people told silly stories which we, because we are so smart, should not take literally or factually but symbolically or fictionally. It is the ancients who knew how to tell a good metaphorical story (a parable, if you prefer) and we moderns who are silly enough to take them factually."[45] We sell ourselves and others short when we think that non-Western people were duped by magic or didn't understand that their rituals served larger societal and symbolic purposes. Ritualistically stabbing your enemy by sticking your knife into a picture of them doesn't mean you don't know how to kill someone.[46] Prehistoric people knew how to *literally* kill their enemies, and so they knew the difference between actually describing or manipulating the physical world, on the one hand, and engaging in myth and rituals, on the other. Myth and rituals couldn't sensibly have been literal to them.

The real mystery is why the usual story wants us to think that myth and ritual are literal, and why technoscience needs to replace them with true science and effective technology. Thanks to its assumption of progressing continuity, the usual story suggests that *non-material* satisfactions (i.e., myths and rituals) should be understood as attempts at *material* satisfactions (i.e., science and technology). Even more oddly, the usual story suggests that *material* satisfactions (i.e., modern technoscience) are themselves *non-material* satisfactions—i.e., the most meaningful story of truth and freedom ever told. Tantalizingly, Gray claims that

> [s]cience hasn't enabled us to dispense with myths. Instead it has become a vehicle for myths—chief among them, the myth of salvation through science. Many of the people who scoff at religion are sublimely confident that, by using

> science, humanity can march onwards to a better world.... Unbelievers in religion who think science can save the world are possessed by a fantasy that's far more childish than any myth. The idea that humans will rise from the dead may be incredible, but no more so than the notion that "humanity" can use science to remake the world.[47]

Perhaps the reason why Gray thinks modern technoscience is a worse myth than 'religious' ones is that religions at least know that they're religious. Like Lewontin's argument about the social function of science (see Chapter Two), the usual story about technoscience may be a myth that mistakenly thinks it isn't actually a myth.

We started this chapter with the idea that myth and science are different things, and that rituals and technologies are different things, but now they don't look so different, at least not if the usual story about Western technoscience is itself a myth! Does this mean that there is progressing continuity which leads from prehistory into modern Western technoscience after all? Not exactly. Myth and ritual do not transform over time into more objectively true and effective forms of science and technology, even though that's what the usual story says. Rather, cultural changes occur which eventually result in a form of rationality that sees itself as different and better than myth and ritual, even though it's still covertly mythological and ceremonial. This form of rationality is the usual story, and its own confusion about itself—and that which it sees as other than itself—is why it both denies that it's superstitious *and* claims to replace magic with better, technological magic. Myth and ritual *are* different from the usual story because they *aren't* confused about what they are.

Indigeneity and STS

We've been examining myth and ritual in the section of this book devoted to prehistory, even though we've discovered that myth and ritual are very much a feature of contemporary society, modern Western technoscience included. So just because we say something's 'prehistoric' doesn't make it out of date or irrelevant to our lives right now. In surprising ways, 'prehistory' is with us right now. We've also occasionally noted that Indigenous cultures provide a wealth of examples of myth and ceremony, but likewise they are not prehistoric people nor out of date and irrelevant to contemporary life. Inversely, the "colonial assumptions" which subjugate and oppress Indigenous people the world over "are by no means only a feature of the past."[48] It behooves us, therefore, to carefully consider the relationship between Indigeneity and modern Western technoscience, although this topic, along with gender, sexuality, race, and colonialism, is still at the periphery of STS.[49]

First, we should remember that categories like 'myth' and 'magic' are Western ones. 'Myth' comes from a Greek word for oral speech, while 'magic' comes from a Persian word for a member of the learned or priestly class (a 'mage'). For thousands of years, epistemological differences between myth, magic, writing, and science have been weaponized by Eurasian empires, manufacturing a hierarchy of value with (eventually) modern colonial science and technology on top and the supposed prehistoric superstition of illiterate savages at the bottom.[50] Western technoscientific rationality sees itself as different than myth and magic because it's supposedly better. Western notions of myth or ceremony are themselves problematic constructions of empire. Therefore, Billy Frank, Jr., late Chairman of the Northwest Indian Fisheries Commission and a member of the Nisqually people of the South Coast Salish, pointedly said that

> I don't believe in magic. I believe in the sun and the stars, the water, the tides, the floods, the owls, the hawks flying, the river running, the wind talking. They're measurements. They tell us how healthy things are. How healthy we are. Because we and they are the same.[51]

To identify Indigenous knowledge with myth or magic within the Western framework is to denigrate it, and so Frank's statement rejects that categorization. Yet he combines what we might call scientific concerns (e.g., measuring the sun or water) with what we might call myth (e.g., wind talking, being one with hawks or rivers), a combination which is anathema to Western technoscience's self-perception. In this way, Indigenous knowledge claims can subvert the categories of colonialism embedded in the usual story.

Another problem with naively associating Indigeneity with mythological forms of expression is that Western cultural frameworks relegate both to the past (e.g., prehistory). Modernity thinks of itself as the only legitimate occupant of the present; the 'out of date' is out of place in the contemporary world. This presumption results from the fact that, according to Wendat scholar Georges E. Sioui, "modern people have long since lost all notion of the intellectual and spiritual tools with which they could have preserved a dignified image of their past...."[52] The ideology of progress is deeply embedded in modern Western culture (see Chapter Ten), and it's where the notion of 'out of date' comes from: the past is automatically irrelevant to the present, and also terrible. To think otherwise about the past is to romanticize it (see Chapter Twelve) or be nostalgic about a lost and irrecoverable time.[53] When Indigenous people are seen in this way, they're either ignoble or noble savages, but either way, irrelevant and ill-suited for the modern present and its glorious future.

But this is false. Neither Indigenous people nor myth exist only in the past, and neither myth nor traditional Indigenous knowledge are automatically false. The past *and* the present can be nonmodern,[54] and nonmodern, Indigenous, and nonliteral epistemologies can be an important counterpoint to the hegemony of Western knowledge systems. Indigenous oral traditions

> document a rich and lengthy history, full of momentous events, migrations and battles, struggles and victories, heroes and sages. Retelling stories generation after generation allows people to engage in a dialogue with their ancestors. Myths allow people to come to terms with their environment by telling of powerful transformers, whose ancient actions resulted in the present appearance of the land and animals. Myths also stress the fundamental unity

> of humans, animals, and supernatural beings, in a manner that allows people to account for the natural phenomena in their world.[55]

Modern science and technology have much to learn from Indigenous knowledge. For example, Potawatomi botanist Robin Wall Kimmerer includes *beauty* in her scientific study of plants, whereas her Western-trained professors could not accept that as a legitimate part of their science.[56] In fact, Western science and technology learned—or 'extracted'—the knowledge European empires needed to survive and eventually conquer their 'New World' from the Indigenous people who lived there.[57] Inversely, Indigenous people today use modern Western technoscience to survive and flourish in the settler colonial societies they find themselves in.[58] Thus "colonized nations and their indigenes have never been the passive victims of colonial rule but rather vibrant actors with agency, active engagement, and resistance, albeit in situations with grossly unequal power."[59]

At one time, all knowledge and action was unabashedly mythological and ceremonial, before it was lost in the West "because of the constraints that led to the development of their present type of civilization."[60] What happened to lead the countries of Europe to abandon myth and ritual in favor of a technoscientific rationality that would seek mastery over all peoples and the Earth itself? You might find it very difficult to believe that myths are not supposed to be descriptive and that rituals are not supposed to be technological. After all, when someone prays to a god, don't they think—or at least hope—that the god will do something in response to the prayer? At the very least, the usual story makes it tough to imagine a non-material satisfaction (expressing hope to a god) as not also a material satisfaction (getting the god to fix a part of the world for you). We should be able to understand that prayers are not technologies. Even so, some people think that they are: some religious people think that prayers can replace medicine, while some non-religious people think that prayers are mistaken attempts at replacing medicine. This conflation has to come from somewhere, right?

Yes. Something had to *change* in order to move away from a mythological and ceremonial culture towards a form of technoscientific rationality that at least *thinks* it is neither mythological nor ceremonial. Something had to change for magic to be seen as a kind of science or technology (e.g., see 'natural magic' in Chapter Eight). Something had to change for symbolic and experiential speech-acts to give way to attempts at *literal* models of the world as it 'actually is.' These changes were cultural. What's more, I am going to argue that these changes were technological. I am going to suggest that the technologies of agriculture (Chapter Five) and writing (Chapter Six) set the stage for a unique cultural perspective to take shape, a perspective which views myth and ritual as the *enemy* of technoscientific rationality.

Notes

1 Martin Fichman, *Science, Technology, and Society: A Historical Perspective* (Kendall/Hunt, 1993), 3.

2 Ludwig Wittgenstein, *Remarks on Frazer's Golden Bough*, ed. Rush Rhees, trans. A.C. Miles (Brynmill, 1979), 5e.

3 George Grant, *Technology and Justice* (University of Notre Dame Press, 1986), 13.

4 Fichman, *Science, Technology, and Society*, 3.

5 Silvia Sebastiani, "A 'Monster with Human Visage': The Orangutan, Savagery, and the Borders of Humanity in the Global Enlightenment," *History of the Human Sciences* 32, no. 4 (2019): 80–99.

6 Paulette F.C. Steeves, *The Indigenous Paleolithic of the Western Hemisphere* (University of Nebraska Press, 2021), xxiii.

7 Steeves, *Indigenous Paleolithic*, xx.

8 Grant, *Technology and Justice*, 12.

9 Arthur C. Clarke, "Clarke's Third Law on UFO's," *Science* 159, no. 2812 (1968): 255.

10 I.e., The Discovery Channel's "MythBusters" (2002–16).

11 *Canadian Oxford Dictionary* (2001), s.v. "Myth."

12 Robert A. Segal, *Myth: A Very Short Introduction* (Oxford University Press, 2004), 5.

13 While there are a number of very precise distinctions between myths, legends, folktales, oral traditions, fiction, ceremonies, rituals, magic, and religion, I will err on the side of generality and treat them all as functionally equivalent, or at least as closely enough related to speak of them all as a coherent whole. From here on in, I will for the most part use the term "myth" as inclusive of ritual, even though the former is a primarily intellectual phenomenon while the latter is primarily an action.

14 Segal, *Myth*, 6.

15 Wittgenstein, *Remarks*, 4e.

16 Wittgenstein, *Remarks*, 4e.

17 Segal, *Myth*, 3.

18 Segal, *Myth*, 11.

19 Segal, *Myth*, 13.

20 Segal, *Myth*, 24.

21 David C. Lindberg, *The Beginnings of Western Science: The European Scientific Tradition in Philosophical, Religious, and Institutional Context, 600 B.C. to A.D. 1450* (University of Chicago Press, 1992), 5.

22 Wittgenstein, *Remarks*, 12e.

23 R.C. Lewontin, *Biology as Ideology: The Doctrine of DNA* (Anansi, 1991), 5.

24 Wittgenstein, *Remarks*, 1e.

25 Wittgenstein, *Remarks*, 3e.

26 Opening narration for the Edmonton Oilers at Roger's Place, transcribed by the author, Edmonton, Alberta, 25 January 2017.

27 Lindberg, *Beginnings of Western Science*, 6.

28 Lindberg, *Beginnings of Western Science*, 6.

29 Lindberg, *Beginnings of Western Science*, 11.

30 John Gray, "Can Religion Tell Us More Than Science?" *BBC News*, 16 September 2011, http://www.bbc.com/news/magazine-14944470.

31 Cf. Thomas Nagel, *The View from Nowhere* (Oxford University Press, 1986).

32 Lindberg, *Beginnings of Western Science*, 7.

33 Lindberg, *Beginnings of Western Science*, 9.

34 Lindberg, *Beginnings of Western Science*, 7.

35 Lindberg, *Beginnings of Western Science*, 7.
36 Lindberg, *Beginnings of Western Science*, 8.
37 Wittgenstein, *Remarks*, 4e.
38 Wittgenstein, *Remarks*, 6e–7e.
39 www.letsfcancer.com, accessed 21 June 2024.
40 Wittgenstein, *Remarks*, 4e.
41 Wittgenstein, *Remarks*, 7e.
42 Shawn Wilson, *Research Is Ceremony: Indigenous Research Methods* (Fernwood, 2008), 11.
43 Wittgenstein, *Remarks*, 12e.
44 Wittgenstein, *Remarks*, 1e.
45 John Dominic Crossan (with Richard G. Watts), *Who Is Jesus? Answers to Your Questions about the Historical Jesus* (HarperCollins, 1996), 80.
46 Wittgenstein, *Remarks*, 4e.
47 Gray, "Religion."
48 Banu Subramaniam et al., "Feminism, Postcolonialism, Technoscience," in *The Handbook of Science and Technology Studies*, 4th ed., ed. Ulrike Felt et al. (MIT Press, 2016), 414.
49 Subramaniam et al., "Feminism, Postcolonialism, Technoscience," 422.
50 James C. Scott, *Against the Grain: A Deep History of the Earliest States* (Yale University Press, 2017), 9–10.
51 Billy Frank, Jr., quoted in Lisa Pemberton, "Refuge That Served as Billy Frank Jr.'s Medicine Now Named for Him," *Bellingham Herald*, 20 July 2016.
52 Georges E. Sioui, *An Amerindian Autohistory*, trans. Sheila Fischman (McGill-Queen's University Press, 1992), 13.
53 Christopher Lasch, *The True and Only Heaven: Progress and Its Critics* (W.W. Norton, 1991), 83.
54 Subramaniam et al., "Feminism, Postcolonialism, Technoscience," 419.
55 Alan D. McMillan and Eldon Yellowhorn, *First Peoples in Canada*, 3rd ed. (Douglas and McIntyre, 2004), 23.
56 Robin Wall Kimmerer, *Braiding Sweetgrass: Indigenous Wisdom, Scientific Knowledge, and the Teachings of Plants* (Milkweed Editions, 2013), 41.
57 Subramaniam et al., "Feminism, Postcolonialism, Technoscience," 417–18.
58 Subramaniam et al., "Feminism, Postcolonialism, Technoscience," 421.
59 Subramaniam et al., "Feminism, Postcolonialism, Technoscience," 423.
60 Sioui, *Amerindian Autohistory*, 13.

Chapter Five

Agriculture

HOW ESSENTIAL IS THE TECHNOSCIENTIFIC INVESTIGATION OF NATURE to the human species? Have human beings always wanted—and always tried—to achieve the mastery of nature that contemporary societies now possess? Was myth always trying to be scientific, and were rituals always misguided attempts at technology? In Chapter Three, George Grant claimed that modern Western technoscience is a unique social phenomenon and that its characteristic way of thinking about the world (i.e., technoscientific rationality) is *not* characteristic of non-Western and non-modern societies.[1] In Chapter Four, we saw that it was an anthropological fact that many societies existed (and may still exist) which simply did not place great value on material progress.[2] Not only were myth and ritual dominant features of prehistoric societies, modern societies still retain many aspects of myth and ritual despite being technoscientific. We also saw that myth (inclusive of ritual) does not appear concerned with exerting or gaining material leverage over the world. Unlike most modern people, prehistoric people knew—and Indigenous peoples still know—the difference between literal descriptions/manipulations of the physical world (on the one hand) and symbolic speech/acts about the meaning and significance of the world we experience (on the other hand). It is only when these two levels of 'explanation' (or expression) are conflated that myth appears as a competitor with technoscience. Myth is not science and ritual is not technology, and that's fine because they're not supposed to be. Myth and ritual are not supposed to be 'literally' true or materially effective, but rather true and effective in a non-material way, perhaps experientially or spiritually. Thus, in the words of the historian of science David Lindberg,

> it is wasted effort, contributing absolutely nothing to the cause of understanding, to spend time wishing that preliterate people would employ (or had employed) a conception and criteria [sic] of knowledge that they have (or had) never encountered—a conception, in the case of prehistoric people, that was not invented until centuries later. We make no progress by assuming that preliterate people were trying, but failing, to live up to our conceptions of knowledge and truth. It requires only a moment of reflection to realize that they must have been operating within quite a different linguistic and conceptual world, and with different purposes; and it is in light of these that their achievements must be judged.[3]

Contrary to the usual story, therefore, simply being unscientific in the modern sense does not make someone irrational; rather, it is irrational and unscientific to think that technoscience is the only way to think and act rationally.

Myths and rituals are therefore neither theories nor methods—they neither attempt to offer world-picturing representations of facts nor do they attempt to exert direct leverage over material things. What's more, myth and ritual are characteristic of human beings for as far back as we can study human beings, whereas technoscience is at best characteristic for only the last one-quarter percent of human history (inclusive of prehistory). Therefore, if anything is characteristic of 'the human spirit,' it would be myth and ritual, not technoscience. *A literal understanding of physical mechanisms for the purposes of improved material leverage is not a universal characteristic of the human condition.* The question before us, then, is how technoscientific rationality came to be, if it could not have been a simple outgrowth of myth and ritual. There must have been some contingent social factors which fostered the growth of this unique way of thinking and acting. In this chapter I will argue that the *agricultural revolution* was a necessary (if not sufficient) condition for the birth of technoscientific rationality.

The agricultural revolution was a technological change: it marked a change in the way that human beings interacted with the material world. Because we are studying technology (in addition to science) as a social phenomenon, we should expect technological changes to have social implications. STS scholar Martin Fichman claims that the "agricultural revolution ... radically altered the course of human cultural evolution."[4] But what about other massive technological changes, such as the advent of tool use or the domestication of fire? In these latter cases, the technological innovation comes *before* the birth of our species *homo sapiens*. Tool use among hominids arises approximately 2.4 million years ago, during the Oldowan period, when *homo habilis* appears to have used rocks to crack open scavenged femur bones to eat the protein-rich marrow which no other animals could access. Fire appears to have been harnessed by *homo erectus* about 500,000 years ago. Our species—*homo sapiens*—only appears around 300,000 years ago, such that we evolved as human beings with tools and fire already in play. Those innovations had significant impacts on the hominid populations that experienced them, of course, but by the time we show up as a species, tools and fire were already part of what it means to be human.

Agriculture, however, only arises at most 10,000 years ago. This places it in the last 3 percent of our history as a species. Ninety-seven percent of the human story had already been told before agriculture appears and, as a result, it is unlikely that it would be innate to human biology or culture. Indeed, if agriculture "radically altered" the course of human culture, then we *cannot* say that it was in some sense inevitable or preordained. Even if we speak of "cultural evolution," we must do so in the same sense as biological evolution: neither of them move inexorably towards any sort of predetermined goal. As a technological and thus social factor, agriculture is *not* a force of nature. It was and remains "a culturally and historically contingent moment"[5] which could have been otherwise. This is what contingency means: it is the opposite of necessity, something which did not have to happen. Social changes, technological changes, cultural 'evolution'—all of these depend on human thinking and action which is itself remarkably indeterminate or even fickle. Even though 'cultural evolution' (if we should call it that) is not a consciously directed process, it is ultimately the aggregate result of masses of people making seemingly insignificant choices. Some people (in Mesopotamia, China, and Mesoamerica) at some times (roughly 10,000 years ago) made choices about how they were going to live in the world, and—unbeknownst to them—the end result of those choices was surprisingly revolutionary.

Contingency: that which did not have to happen; caused by historical or social human action rather than natural forces; the opposite of necessity.

Foraging

Just as we tried to gain an initial understanding of the uniqueness of technoscience by contrasting it with myth and ritual, in this chapter we will try to understand the uniqueness of agriculture by contrasting it with its opposite: *foraging*. Foraging is a way of procuring food by finding where it already is. I will use the term to encapsulate both hunting (finding primarily land animals), gathering (finding primarily plants), and fishing (finding primarily aquatic animals). The animals and plants that foragers seek are characteristically wild, meaning that these organisms are in charge of where they will be at any given moment. A wild blueberry bush will grow where it grows because that's where its seeds fell after being deposited by the wind or some animal's feces, not because a gardener decided to plant it there. A wild sheep will roam where it roams because of what the wild sheep is interested in doing, not because a shepherd has ensured that the sheep will be where the shepherd wants it to be by means of fences, dogs, or crooks. Foragers, then, have to go where the blueberries and sheep are, because such food is wild. Most animals are foragers (some insects are

Foraging: a subsistence practice reliant on locating wild animals and plants for food and other resources.

cultivators or herders, like ultrasocial species of ants and termites),[6] and all humans were foragers too, up until about 10,000 years ago. Like tools and fire, human beings evolved with the technique of foraging already in play, living in an environment we would now call 'the wilderness.' Importantly, this wild way of life was very rewarding:

> With relatively simple technology—wood, bone, stone, fibers—[foragers] were able to meet their material needs without a great expenditure of energy, leading the American anthropologist and social critic Marshall Sahlins to call them, in another famous phrase, "the original affluent society." Most striking, the hunter-gatherers have demonstrated the remarkable ability to survive and thrive for long periods—in some cases thousands of years—without destroying their environment.[7]

As a form of subsistence, foraging is very successful, a context within which human beings flourish.

While anthropologists point out that there is considerable diversity and variability among forager cultures, there are also many commonalities that go beyond mere subsistence practice. That is to say, there are correlations to be drawn (not without exceptions) between subsistence foraging and other cultural elements, such as social organization and cosmology; foraging is a technique with social implications. One common impact of foraging has to do specifically with hunting, and it can be applicable to ancient as well as contemporary hunting. Because hunted animals are wild, hunters by definition do not know where the prey animal is. In the (unfortunately gendered) language of José Ortega y Gasset, the hunter

> does not believe that he knows where the critical moment is going to occur. He does not look tranquilly in one determined direction, sure beforehand that the game will pass in front of him. The hunter knows that he does not know what is going to happen, and this is one of the greatest attractions of his occupation.[8]

Like wild animals themselves, hunters have to be continually aware of everything, of the possibility that the unexpected could happen at any moment at any point on the horizon. To narrow this vast range of possibilities down, hunters must know a lot about animal behavior and environmental conditions to be able to have even a hope of predicting where the desired animal might be. This broad, semi-focused awareness requires continual preparedness, simply because the hunter is not in control of the animal or the environmental situation.

The hunter's lack of control may sound like a terrifyingly insecure way to get food, but Ortega argues that it requires a very satisfying form of intellectual activity: *alertness*. Alertness for Ortega is not a transcendent form of intellection (like Greek *theoria*) where an ideally rational spectator hopes to see all things as they really are in truth (see Chapter Two), but is rather "truly think[ing]" that does not simply accept "what habit, tradition, the commonplace, and mental inertia would make one assume."[9] That is to say, alertness is critical thinking. This form of thinking requires *immanence* in the world—surrounded by the subjective appearances of nature or the jungle of ideas when we need to discover an answer—through which we seek a path forward that hopefully leads to the moment of truth. Neither the hunter nor the philosopher is in control of the appearance and capture of the intended object, but neither are they willing to forego the hunt for the promise of control and a god's eye view of things.[10] Hunting, like myth, is an experiential and non-theoretical approach to the world which is nevertheless rational and intellectually demanding. It does not see its lack of guarantees or control of the world as a problem, but rather as definitive of itself.

The relation between alertness and the surrender of control is connected to another commonality across forager cultures: "the *giving environment*, the idea that the land around [foragers] is their spiritual home and the source of all good things."[11] Because the hunter is not in control of the animal's chance appearance, experientially the animal appears as a *gift* to the hunter by forces beyond the hunter's control. "Thus in gather-hunter or foraging societies, animals and nonhuman nature are seen as the source of the gift of life," T.R. Kover argues, which in turn brings "the fundamental recognition that the gift is undeserved and that we are indebted to, and dependent on, the sources of this gift."[12] The perception of animals giving themselves to hunters is a meaningful signification which is unsurprisingly conveyed mythologically and ritualistically:

> [T]he ritual preparations of the hunt among many hunter-gatherers emphasize not only the importance of conveying the appropriate attitude of respect to the prey, but also the prevalent belief that the prey itself decides to give itself to the hunter who demonstrates appropriate skill and respect.[13]

Many forager cultures, past and present, in fact have totemic treaties of affiliation between themselves and particularly significant species of animals, mythologically political agreements to exchange ritual respect for hunting success.[14]

In such a social context, the lack of control over animals and the larger nonhuman world does not need to be seen as a source of anxiety. If the world itself mythically gives sustenance and life to human beings who respect it, then there simply isn't any need to increase one's knowledge and control over it. If the killing and eating of other animals is—through the proper ritual channels—experienced as a gift, increasingly powerful tools are unnecessary for securing a consistently successful harvest. If being immersed in the wild world is intellectually stimulating and satisfying, then a theoretical and detached form of knowing which eliminates hunting per se is undesirable. If foraging is a stable and sustainable sociological platform for long-term human flourishing, it's unclear why anyone would want to replace it with something 'bigger and better.' Without misrepresenting foragers as 'noble savages' but rather relying on scientific facts from archaeology and anthropology, we can see that—like myth and ritual—hunting and gathering do not share with technoscience an interest in increasing our material leverage over the world or gaining a divinely correct picture of the world's physical operations. And yet, agriculture happened.

Domestication

There is no consensus as to why some human beings in some parts of the world supplanted foraging with agriculture. From a minute analysis of the stone tools of the transitional period, archaeologist Jacques Cauvin hypothesizes that plants and animals were first domesticated in bull-worshipping cultures obsessed with weapons, as if agriculture was the result of a power trip.[15] Regardless, once agriculture was in place, agrarian societies found they had massive military advantages over their forager neighbors, which explains both the retention of agricultural practices and their euphemistic 'spread' across the rest of the globe (even though the earliest agriculture did *not* produce food yields greater than that of foraging).[16] For our purposes, it is enough to recognize that agriculture was a cultural and historical contingency which had certain social effects down the line, consequences that helped facilitate technoscientific thinking, even if we don't know for sure why some people decided to pursue this way of life.

> **Domestication:** the process or effect of making a nonhuman lifeform dependent on humans for its reproduction and nourishment; usually reliant on selective breeding resulting in genetic divergence from wild ancestors.

Agriculture is a technological innovation, but it comprises a number of distinct tool sets and techniques. Key among these is the technique of *domestication*. Domestication is the process and effect of making a nonhuman lifeform (either plant or animal) dependent on humans for its reproduction and nourishment. Literally, the word means 'house-training' (cf. the English words 'domestic' and 'domicile'), which is itself etymologically related to the Latin word for ruler or lord (*dominus*, cf. the English words 'domination' or 'dominion'). Most domestication occurs genetically through selective breeding, and generally the selection adapts plants and animals to human *sedentism*—staying in one place more or less year round. Foragers are typically nomadic, whereas farmers are especially sedentary, remaining in place to tend stationary crops and penned animals. Domestication was thus a prerequisite for phenomena such as bread, cows, and cities—all of which we now take for granted.

> **Agriculture:** using domesticated organisms to increase the amount of food that the environment can produce. Technically, agriculture only refers to plough farming, but for simplicity's sake I also include horticulture (gardening) and pastoralism (herding or ranching).

In addition to these changes in material culture, domestication also implied "a series of complex thought processes and cultural innovations" which set it apart from foraging.[17] First, domestication requires treating lifeforms as a "standing reserve."[18] This means that living things are put into storage so that their energies can be drawn upon at will. Thus, while all human cultures would have stored food (such as dried berries or rendered fat), agri-cultures store seeds not only for food in the present but also for growing living plants in the future. Similarly, domestic animals are not all killed and eaten as soon as they are brought under control, but rather killed at the time deemed right by the owner, or kept alive as beasts of burden or for breeding stock. Paul Shepard remarks that with agriculture, living animals and even "the land itself becomes a tool, an instrument of production,"[19] whereas foragers only make tools from dead or inorganic matter. Thus with domestication we can see the significance of the English term "livestock"—economic investments that are biologically alive.

Second, domestication requires a different kind of knowledge about animals and the environment than we saw with hunting. While hunters need to know a lot about the seasons, weather, and reproductive cycles of their prey, they do not need to know exactly when seeds should be planted or when cattle should be inseminated—whereas agriculturalists do. Domestication requires precise calendar knowledge so that the sexual reproduction of plants and animals can be more efficiently manipulated. This may be why astronomy/astrology was such a prominent concern among early Mesopotamian and Egyptian civilizations (i.e., city-based agrarian cultures). Furthermore, the control over organisms' reproductive cycles can lead, in some cases, to a lifeform's complete dependence on human being for reproduction.

My cousin helping a cow give birth on his ranch.

Domesticated bread wheat does not spontaneously fall from the stalk, but must be threshed to loosen the seeds. Green peas, meanwhile, were artificially selected for pods which did not open to distribute seeds. In both cases, these cultivars will go extinct if humans do not collect the seeds and plant them, because the plants themselves cannot do that. Similarly, some dogs cannot reproduce without human intervention: British bulldogs can hardly give birth without caesarian section, while Irish wolfhounds require the male to be manually stimulated for coitus to occur. Domestication involves invasive interventions into the sex lives of plants and animals, which is likely to produce a different perspective on the world than foraging, which allows other organisms to be in charge of their own reproduction.

Finally, because domestication is an attempt to store life as a standing reserve, it also requires risking the delayed gratification of eating in the hopes of generating a surplus nearly a year into the future. Early farmers "made the seemingly senseless choice of literally taking the bread from their mouths and burying it, hoping to multiply their agricultural 'investment' in a few months."[20] Rather than a foraging universe which gives forth sustenance (provided you show proper respect), agricultural food becomes associated with a risky investment of labor that entitles the laborer/owner to future profits. Each of these three intellectual impacts of domestication—standing reserve, reproductive control, and investment multiplication—is a form of practical knowledge that *increases leverage over the material world*. But that's only the beginning.

Peasantry and Bonded Labor

Domestication also had effects on human social structure. Because the "agricultural revolution provided Neolithic farmers with harvests that had previously been inconceivable," this led to "a communal food surplus which greatly exceeded the subsistence of earlier communities. This surplus permitted the feeding of a (gradually) increasing population."[21] Of course, as human populations grew, so did the need for more agricultural production. These two aspects formed a positive feedback loop in perhaps the first instance of the "treadmill of production."[22] However, this feedback loop made lapses in food production catastrophic. Populations could grow past the point where they could be sustained by foraging if agricultural crops failed. If the environment (weather, soil, vermin etc.) did not cooperate in precisely the right way to make agriculture productive, famine was inevitable. This marked a departure from the gifting cosmos and a shift towards

> an obsession with the great round of the year, perhaps because of the increasing uncertainty of the yield and because the farmer had fewer alternatives than his foraging ancestors. With agriculture it was likely to be boom or bust, and in the bad years nature seemed to withhold that to which the farmer, after all his labor, felt he had a right.[23]

From the agrarian perspective, the only practical solution was expansion and intensification of agricultural land holdings, including the invention of irrigation technology.

On average, however, population growth happened, and it allowed for sharply defined social classes to emerge: the maintenance of a "communal food surplus ... allowed some members of the community to be freed from primary production and used for specialized jobs, such as making tools and implements."[24] But more striking than an artisan class was the emergence of wealthy ruling elites, to whom the collected surpluses were upwardly funneled. Unlike foraging (which fosters largely egalitarian social structures), agriculture requires a large investment of time into difficult physical labor, which the wealthy were understandably keen to avoid. Thus the vast majority of agricultural populations were comprised of what we would now call peasants, if not slaves. This large class of manual laborers engaged in repetitive tasks and drudgery to produce wealth concentration for the fortunate few who owned the land. Moreover, as "the diversity of foods diminished—the wild alternatives become scarcer and more distant from villages—the danger of malnutrition increased."[25] Archaeological examination of skeletal records shows a *decrease in human health and lifespan* with the adoption of agriculture, except for those (wealthy) skeletons buried closest to the granaries. Human flourishing was far less equally distributed in agricultural societies than in foraging ones.

Finally, increasing population densities in both cities and the rural countryside led to the creation of a bureaucratic or managerial class. Large numbers of enslaved peasants require organization and co-ordination to effectively achieve the goals of the state, both for food production and the construction of public works. Furthermore, because famines are inevitable in a boom and bust cycle, it is imperative—from the perspective of the non-peasant classes—that food surpluses be stored away from their sites of production, lest the laborers consume the last reserves while the elites starve. The managerial classes thus relocated, hoarded, and defended food (with military strength, as necessary) at centralized locations safe from lower-class rioters. The supposed inevitability of 'death and taxes' may in fact stem from this agrarian and especially civilized context.

Civilization: a form of sedentary human culture characterized by urban centers; from the Latin *civitas* ("city").

Attempted Mastery of Nature

With the domestication of animals and plants, the sociological pressures inherent in agriculture resulted in enforced social hierarchies that kept food and wealth away from the peasant masses who were under the control of elites. As with nonhuman animals and plants, lower-class human beings were subjected to the kind of practical knowledge that exerts increasingly effective leverage over matter. Slaves, peasants, laborers (etc.) were treated and perceived much the same as domesticated livestock. Thus, in Fichman's words, the "[agri]cultural complex that associated humans, plants, and animal can be regarded as the starting point of that attempted 'mastery of nature' which became one of the dominant goals of subsequent human social and technological evolution."[26] The meaning and significance of agriculture is, therefore, not that 'nature' should be better respected, but rather that it should be better manipulated: "The entire system [of agriculture] depends on the subjugation of nature, and the domination and manipulation of living creatures."[27] With agriculture, nature's bounty is no longer a gift but an extraction of dues. Those humans most closely associated with dirt—unfree laborers—are subject to the same extraction of dues (in the form of taxes) as are the fields and herds.

Unsurprisingly, however, these 'natural' constituencies are recalcitrant to such extractions. Peasants no more want to surrender their food surpluses than grasslands 'want' to be tilled or deer 'want' to be caged. As Kover argues, agriculture "depends on a decisive separation between the natural and the human world, a state in which the [latter] is seen as completely compliant with human ends and needs and the [former] is seen as defiant and antagonistic to the natural world."[28] The need for control implicit in agriculture thus generates an *oppositional framework* for the cosmos in place of the generous foraging cosmos.

In the first place, agriculture creates a clear visual impression that it is distinct and separate from wild nature. The boundary between a garden or a tilled field is obvious in comparison to a forest or a grassland, while the boundaries of pastures are clearly marked by fences or walls. These agrarian spaces are fabrications of human hands, places where domestication is supposed to rule. While forager camps are also distinguishable from the environing wilderness, their way of life creates no clearly delineated territory of human reorganization outside of which lies an unproductive wasteland.

Secondly, the visual division between agrarian settlement and untamed wilderness is seen—from within the agricultural perspective—as oppositional. This is because wild nature resists agrarian reordering:

> [N]ot only does the wild serve no discernible advantage for the farmer..., but it seems to actively hinder and undermine their aims. Wild flora or weeds usurp the fecundity of the soil, and wild fauna preys on domestic crops solely reserved for humans.... [F]ailure to protect against these depredations can spell the difference between success or outright disaster. Thus, with agriculture, the wild becomes the enemy of the tame as its purposes seems to be constantly working to undermine the domestic human order.[29]

'Nature' becomes divided against itself in the agrarian context, between the helpful and productive rural nature that is under sedentary human control, and the antagonistic or useless wild nature that is not (yet) under human control. Agricultural myths reflect this denigration: "wild or untamed nature is frequently seen in agrarian cosmologies as a source of disorder, malevolence and evil, and is often personified by demons, ghosts and monsters."[30] In the *Enuma Elish*, for example, the Mesopotamian god Marduk created the world we live in out of the corpse of Tiamat, a female dragon of the sea, after smashing her skull with a club.

> **Recalcitrance:** a strong innate resistance to external influence or manipulation.

Agriculture thus encourages the view that nonhuman nature is naturally disorderly and must be socially constrained.

This is also the case for human nature. Inasmuch as the peasant classes are more closely connected to the physical world (by virtue of their manual labor), they stand in need of more social control than the superior social classes who see themselves as rising above the muck. This is especially true as rising population densities contribute to epidemics or social unrest among the masses. The human body itself can become associated with unruly nature, a source of uncivilized or animal urges that need to be controlled if people are going to live in dense sedentary conditions. This is why the phrase 'going wild' isn't supposed to sound decent or polite when applied to human beings.

Beyond peasants and human bodies, however, are the recalcitrant human beings from neighboring cultures, who invariably clash over land and resources as agrarian expansion ratchets up. Population growth requires converting more land to farms, and eventually your society will come into conflict with neighboring societies who also need the same land. Famines, meanwhile, make the capture of enemy storehouses incredibly lucrative, while undesirable agricultural labor makes the acquisition of slaves quite desirable. Indeed, capturing slaves to maintain a bonded workforce adequate for economic output was the main motivation for warfare in early Mesopotamia, not territorial expansion.[31] Finally, high population densities allow for large armies, capable of overwhelming forager societies and conquering neighboring agrarian societies. Thus, "anthropologists and archaeologists have noted that the transition from foraging to farming often appears to be accompanied by an intensification of intercommunal hostility and warfare." Oppositional agricultural myths not only view wild nature as chaotic and evil, but also view "humans outside the boundaries of one's community ... as chaotic and evil."[32]

Human nature which threatens your culture's precarious existence comes to be seen as wild, outside the limits of the moral order, savage, barbaric, subhuman, animal, or verminous: "a similar conceptual and symbolic motif is found throughout [both] segmentary tribal societies [and] the great archaic civilizations: the boundaries of the polity or state define the ordered world, and the world beyond its frontiers is identified with hostile, malevolent forces and its inhabitants equated with wild animals."[33] Human beings can be just as threatening as wild nature is, and the only way to contain that threat is to domesticate it, i.e., to bring it under the lordship of the sedentary domicile at the heart of one's own agricultural civilization.

The Inadequacy of Myth?

In Chapter Three, we noted George Grant's claim that the will to power embodied in modern Western technoscience is historically and culturally unique. In Chapter Four, we saw that myth and ritual are not well suited for descriptions of facts about the physical world that could lead to material leverage or power. In this chapter, we've seen that foraging—itself a technique—also does not lend itself to the idea that humans must increase their power over the material world. Rather, foraging tends to generate myths and rituals which see all of the wild natural world in terms of kinship and gift. In terms of non-material satisfaction, foraging and myth appear well suited to each other.

With the advent of agriculture, however, we've seen a number of shifts. Agriculture tends to generate the view that life itself requires the manipulation and control of recalcitrant living beings, human or otherwise. If and when such lifeforms resist agrarian manipulation and control, they are deemed evil. Moreover, we've seen that myth can and has been employed to express these agrarian meanings. The question I want to pose at this juncture is whether agriculture will place strains on the psychological resources myth can offer, strains which might eventually result in the rejection of myth by critics. Agriculture creates a precarious social reality: large populations dependent on increasing control over land, food production, and the populations themselves. Within this context, *if you lose control of the world, your world will end.* Myth may not be a speech-act suited for providing explanations which increase control of the world, but might agriculture create an expectation that myth *should* contribute to the project of controlling the world?

Consider the need in agrarian societies for exerting social control over large populations of potentially riotous workers. If, as we saw in Chapter Four, one of the functions of myth is to provide social cohesion, we would expect agrarian myths to contribute to the social cohesion of large masses of laborers. It would not be surprising if myths were intentionally manipulated by social elites to solidify the legitimacy of their claims to the majority of the society's wealth, for example by mythologically making the aristocracy divine. But should this social order come under scrutiny by critics who see it as oppressive and unjust, myth could also be seen as facilitating social oppression (thus guilty by association) and repudiated by such critics. Even so, that negative assessment would not alone lead anyone to expect myth to function as a literal description of facts.

But when it comes to earth control, rather than social control, a desire for more effective knowledge than myth might reasonably arise. If the very existence of one's society is dependent on precarious human manipulations of recalcitrant lifeforms and capricious environmental factors, the expressions of a culture's desires for successful harvests might well take on a desperate character. Mother Earth *really* needs to be convinced to yield her fertility to human insemination, and in such contexts myths and rituals could feasibly transfer over from symbolic gestures to literal attempts at—or methods for—ensuring agriculture success. The usual story's view of myth and ritual as literal attempts at physical description and manipulation may originate here, in agriculture. This is speculation on my part, however, and no more can be said about it here.

Even if we continue to maintain myth as separate from scientific descriptions and ritual as separate from technological manipulation, the fact is that *agriculture makes people think that the world must be controlled technologically if humans are going to survive and flourish.* This way of thinking was not true for 97 percent of human (pre) history, but agriculture was a social construction that made it at least plausible, if not universally true. Agriculture, then, is the first step towards understanding the will to power encapsulated in modern Western technoscience, and it may have helped to make myth an enemy of technoscientific rationality. But the decisive break from myth requires another technological change: the invention of *writing*. We have already contrasted mythological expressions with 'literal' explanations, and in the next chapter we shall see why. Technoscience is not only characteristically agricultural, but also characteristically *lettered*.

Notes

1 George Grant, *Technology and Justice* (University of Notre Dame Press, 1986), 12–13.
2 Martin Fichman, *Science, Technology, and Society: A Historical Perspective* (Kendall/Hunt, 1993), 3.
3 David C. Lindberg, *The Beginnings of Western Science: The European Scientific Tradition in Philosophical, Religious, and Institutional Context, 600 B.C. to A.D. 1450* (University of Chicago Press, 1992), 10–11.
4 Fichman, *Science, Technology, and Society*, 4.
5 Tihamer R. Kover, "The Domestic Order and Its Feral Threat: The Intellectual Heritage of the Neolithic Landscape," in *Nature, Space and the Sacred: Transdisciplinary Perspectives*, ed. Sigurd Bergmann et al. (Ashgate, 2009), 235.
6 John Gowdy, *Ultrasocial: The Evolution of Human Nature and the Quest for a Sustainable Future* (Cambridge University Press, 2021), 18–24.
7 Richard B. Lee and Richard Daly, "Foragers and Others," in *The Cambridge Encyclopedia of Hunters and Gatherers*, ed. Richard B. Lee and Richard Daly (Cambridge University Press, 1999), 1.
8 José Ortega y Gasset, *Meditations on Hunting*, trans. Howard B. Wescott (Wilderness Adventures Press, 1995), 138.
9 Ortega, *Meditations on Hunting*, 140.
10 At least Ortega's ideal critical philosopher is not willing to do this. Many philosophers of the Western tradition have, in fact, desired to know things *sub specie aeternitatis*—"from the perspective of eternity." This desire to know reality in this way has had an influence on Western science's own self-understanding, as we will see in subsequent chapters.
11 Lee and Daly, "Foragers and Others," 4.
12 Kover, "Domestic Order," 238.
13 Kover, "Domestic Order," 237.
14 Paul Shepard, *Nature and Madness* (University of Georgia Press, 1982), 34.
15 Jacques Cauvin, *The Birth of the Gods and the Origins of Agriculture*, trans. Trevor Watkins (Cambridge University Press, 2000), 121–34.
16 Samuel Bowles, "Cultivation of Cereals by the First Farmers Was Not More Productive Than Foraging," *Proceedings of the National Academy of Sciences* 108, no. 12 (2011): 4760–65.
17 Fichman, *Science, Technology, and Society*, 4.
18 Martin Heidegger, "The Question concerning Technology," in *The Question concerning Technology and Other Essays*, trans. William Lovitt (Harper & Row, 1977), 17. However, Heidegger himself didn't think that traditional agriculture was an example of standing reserve. He reserved that categorization for industrialization.
19 Shepard, *Nature and Madness*, 36.
20 Fichman, *Science, Technology, and Society*, 4.
21 Fichman, *Science, Technology, and Society*, 4.

22 Allan Schnaiberg, *The Environment: From Surplus to Scarcity* (Oxford University Press, 1980), 227–31.
23 Shepard, *Nature and Madness*, 30.
24 Fichman, *Science, Technology, and Society*, 4.
25 Shepard, *Nature and Madness*, 32.
26 Fichman, *Science, Technology, and Society*, 5.
27 James Serpell, *In the Company of Animals: A Study of Human-Animal Relationships* (Basil Blackwell, 1986), 175.
28 Kover, "Domestic Order," 236.
29 Kover, "Domestic Order," 240.
30 Kover, "Domestic Order," 240.
31 James C. Scott, *Against the Grain: A Deep History of the Earliest States* (Yale University Press, 2017), 171ff.
32 T.R. Kover, "The Domestic Order and Its Feral Threat: Paul Shepard on the Intellectual Heritage of the Neolithic Landscape," unpublished draft (2007), reproduced by permission of the author.
33 Kover, "Domestic Order" (2007).

Chapter Six

Writing

FOR THE OVERWHELMING MAJORITY OF THE HUMAN STORY, WHAT WE now call technoscientific rationality was atypical. Instead, myth and ritual were dominant forms of thought and action, neither of which encouraged the idea that technologies should get progressively more powerful, nor that human practical knowledge of the world was inadequate or inaccurate and should be improved. One of the intellectual impacts of agriculture, however, was to create the impression that the wild, untamed world is opposed to human endeavor and as such must be controlled technologically lest catastrophe strike. In this social context, technological progress is seen as inevitable and innate to human nature. The usual story about science and technology, then, is characteristically agricultural.

But the usual story also holds that technological progress goes hand-in-hand with scientific progress. That is, without scientific knowledge, technology won't progress. However, what we now call scientific thinking is also historically atypical. Scientific knowledge (according to the usual story) is transcendent and universally true. Unlike myth, it offers a systematic and theoretical account of physical facts which is supposed to be *literally* true. But this notion of demythologized scientific rationality occurs late in human history, and for that to happen, certain social and intellectual changes also had to occur. According to Martin Fichman, "the invention

> **Conventional Definition of Literal:** an exact, non-symbolic correspondence between an idea or concept and a real thing.

of writing ... was absolutely essential for the growth of abstract science and its transmission."[1] In this chapter, therefore, we will explore how scientific thinking is a specific product of *literate* cultures. Indeed, due to an ironic twist in the English language, the word "literal"—which we use for an exact, non-symbolic correspondence between a concept and a reality—has the same etymological root as the words "letter" and "literary." It is thus no coincidence that we say scientific rationality offers a *literal* account of facts. The very notion of literal correspondence between thoughts and things comes from writing.

Writing is an information technology. As we saw with agriculture, technologies have impacts on the way human societies view the world. Technologies also create new contexts within which other technologies are invented that otherwise wouldn't have a purpose. In the case of writing, it is a technology that emerges independently only in agricultural societies. In Chapter Five, we saw that agriculture depends on a very controlling manipulation of the physical world. Systematic attention must be given to the weather and the seasons so that planting or harvesting (etc.) can be done at the correct time each year. We also saw that acquisition and defense of territory is crucial to agricultural systems of survival: "an agricultural way of living [requires] that serious attention ... be given to geographic boundaries and property divisions,"[2] lest your areas for cultivation and ranching shrink while your neighbors' suspiciously grow. Thirdly, the manpower needs of agricultural states required careful management of human females, especially slaves, for their reproductive capacities were the basis of the economic output that provided wealth for the elites.[3] In all three cases—cosmologically, terrestrially, and reproductively—agriculture creates "the need to devise ... an established and relatively stable unit ... of measurement."[4] There must be an authoritative and standardized way to plan the farming year, resolve territorial disputes, and determine how much fertility is needed to keep the economy going, if humans are to exert efficient control over the world. In this context, information is certainly power.

However, the standards of measurement devised by agrarian civilizations are unavoidably conventional in character. When one moment stops and the next starts will ultimately be arbitrary, just as the boundary between your strip of riverfront property and that of your neighbor's will be. Markers that are placed—such as a surveyor's stake or setting the start of each day at 12:00 AM—are ultimately human constructions, not objective indications given to people by the universe itself. Information may be power, but that information—like the exercise of power itself—is the result of an historically contingent sedentary human culture imposing itself onto a world that is generally resistant to such human manipulation and control. Like these conventional markers, writing is an invention which codifies information important to the purposes of agrarian societies, standardizing and stabilizing data in objective records just as a silo stores and protects the harvest for the future. Writing is the granary of the mind.

Administration

The codification of agrarian information is an exigency of the bureaucratic classes of highly populous and stratified agricultural civilizations. Thus, "anyone hoping to discover how Sumerians of 3000 BCE thought and felt is in for a disappointment. Instead, the first Sumerian texts are emotionless ... clerical records of goods paid in, workers given rations, and agricultural products distributed."[5] Writing was not invented for the recording of myths or poetry. It wasn't even invented to record human speech![6] It was a tool for the efficient management of human and nonhuman resources. Precise meteorological observations are difficult to memorize, and thus some of our "oldest written records, going back more than 4,000 years, are astronomical in character."[7] Precise information about property boundaries is also very difficult to commit to memory, and even if it were easy to remember, it would not be a stable enough method of storage to be authoritative in disputes. Managing reproductive services to produce a surplus for expropriation involves a lot more information than human memory can store.

Therefore, if your job as a member of the managerial middle class is to keep track of how many goats each of a thousand farmers is supposed to surrender to the crown each year, you might eventually decide to try writing it all down. Writing is invented to alleviate some of the mental pressure of ruling and taxing masses of peasants or slaves: the "recording and transmission [of] information" is a "crucial form of urban management."[8] Writing helps to exert more efficient control over the heavens, the earth, and the various beasts that toil therein by keeping permanent lists of objects and numbers. Information is power over people no less than dirt and weather.

Originally, then, writing was mathematics, and mathematics was for empire.[9] From the celestial observations of priest-astrologers to filing your taxes, writing is about breaking complicated things down into more easily manageable units—i.e., *quantification*. The reason why we still have 60 minutes in an hour or 360° in a circle is because the Mesopotamians found that base-60 numbers work much better for fractions than base-10 numbers (which they also used). And of course, modern technoscience would be impossible without mathematics.

Quantification, in turn, assists with *prediction*: Egyptian and Mesopotamian scholars invented mathematical notation which allowed them to predict new moons and lunar eclipses, "not by the use of geometric models, as Greek astronomers would [later] do, but simply through the use of numerical methods that extrapolated past observations into the future."[10] While prediction isn't always an administrative need, it is relevant to later developments in scientific method.

Thirdly and most obviously, written lists were forms of (hopefully) *permanent* storage. Not only is it difficult to memorize lists, memory also fades and is restricted to the mind of the memorizer. Writing—e.g., on soft clay tablets that were later baked hard—does not decay nearly so quickly. Writing can even survive the death of the author, allowing the dead to communicate with the living (but not the other way around). Writing is the literal embodiment of speech into a tactile, solid object. This permanence contributed to the fact that the "first ancient writings were all sacred. Scribes worked in the temples or in the courts of deified rulers. In Egypt, the scribes worked in the tombs where the written word perpetuated the spoken word and made it 'eternally' effective."[11] Each of these three characteristics—quantification, prediction, and permanence—are all key components of our current understanding of technoscientific rationality. Science is mathematical, provides causal predictions of events, and is supposed to be objectively or universally true.

An example of hieroglyphic writing.

Sociologically speaking, however, it is important to note that early writing was not distributed equally across society. This is readily surmised from its purpose as an administrative tool of the bureaucratic elites: "Early writing served the needs of [complex and centralized] political institutions (such as record keeping and royal propaganda), and the users were full-time bureaucrats nourished by stored food surpluses grown by food producing peasants."[12] Early writing was very difficult to learn, as there could be thousands of symbols to memorize connected by a mysterious system of grammatical shorthand. Hieroglyphic (Egyptian) and cuneiform (Mesopotamian) scripts were understandably "accessible to only the (few) scribes and priests" who underwent the extensive training to use them.[13] If information is power, scribes (and their deified employers) had a monopoly on that power while the lower classes were disenfranchised from it. Writing was a tool to maintain that disenfranchisement, and it is fair to say that the control exerted over the majority of human beings in archaic civilizations was oppressive. In the clearly value-laden language of anthropologist Claude Lévi-Strauss, "ancient writing's main function was 'to facilitate the enslavement of other human beings.'"[14]

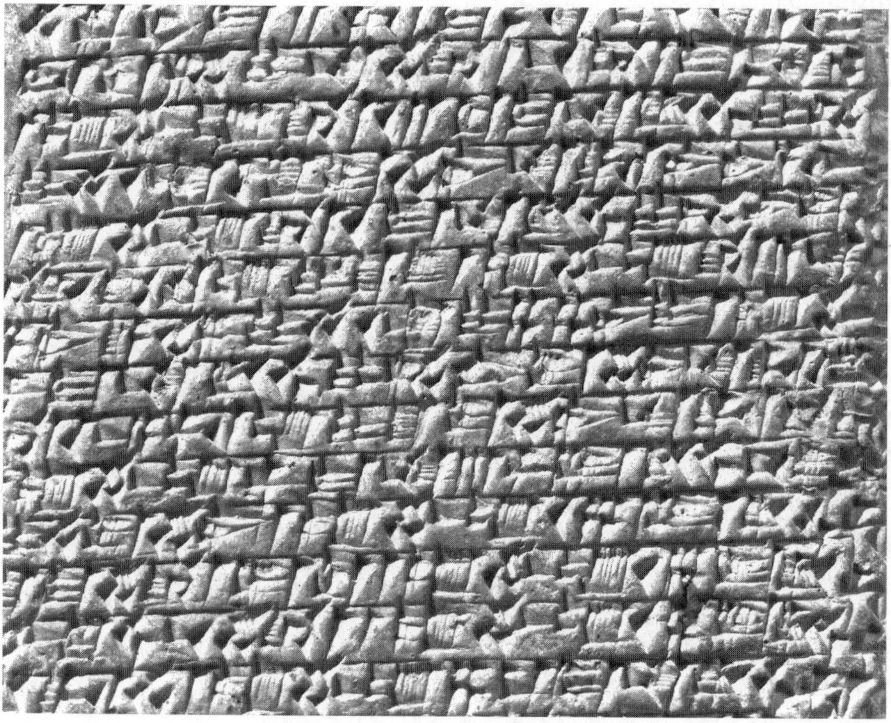

An example of cuneiform writing.

Technologies do not always turn out the way they were intended to, however. Writing could be turned against the hierarchical social powers if it could be used to spread counter-propaganda or facilitate secret plots among those very same rebellious peasants that writing was supposed to control. However, for these subversive purposes, writing needed to be both more expressive than mathematical, and be easier for the masses to learn. Even if you are the elite minister of propaganda, it's not easy to record ideas like "the king is awesome" if your scripts are essentially symbolic systems for counting jars of olive oil. Alphabets—which "apparently arose only once in human history: among speakers of Semitic languages, in the area from modern Syria to the Sinai"[15]—overcame this limitation by being phonetic rather than logographic. That is, alphabets are a symbolic convention for representing *sounds* rather than words or syllables; they break words down into even smaller and more abstract pieces: consonants and eventually vowels. This makes the script more expressive (you don't have to symbolize concepts like "awesomeness," you just spell it) and easier to learn, because the total number of symbols to learn is usually under 30. Alphabetic writing is an information technology even peasants could access, and it can be turned to purposes like writing poetry or expressing ideas like "the king is *not* awesome"—which you might be inclined to express if you're an unfree laborer. The scribal monopoly on writing power can be broken and the tool itself can be turned against the purpose it was originally designed for, making the masses more difficult to 'manage.' Whoops!

The Logical Cosmos

Writing, therefore, is a social convention of agricultural and especially stratified civilizations which was designed to serve the purpose of controlling agrarian laborers (who in turn exerted control over nature) but could also be turned against itself, especially with the advent of alphabets. (The possibility of unintentional technological subversion is important to remember about *all* technologies.) Secondly, writing also had important *intellectual* impacts arising out of its use to exert power over a world recalcitrant to agriculture and oppression. As we saw above, writing abstracted human language in order to quantify various particles of matter (be it corn or people), predicted events, and (hopefully) overcame human limitations by recording hard-to-remember data in static objects. These three aspects—abstraction, prediction, and objectivity—are almost enough to define science! Indeed, this is where we get one of our words for 'science.' As we saw in Chapter Three, scientific fields like biology contain the root word *logos*. Logos isn't just the Greek word for "word," but rather for *rationally organized and principled* words. The logos is the eternal word, baked into reality and written in stone, which organizes and manages the entire universe. Science *is* logos *is* writing. How did this epistemological framework come to be?

According to David Lindberg, writing

> had the revolutionary effect of opening knowledge claims to the possibility of inspection, comparison, and criticism. Presented with a written account of events, we can compare it with other (including older) written accounts of the same events, to a degree unthinkable within an exclusively oral culture. Such comparison encourages skepticism and, in antiquity, helped to create the distinction between truth, on the one hand, and myth or legend, on the other; that distinction, in turn, called for the formulation of criteria by which truthfullness could be ascertained; and out of the effort to formulate suitable criteria emerged rules of reasoning, which offered a foundation for serious philosophical activity.[16]

These rules of reasoning we now call "logic" (etymologically derived from the Greek word *logos*). What we now call science is the application of such logic to nature: aka natural philosophy (see Chapter Two). As we shall see below, logic is the set of rules for writing rationally, communicating literally and without contradiction (unlike myth), and thinking abstractly and objectively (not subjectively). Science *thinks* the way writing *operates*, and the physical universe operates logically (if science is to understand it at all). The cosmos has to be a written place.

Of course, the intellectual impacts of writing were not part of the intention behind the design of writing. Still, these impacts flow from its structure. As a convention for efficient management, writing has to ignore much of the variation in actual speech. Spoken words are expressed at certain volumes, pitches, and speeds with differences in emphasis, grouping, and pronunciation. While writing carries with it certain textual markers of individuality (e.g., you might not have written this chapter the same way I did), many other particularities of an individual's speech are not conveyed by writing. Auditory and temporal data (e.g., volume and speed) are *abstracted* away by writing because they are deemed insignificant. The sound of my voice is the last thing you need to think about when reading this book! (The fact that this book is typed out strips even more of my individuality from the text, but surely it is a relief that you don't have to read my idiosyncratic handwriting.) Writing abstracts away from particulars towards what are seen as the essentials of meaning, because this assists with sorting and quantification. If you're making a list of pigs, it doesn't matter how everyone sounds when they say the word "pig." This process of abstraction goes on to encourage us to think that we are boiling things down to 'the basics,' as if we're finally getting to the truth or the core of things. Particularities don't matter as much as the 'essence' of what is being said; quantities matter more than qualities. Writing encourages us to think that the truth is abstract. It's no coincidence, then, that science is highly abstract.

There is more to writing than abstraction, however. Writing *orders* abstractions. Regardless of your assessment of my chapter's argument, you'll at least notice that the letters are not randomly scattered across the page. The letters are rather consistently and grammatically grouped into words, phrases, and sentences, and these are all evenly spaced along lines starting at the left and proceeding to the right. While speech can seem fragmented or tangential, writing at least looks orderly. Indeed, writing is a visual and tactile translation of an auditory experience. While sounds can seem ephemeral, writing converts mere words into solid things you can *see* and *grasp*—two words we consistently use as metaphors for comprehension and understanding. By comparison, the wilderness outside agricultural societies may have looked disorderly and uncooperative, but a plowed field clearly marked and geometrically organized that world by the straight lines of its furrows. (Early Greek writing was called "boustrophedon" because it undulated back and forth, from right to left and then left to right, following the pattern an ox would take while plowing a field.) Like agriculture's reorganization of the natural world, writing reorganizes the confusing world of thoughts and sounds along solid lines. The two worlds of thought and matter mirror each other; the truth in both is literally a *linear* order.

> **Boustrophedon:** a style of writing where the text alternates direction on each line, from left to right and then right to left and back again, as if the page is ploughed like a field.

Furthermore, this organized abstraction is supposed to go beyond mere human convention. While in theory we know that property boundaries and agrarian social structures are contingent arrangements, we know in practice that if our agricultural control systems fail, the 'world' as we know it will catastrophically come to an end. Therefore, agrarian cosmology (i.e.,

> **Cosmology:** the science of the universe as a whole; literally, 'cosmos-logic.'

the scientific logic of the cosmos) is not experienced as a human convention or a contingent social construction. It is rather seen as the true, though hidden, nature of things. The abstract truth transcends the way people may perceive things, just as writing transcends the ability of people to remember things. Writing transcends a lot of things! It can be transmitted through time, whereas oral speech (until the invention of the phonograph—literally, "sound writing") must be experienced in real time. Writing can also be transmitted across space, whereas oral speech (until the invention of the telephone—literally, "far sound") is limited to how far people can shout and hear.

Finally, the permanence of writing overcomes the fluidity of oral speech; to get a promise or contract 'in writing' is supposed to make something official. Permanent records make conventions *exist* in the physical world. Because writing is embedded in a stable object instead of a human mind, it takes on a (supposed) character of objectivity, as it appears to be immune to certain human limitations. Even though it is a social convention, writing can take on a perceived character of natural solidity and authority, as if written words are more 'real' or trustworthy than spoken ones. While the human scale is transient and limited, writing (supposedly) sees through the tangles of uncertainty to the bare bones of truth beneath the way things appear. It is no wonder that early writing gave words the air of divinity and eternity. Therefore, thanks to writing, the nature of the universe as abstract essence, linear order, and unchanging permanence is not perceived as a human construct but rather as an objective reality that overcomes human subjectivity—even though this writing-centered vision of the cosmos is literally modeled after a human construct.

Intellectual Mastery

Writing is a tool for reorganizing the world, and it re-images the world as a written place. But on first glance, the world doesn't necessarily look organized—especially if it lies outside the sphere of domestication and civilization. In the raw, the physical world appears chaotic and evil (to an agriculturalist)! However, as Lindberg pointed out earlier, logic encourages skepticism, not only between different written accounts of the same event, but also between written accounts and unrecorded claims. The way the world initially appears—especially how agrarian cultures experience it—is a disorganized mess compared with what agriculture wants to do with it, and compared to the way that writing imagines the truth of all things. The truth, therefore, must be *beyond* the way things look. Just as a skeptical mind knows that a stick is not actually bent when it is half submerged in water, the cultural mind-set of writing thus looks at the universe with a skeptical eye, eager to discover (i.e., uncover) the abstract and unchanging order which supposedly lies beneath the apparently chaotic fluctuations of individual things. 'Reality' is not necessarily the way things appear in experience, therefore. The truly rational thinker—the scientist—will go beyond the way things look to how they *really are* in actuality. Logic will uncover the ultimately logical character of the cosmos. This is the usual story.

Therefore, while agriculture attempts to master the world with technology, writing creates the idea that the world can be *intellectually mastered*. The world *can* be understood, grasped by literate, lettered, literal thinking. The written word "revealed the world, line by line, page by page, to be a serious, coherent place, capable of management by reason, and of improvement by logical and relevant criticism."[17] The wild chaos of nature is not the ultimate truth, it is only the way things appear experientially to certain forms of culture. The hidden truth of things is logic—the written word, abstract reason—so much so that the classical Greeks eventually referred to the central organizing principle of reality as the divine Logos, which is why early Christians identified Jesus Christ as the Logos.[18] In this societal context, myth looks decidedly bad. Myth does not provide literal descriptions of the hidden realities behind material facts, whereas logic reveals the quantities, causes, and unchanging essences of all things. Myth can be used to entrench oppressive social powers, whereas logic can be used to unmask propaganda as unfounded or irrational. Thanks to logic, the human mind can overcome the powers that be, either political tyrants or the tyranny of the natural world. Logic can overcome these obstacles because it sees itself as finally having access to *the literal truth*: the direct correspondence between a mental conception and an external reality. Truth, on this conception, is when mental pictures of the world precisely and accurately correspond to the actual facts out there. Truth is when we see the way things *really* are. Myths never correspond to literal facts. Myths do not offer intellectual mastery of reality. Therefore, myths are false. The systematic application of logic to matter will reveal the literal truth of all things. The usual story, then, is characteristically literalistic.

The Correspondence Theory of Truth: truth is when a mental conception, idea, or world-picture directly, precisely, and accurately corresponds to actual facts in the world.

The Greek Miracle

The first fully alphabetic script we have written records of was Greek; they appropriated the Phoenician alphabet, using the unpronounceable (to them) Phoenician consonants as Greek vowels. This alphabetical system of writing was easy to learn, and it spread widely. In addition, the Phoenicians and the Greeks were both maritime trading cultures, sailing and colonizing much of the Mediterranean basin. The islands of the Aegean Sea in particular had limited natural resources, which

> imposed frugality on their inhabitants, making them ever more inventive. At the same time the continuous quest for means of subsistence and resources impelled them into adventurous voyages far away from their homeland, bringing them into contact with other societies and cultures. The goods and ideas brought back home by these intrepid sailors and barterers were quickly adapted to their

> own needs. They lost their foreign character and were fully assimilated by the island communities.... [T]he Aegean with its island groups was for millennia the crossroads where cultures of three different continents met. The Aegean merchant and mariner was at the centre of this meeting, if not its instigator. So he had the opportunity of comparing and assessing different ideas and of forming his own opinion.[19]

Moreover, the Greek merchant class also developed heavy infantry tactics (a recent innovation in the ancient world), and as a result their bronze-clad citizen-soldier *hoplites* were able to militarily overthrow local monarchs, finding themselves in charge of comparatively democratic city-states. As such, they found themselves in need of new ideas for laws and ways of generating political agreement.

A Greek hoplite warrior.

Fully alphabetic writing was thus eagerly learned by a populace with newfound political freedoms and responsibilities. In such a social context, adding the component of widespread literacy had a revolutionary effect. The "Greek miracle" was "the spectacular development of philosophy and science" beginning in the sixth century BCE.[20] While the Greeks famously had many myths, their natural philosophers relegated "the gods ... to the background and [made] an attempt ... to explain the world in natural, rather than supernatural terms."[21] That is, because these thinkers clearly repudiated myth in favor of the writing-centered (or "logocentric") model of reasoning, they "separate[d] the natural world from the supernatural world" so that their vision of the universe was governed by logic, not the whims of the gods.[22] A characteristically Greek response to mythological or theological explanation, therefore, would be to ask "what's the *true* explanation?" (even though their answers to such questions might still strike us as mythological or untrue). The point is that the Greeks were, astoundingly, the world's first scientists. Therefore, the usual story's repudiation of myth is also characteristically Greek.

We began the 'prehistory' section of this book with the point that 'history' only technically begins with the advent of written records. Therefore, now that we've covered the invention of writing, it is clearly time for our discussion to move on to a new phase. We also began this section of the book with a concern about "the startling novelty of the modern enterprise."[23] We already 'knew' (thanks to the usual story) that modern Western technoscience seeks to achieve ever more powerful mastery over the physical world. Our question as STS inquirers was whether that mind-set has always characterized human beings, or whether it was an historically and socially contingent idea which was not and could not be innate to human nature. We then saw, in Chapter Four, that to understand nature symbolically and experientially was overwhelmingly dominant across human cultures. Moreover, we saw that mythological and ceremonial understanding did not lead to the conclusion that the physical universe needed to be mastered by human beings. Rather, in Chapter Five we saw that agriculture makes us think that the world *must* be controlled *technologically*. Now in this chapter, we have seen that writing makes us think that the world *can* be understood *scientifically*.

Unlike agriculture or writing, science does not appear to be a technology itself. Science isn't a thing, like a "person, an object, or an event. It is an idea." And what is the scientific idea? It is "the idea that humans can understand the physical world."[24] *Scientific rationality is the intellectual mastery of the knowable universe.* I have argued that this form of rationality is a result of (at least) the impact of writing technology. While this form of rationality is incubated in a larger agrarian setting—wherein the recalcitrance of the physical world requires mastery—intellectual mastery is not necessarily the same thing as technological mastery of physical things. Just as science can be pursued in a purely theoretical way (see Chapter Two), the supposed ability to use logic to find the hidden reality behind chaotic appearances doesn't mean we will gain thereby more effective tools for controlling objects. If we put these two things together, however, we will get the notion of *technoscience*: intellectual mastery *and* technological control rolled into one (see Chapter Three). But just as a yearning for technological control is a social phenomenon—and just as a belief in intellectual mastery is a social phenomenon—so is the rolling together of intellectual and technological mastery a social phenomenon. To tell that part of the story, we will have to examine the premodern Greeks, Romans, Goths, Arabs, and medieval Christians.

Notes

1 Martin Fichman, *Science, Technology, and Society: A Historical Perspective* (Kendall/Hunt, 1993), 6.

2 Philip Wheelwright, ed., *The Presocratics* (Prentice Hall, 1997), 6.

3 James C. Scott, *Against the Grain: A Deep History of the Earliest States* (Yale University Press, 2017), 150–82.
4 Wheelwright, *The Presocratics*, 6.
5 Jared Diamond, *Guns, Germs and Steel: A Short History of Everybody for the Last 13,000 Years* (Random House, 1997), 234.
6 Scott, *Against the Grain*, 145.
7 David C. Lindberg, *The Beginnings of Western Science: The European Scientific Tradition in Philosophical, Religious, and Institutional Context, 600 B.C. to A.D. 1450* (University of Chicago Press, 1992), 16.
8 Fichman, *Science, Technology, and Society*, 6.
9 John Gowdy, *Ultrasocial: The Evolution of Human Nature and the Quest for a Sustainable Future* (Cambridge University Press, 2021), 105.
10 Lindberg, *Beginnings of Western Science*, 18.
11 Fichman, *Science, Technology, and Society*, 6.
12 Diamond, *Guns, Germs and Steel*, 236.
13 Fichman, *Science, Technology, and Society*, 6.
14 Claude Lévi-Strauss, quoted in Diamond, *Guns, Germs and Steel*, 235.
15 Diamond, *Guns, Germs and Steel*, 226.
16 Lindberg, *Beginnings of Western Science*, 12.
17 Neil Postman, *Amusing Ourselves to Death: Public Discourse in the Age of Show Business* (Penguin, 1985), 62.
18 John 1:1–18.
19 *Archaeological Atlas of the Aegean: From Prehistoric Times to Late Antiquity* (Ministry of the Aegean–University of Athens, 1999), 12.
20 Lindberg, *Beginnings of Western Science*, 13.
21 Fichman, *Science, Technology, and Society*, 7.
22 Andrew Ede and Lesley B. Cormack, *A History of Science in Society: From Philosophy to Utility*, 3rd ed. (University of Toronto Press, 2017), 3.
23 George Grant, *Technology and Justice* (University of Notre Dame Press, 1986), 13.
24 Ede and Cormack, *History of Science*, xii.

III

Premodernity

Chapter Seven

The Classical West

GREECE AND ROME

WE HAVE SEEN THAT MYTHOLOGICAL FORMS OF THINKING ABOUT THE world were increasingly rejected by agricultural and literate societies, because such societies saw little benefit to stories or symbolic actions that did not contribute to the technological and scientific mastery of the world. This was nowhere more true than in ancient Greece, where "the remarkable achievement of separating the investigation of the laws of Nature from any religious questions of the relationship between humans and the gods" began to blossom.[1] As the world's first widely literate society, the Greeks were confident that the use of abstract, linear, static, and skeptical rationality would reveal the truth about the universe—that science (as *epistemē*, see Chapter Two) would provide them with the *logos*, the picture of the world that correctly corresponds to the facts (see Chapter Six).

However, we also saw that faith in the power of human reason to intellectually comprehend a logical cosmos does not necessarily translate into practical knowledge that will amplify our technical powers, nor translate into a desire for such practical knowledge. Knowing the *logos* of all things does not necessarily mean one wants or will get *technē* as well. Technē and logos can be quite independent of each other. Rather, as we saw in Chapter Three, it is *modern* civilization that "is distinguished from all previous civilisations because our activities of knowing [logos] and making [technē] have been brought together in a way which does not allow the once-clear distinguishing of them."[2] Modern Western cultures view science and technology as mutually reinforcing and integrated, such that 'useless science' is about as nonsensical to our ears as 'purely theoretical technology' is. Yet our very word 'technology' combines the Greek words for theoretical science and practical crafts (see Chapter Three), which is why we frequently refer to 'technoscience' in this book. Both prehistoric agriculture and literalism involve desires for mastery—technological mastery of the physical world in agricultural societies and intellectual mastery of the physical world in literate societies—but the systematic combination of these two lies in modernity. If we are going to at least consider how these two aspects might have come together, we will have to investigate *premodern* societies—specifically the classical and medieval cultures of the Western world, wherein the eventual unification of technē and logos took shape. In this chapter, we will examine the different but related societies of classical Greece and Rome.

Greek Natural Philosophy

My students (which may include you) are often disappointed or perplexed to learn that there is a philosophical component to an STS course. The only consolation I offer them is that there is *always* a philosophical component to science and technology, and it's only the very recent invention of the word 'scientist' that unhelpfully blinds us to that fact. As we saw in Chapter Two, the term 'scientist' was invented in 1833, and so a great many scientists of the recent past (e.g., Sir Isaac Newton) identified themselves as natural philosophers. We also saw that natural philosophy simply means 'systematic theoretical inquiry into nature,' and as such we should try to be less squeamish about philosophy.[3] Philosophy is simply "an unusually disciplined and persistent attempt to understand ourselves and our place in the world" characterized by a "relentless use of reason."[4] Philosophy is the transliteration of the Greek word for "the love of wisdom." In these respects, then, there's not a terrific amount of difference between Western philosophy and Western science. Indeed, the very first scientists were also the first Western philosophers. They were Greeks.

The first recorded natural philosopher was a man named Thales, who flourished around 585 BCE, predating the Buddha. We don't have access to any of his own writings, but ancient authorities all attest that his big idea was that *everything is made out of water*. It's beside the point that we now think this claim is false. What matters is rather that Thales proposed it at all! For the first time in recorded history, we see somebody proposing a description of the ultimate nature of all things which was supposed to be literally reducible to something utterly mundane and boring: water. In an astounding feat of original thinking—saying what no-one had said before—Thales proposed a naturalistic or materialistic explanation of the world.[5] His watery worldview was not supposed to be a myth.

Clearly, nobody thought the world actually looked like it was made out of water. Experientially, the world does not appear watery. Rather, as we saw in Chapter Six, the unifying feature of the world (in this case, water) is supposed to lie *behind* or *beyond* the way things look: appearances do not equal reality. "Philosophy," says W.K.C. Guthrie, "started in the faith that beneath this apparent chaos there exists a hidden permanence and unity, discernible, if not by sense, then by the mind."[6] Even in terms of modern science, the world doesn't look like it's made out of (sub)atomic particles, but we reason that it must be. The logos, therefore, transcends what things look like.

Now to say that the world is made out of a natural element is a theoretical rather than a practical claim. You are not going to get rich or design better ploughs by 'knowing' that everything is made out of water. According to Aristotle, other Greeks mocked Thales for how useless his natural philosophy was, especially in terms of how economically ineffective it was (Thales was financially poor). His watery logos was not a techno-logos, because its supposed intellectual mastery did not give him practical mastery over nature or money. But Thales rose to this challenge:

> [H]e drew upon his knowledge of the heavenly bodies to predict a large olive crop, and collecting some money while it was still winter he bought up all the olive presses in Miletus and Chius, securing them by partial payments very cheaply because of the absence of competing bids. When the proper time arrived there was a sudden demand for olive presses, which he then rented out on his own terms, making large profits for himself.[7]

Thales was able to use his scientific knowledge to predict agricultural conditions and make a killing with financial speculation, but he did this only to prove that his knowledge wasn't necessarily useless, not to indicate a new practical direction for his philosophical investigations. For Thales, theoretical science was valuable apart from any economic or practical value it might be turned towards, and this non-pragmatic attitude towards science characterized Greek scientists in general.

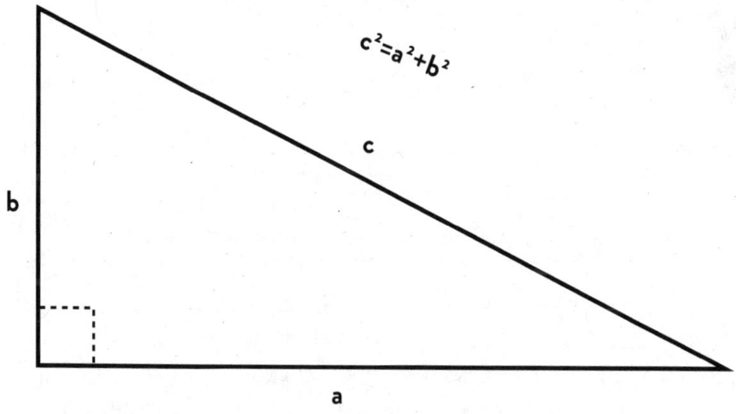

The Pythagorean theorem.

The Pythagoreans, a Greek religious sect that followed the teachings of the semi-legendary Pythagoras, believed that *number* (rather than a material element, water) was the ultimate nature of all things. They discovered the mathematical relations present in musical harmony (like the relation between the pitch of a note and the length of a harp string) and the Pythagorean theorem (which was useful in architecture), but they were more interested in worship and ritual action than using their musical or geometrical knowledge for any practical benefit. Mathematics was divine, not a tool of the ruling class. Other Greek natural philosophers also made startling scientific discoveries: e.g., all things are made of 'boundlessness' or unformed material/pure matter (Anaximander); qualitative differences in things are the result of quantitative differences in the basic element (Anaximanes); complex life forms biologically evolved out of simpler ones (Empedocles); and all things are made out of *atoms*—the Greek word for 'that which cannot be divided' (Leucippus and Democritus). But in all these cases and more, the natural philosopher's purpose was intellectual mastery alone, with virtually complete disregard for practical or technical applications of scientific knowledge. Classical Greek scientists were remarkably uninterested in combining technē with logos.

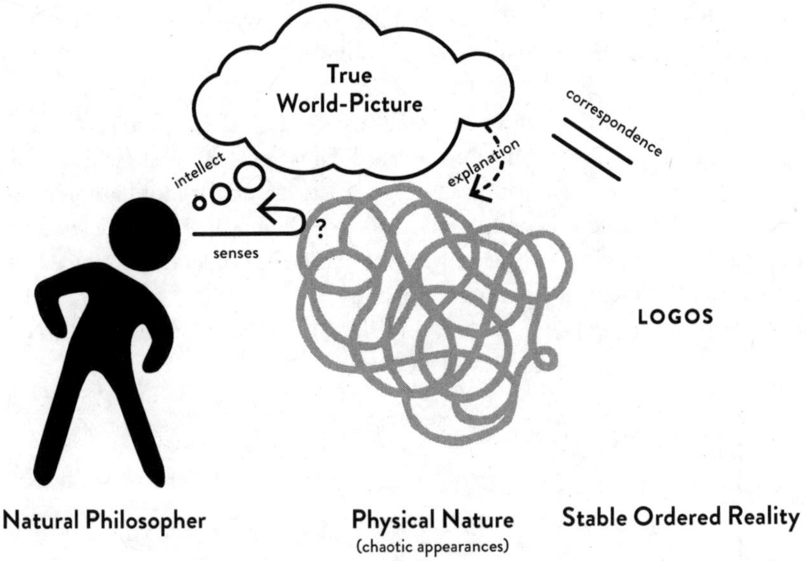

The distinction between appearances and reality.

The enduring gap in the Greek mind between intellectual and technological mastery is related to the distinction between appearances and reality. If the truth about all things is an ordered, logical reality that lies beyond the chaotic appearances perceived by our senses, then a scientist will always be transcending the sensible realm to intellectually access the Stable Ordered Reality which can only be known by the mind. The physical world as it appears to us is not a primary scientific concern; rather, the universal laws of organization behind it are. But practical concerns—like making money, building temples or roads, even going to war—all exist in the realm of appearances, the world we access with our senses. Therefore, the concerns of this world—including economics and technology—are, for a Greek scientist, secondary at best.

This neglect was no more clearly the case than with Socrates, who eschewed natural philosophy altogether, and his student Plato. For them, the visible world was downright untrustworthy, and provided no helpful information about ethics, politics, or life after death. The physical world is constantly in flux, but truth cannot always be changing. Therefore truth exists in an unchanging realm altogether separate from the physical realm, and the job of the philosopher is to get to this realm of intellectual reality the way a prisoner escapes from a dungeon.[8] (In this respect, they had much in common with the earlier natural philosopher Parmenides, who argued that because reality is unchanging, all change or motion is illusory and thus the physical world is illusory and does not actually exist!) True scientists, for Socrates and Plato, don't even bother themselves with natural philosophy, but rather focus on mathematics, formal logic, and the unchanging 'ideas' or 'forms' (cf. Aristotle's formal cause below) of which everything physical is an imperfect copy. Science, here, is about leaving the world behind as much as possible. As such, Socrates and Plato have a reputation as *otherworldly* philosophers. Their intellectual efforts were directed largely towards a world other than the one we experience with our senses.

Aristotelian Science

Plato's student Aristotle, however, disagreed with his teacher's otherworldly interpretation of the appearance-reality distinction. What is the point of science if it doesn't help us understand the world we live in? By comparison to Socrates and Plato, Aristotle was a *thisworldly* philosopher. He was the last of the great Greek scientists, and as such was immensely influential in subsequent Arabic and European science.[9] Even though science must transcend individual physical things and articulate universal, causal explanations, Aristotle insisted that it must do so in a way that does not explain away the physical world. In fact, all knowledge starts with sense experience: "the senses," he said, "give the most authoritative knowledge of particulars."[10]

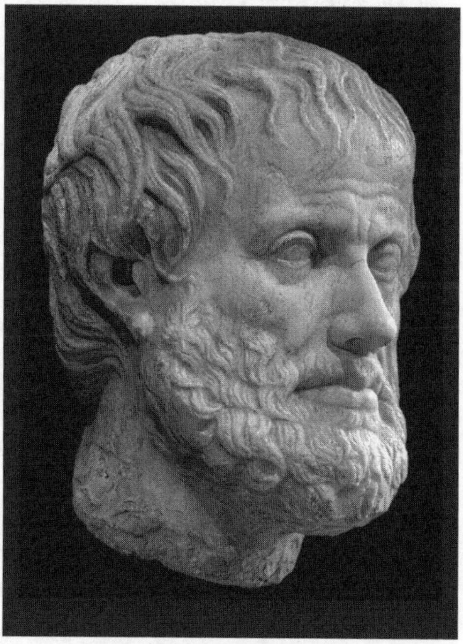
Aristotle (384–322 BCE).

Aristotle was therefore renowned for his interest in biology and his study of natural specimens. As a field researcher, however, he was careful to distinguish his method from experimentation: "to find out what is natural," he argued, "we must study specimens which retain their nature and not those which have been corrupted."[11] An experiment is a highly artificial arrangement, something which would never spontaneously occur in nature. As a corruption of a natural state, Aristotle would deny that experimental science tells us the truth about nature itself: "such tests only showed how the thing being tested acted in the test rather than in nature."[12] As artificial conditions, experiments are *technical* arrangements, not natural ones, another reminder that (unlike in classical Greece) modern science is inextricably bound up with technology. Say what you like about ancient Greek science, but contemporary STS scholars will insist that modern technoscience is *artificial* knowledge.[13] This point will come to bedevil us later, in Chapter Fifteen, when we consider the social construction of scientific knowledge.

If all knowledge begins with sensation, then for Aristotle, human scientists lie on a continuum with nonhuman animals which also possess faculties of sense. Some animals not only sense things but also remember those sensations, and as a result can be taught or trained, like a dog. Aristotle did not think many animals have what he calls "connected experiences," however, whereas humans do. Therefore, humans can learn things through experience, from paddling a whitewater rapid to providing a stylish haircut to baking the perfect bagel. Experience, says Aristotle, is knowledge of individual things (like a stretch of river, a person's hair, or a lump of dough), and for the most part human beings get along just fine with experiential knowledge.

But experiential knowledge is not science. For Aristotle, it's not even technical knowledge! Knowing how to do things from past experience is not the same as knowing how or why those things work. Aristotle categorized experiential knowledge at the level of manual labor, and characterizes manual workers as being "like certain lifeless things which act indeed, but act without knowing what they do, as fire burns,—but while the lifeless things perform each of their functions by a natural tendency, the labourers perform them through habit...."[14] That is to say, the knowledge of habit and experience is suitable for slaves, the same people for whom mathematics was invented to control (see Chapters Five and Six).

To have practical or technical knowledge, by contrast, is to be a master craftsman or artisan who can instruct or teach his slave laborers what to do, how to do it, and why. In our day and age, there is a derogatory saying that "those who can, do, and those who can't, teach." Aristotle would disagree: "in general it is a sign of the man who knows, that he can teach, and therefore we think art more truly knowledge than experience is; for artists can teach, and men of experience cannot."[15] Technē (which is art or craft, as we saw in Chapter Three) is a knowledge that goes beyond mere individual or particular things to a certain class of universals. Rather than simply knowing how to place one brick upon another, the artisan or craftsman can design a wall or even a building, proceeding from a set of abstract plans and directing laborers so that the final product can be realized. Therefore, Aristotle says that

> *knowledge* and *understanding* belong to art [technē] rather than experience, and we suppose artists [craftsmen] to be wiser than men of experience ... and this because the former know the cause, but the latter do not. For men of experience know that the thing is so, but do not know why, while the others know the 'why' and the cause.[16]

Because technē is a form of knowing the causes of things, it is closer to science than simply making practical effects happen.

A catapult was originally a device that could pierce a shield. *Kata* means 'through,' and *pelta* was a type of shield.

Yet being a skilled artisan who knows how to cause general classes of technics like 'temple' or 'catapult' to take shape does not make someone a scientist either. Technical knowledge, though it is knowledge of certain universal patterns, is still only universal within certain categories of effect production. As such, it is ultimately directed towards particulars (rather than towards universals themselves), otherwise it would not be applied or useful knowledge. But the 'higher' sciences—i.e., science per se, or 'pure' theory—are more general. Full-blown sciences are not limited to certain categories of production, nor are they directed towards manipulating physical materials; they are directed towards universal laws and patterns in themselves. For Aristotle, the more general, universal, and simple a knowledge claim can be, the more scientific it is: "all men suppose what is called wisdom to deal with the first causes and the principles of things … and the theoretical kinds of knowledge to be more of the nature of wisdom than the productive. Clearly then wisdom is knowledge about certain causes and principles."[17] The more abstract, eternal, and universal knowledge is, the more scientific it is, and the farther away it is from practical technology.

Universal truths (like the laws of nature discussed in Chapter Two) are 'simple' in the sense that there are fewer universal laws than there are particular things. For example, there is only one law of gravity while there are innumerable physical things which are subject to the law of gravity. It would be 'wiser' to try and understand the law of gravity than to try to figure out an individual and unique reason for why each individual and unique thing falls when dropped. So for Aristotle, the characteristic "of knowing all things must belong to him who has in the highest degree universal knowledge; for he knows in a sense all the subordinate objects."[18] That is why he supposes "that the wise man knows all things, as far as possible, although he has not knowledge of each of them individually."[19] Knowledge of universal principles is therefore better than knowledge of particular things, including how to manipulate them technologically.

The Aristotelian scientist will, when all is said and done, know the universal causes and principles of *everything*. Clearly, this is a monumental task, but scientists should not be daunted by it. After all, "he who can learn things that are difficult, and not easy for a man to know, is wise (sense-perception is common to all, and therefore easy and no mark of wisdom)."[20] Even though, for Aristotle, all knowledge begins with sense-perception, "the most universal [things] are on the whole the hardest for men to know; for they are furthest from the senses."[21] It's so difficult, in fact, that the upper reaches of Aristotelian science are characterized as superhuman: "the possession of [the highest science] might justly be regarded as beyond human power...."[22] The best science goes beyond physics to *metaphysics*, which is where we will find the set of universal laws governing physics itself, the universal laws which make anything (including physics or physical things) possible. For example, how is it possible for causation to occur? Or why is there something rather than nothing? Is there a purpose to existence at all? What is being? These are metaphysical questions. Aristotle goes so far as to say that "this science [i.e., metaphysics] alone is ... most divine."[23] The ideal Greek scientist knows the world the way a god would know the world.

Metaphysics: the ultimate reality of all things, physical or not; the study of the universal laws which make anything possible, physical or not.

Greek Science, Technology, and Social Hierarchy

For Aristotle, metaphysics is the divine science. Technology is also a form of knowing, but it is far down the list of importance, because technologically mucking about with actual physical things is emphatically *not* divine! Aristotle contributes to the usual story here by giving a narrative of progress which starts with the lower arts and terminates in the divine sciences of the leisure class:

> At first he who invented any art that went beyond the common perceptions of man was naturally admired by men, not only because there was something useful in the inventions, but because he was thought wise and superior to the rest.[24]

This is pretty easy for modern readers to relate to. Of course someone who can invent something useful is wiser than and more superior to someone who can't! Aristotle continues:

> But as more arts were invented, and some were directed to the necessities of life, others to its recreation, the inventors of the latter were always regarded as wiser than the inventors of the former, because their branches of knowledge did not aim at utility.[25]

This might be harder to relate to, because aren't useful skills more valuable than merely recreational ones? And yet, mainstream Western culture pays agricultural workers substantially less than workers in sports and entertainment, even though without the former we'd mostly starve to death while without the latter we'd mostly be bored. Contemporary Western cultures certainly grant more economic value (if not prestige) to leisure activities than to labor activities, and in this respect we're not much different from ancient Greeks.

Aristotle concludes that the theoretical sciences (which, unlike the practical sciences, do not aim at producing artifacts) only arose historically when there was an agricultural surplus sufficient for freeing a fortunate few from the life of labor:

> Hence when all such inventions were already established, the sciences which do not aim at giving pleasure or at the necessities of life were discovered, and first in the places where men first began to have leisure. This is why the mathematical arts were founded in Egypt; for there the priestly caste was allowed to be at leisure.[26]

As we saw in the previous chapter, writing was a communication tool invented to facilitate the enslavement of other human beings. These 'managed' peasants were not in any position to pursue a life of scientific research. But the upper classes had both access to writing (and its attendant intellectual effects) and the money and time to make scientific investigation possible. The leisure of Aristotle's scientists was only possible because "a large proportion of the work to keep society going was done by slaves.... [M]ost of the menial positions and even the artisan class were made up of slaves. Those who worked with their hands were at the bottom of the social hierarchy."[27] Therefore, another reason why Aristotle would value theoretical science for its own sake rather than "on account of its results"[28] is because theoretical science is for free men, while technology is for slaves: "as the man is free ... who exists for himself and not another, so we pursue this as the only free science, for it alone exists for itself."[29] Slaves have no choice but to be useful, whereas if you're rich, you can be as useless as you want to be.

The social divide between free and bonded men is why we do not see technē and logos conjoined as technoscience in the ancient Greek context. Menial tasks of production are for subhumans while scientific enlightenment is for men who approach the status of gods. (Aristotle was also explicitly sexist, saying that women were by nature unsuited for scientific rationality; this is why he always refers to 'men' in this chapter.) A person works at a job because they have to, but a free person does whatever they want to. Therefore, science doesn't *have to* have a job or a technological application; science can be pursued for entirely non-utilitarian or non-economic ends. Why should science have a practical use anyway? The activities of utility are activities of necessity, not of freedom. Technology is not liberating: it is suitable for slaves. After all, nobody really wants to work—we only get jobs because we have to! Thus science should be free whenever it can be.

An ancient Greek would therefore think that what we now call the 'Liberal Arts'—philosophy, literature, history, etc.—should be taught in a university's Faculty of *Science*, while useful skills like engineering, dentistry, or welding should be taught in the Faculty of *Arts* (the arts are for artisans, as you'll remember from Chapter Three). Sciences (like philosophy or the humanities) are *not* supposed to be about jobs, whereas arts (like engineering and the trades) are skills used by manual laborers to economically scrape by. A Greek would tell you it's completely foolish to go to university to get a better job, because job training is for slaves whereas education is for free (aka wealthy) citizens who can choose to learn whatever it is that interests them. Therefore, you aren't supposed to feel like you need to 'do anything' with your degree; you're supposed to study STS (or English or art history or whatever) because *you want to and can*. Science isn't about work either, then; it's rather one of the things you could do (if you wanted to) if you won 'the lottery' (i.e., being born into the upper social classes in Greece). This is why it would not make sense to a Greek to combine technē with logos, any more than it would make sense (to them) to combine a slave with a god. This is also why our current attitudes towards education are characteristically modern: we operate on the inverse assumption that knowledge is power and that education is a tool; university is all too often about training for a job and making money, rather than learning and exploring new things simply because knowledge is good in itself.

The Naturalness of Artifice

The Greek association of technē with the lower classes had an ironic implication for the nature of technology, however: it meant technology was *natural*! By contrast, nowadays we tend to think that technology is artificial, and artificiality implies *unnaturalness*. 'Nature' or 'the physical,' as we saw in Chapter Three, means roughly 'that which gives birth to itself,' and clearly artifacts do not give birth to themselves. However, in ancient Greece the issue was more complex than that. Remember that within an agrarian context, monotonous and interminable physical labor seems necessary for human survival. Nature is recalcitrant to sedentary human demands, and if we don't struggle against it technologically, then agricultural civilization collapses. Therefore, the (supposed) 'nature of Nature' *requires* humans to create the technologies by which agrarian civilizations survive: ploughs, oxen, granaries, fences, houses, shovels, etc. So by crafting clothing (etc.) for ourselves, we're only doing what 'nature' forces us to do anyway. Necessity comes from nature, and the laboring and artisanal classes work out of necessity. Leisure and freedom, by contrast, are divine and supernatural (if not technically 'unnatural').

It's like a gift of the gods if you don't have to work for a living. 'Nature' sure won't let you take time off (if you're a farmer).

Craftsmanship is necessary, therefore, because nature (supposedly) requires us to have the tools of agrarian survival and to satisfy our basic animal needs. This is why the Greeks thought that lower-class people—e.g., slaves, foreigners, women—were better suited to technological work: they're barely human anyway. To a Greek mind, the lower down on the social hierarchy, the closer you are to animals, dirt, and nature, while the higher class you are, the closer you are to the gods. Manual laborers are—by 'nature'—suited to 'natural' work, namely technē. The practical arts and crafts are closer to material nature than scientific reasoning is, while the leisure classes are closer to the hidden and divine truths of the universe which are far from the senses. Technology is immanent while science is transcendent, which is why technological artifacts can be thought of as natural.

Therefore, when an artifact is created by a (sub)human artisan, that artifact is designed to imitate nature: "the ancient Greeks did not conceive of technological change and economic production in the modern terms of efficiency and progress. Practical techniques were judged as analogous to and facilitating for our purposes … cosmic and natural processes…."[30] A Greek temple should be as massive and as permanent as a cave in the side of a mountain, so that it will last as long as the natural feature it is inspired by. Roman aqueducts "may seem 'overbuilt' by modern standards, but that is because they are designed not just to carry water but to do so in perpetuity, like rivers or streams."[31] Biodegradable architecture wouldn't have made sense to the Greeks (and the Romans, who we'll examine below) because the logos behind the apparent chaos of natural appearances is eternal and unchangeable. For them, the nature of Nature is permanence, and so natural technologies and artifacts must be in harmony with unchanging and permanent natural laws.

This doesn't mean that the Greeks failed to make a distinction between artifacts and objects of nature. After all, as we saw in Chapter Three, the distinction between physical and artificial things is terminologically Greek. Here the difference is simply between who brought the object into being; physical things are their own agents of actualization whereas technical objects are actualized by external agents. In both cases, however, the objects are to follow their respective cosmic principles. Even when particular tools and techniques are necessitated by an agrarian social context, that social context is assumed to be natural (not socially constructed); humans are (falsely) assumed to be naturally agricultural. Technologies, then, are (for the Greeks) artifacts that humans create in the larger, surrounding context of naturalness. To *cause* something to exist—the way a silversmith might cause a chalice to exist—is to reveal the natural truth about something, to disclose (uncover) the hidden logos of something—as if the chalice was always inside the ingot of silver, waiting to come out.

According to Martin Heidegger, "[w]hat we call cause and the Romans call *causa* is called *aition* by the Greeks, that to which something is indebted."[32] For the Greeks there were four ways in which a thing is caused to be, and the silversmith (the artificer or technician) is only one of these. These are the famous 'four causes' of Aristotle (although he was using concepts from his predecessors):

> (1) the *causa materialis*, the material, the matter out of which, for example, a silver chalice is made; (2) the *causa formalis*, the form, the shape into which the material enters; (3) the *causa finalis*, the end, for example, the sacrificial rite in relation to which the required chalice is determined as to its form and matter; (4) the *causa efficiens*, which brings about the effect that is the finished, actual chalice, in this instance, the silversmith.[33]

The modern technoscientific worldview sees only the agent (the *causa efficiens* or efficient cause) as defining what 'causality' means: "For a long time we have been accustomed to representing the cause as that which brings something about."[34] But an overemphasis on this form of causality leads us to the impression that human technologies are unnatural interventions into the realm of nature.

Aristotle's Four Causes:
- the material cause (matter)
- the formal cause (shape)
- the final cause (purpose)
- the efficient cause (agent)

The classical Greeks, however, saw causation as a fourfold type of indebtedness. An artificial thing like a chalice can only exist on account of the silver material it is made out of; therefore, the "chalice is indebted to, i.e., owes thanks to, the silver...."[35] To identify the material cause is to recognize our indebtedness to the matter for being the base of the technical object. The chalice can also only exist on account of the shape or idea of 'chaliceness,' the abstract model or form which the raw material is formed into. To identify the formal cause is to recognize our indebtedness to the form for being the logical structure of the technical object. Third is the cause most strange to our modern ears: the *telos*, the purpose or goal of the object. But a chalice that is not "circumscribed as [a] sacrificial vessel" would not be a chalice.[36] To a Greek, everything has a purpose; of course a technology must have a purpose to fulfill, otherwise it would be useless! Therefore, to identify the final cause is to recognize our indebtedness to the reason for which a thing is made, without which it would not be made at all. Only when these three causes are in view can the fourth, the agent or efficient cause, be called upon to combine them all into a single thing. The silversmith does not do this automatically, but critically appraises the matter, the form, and the goal in the process of bringing forth the chalice into being. So to identify the material cause is to recognize our indebtedness to the agent for purposefully shaping the matter into the artificial object. For the Greeks, none of these causes—not the matter nor the form nor the purpose nor the agent—are contrary to nature. Technologies are rather imbued with meaning and significance which derives from the larger context of the universe itself (as they saw it). Yet being natural does not make something divine; as we have seen, for the Greeks that was something else entirely.

Roman Power

Because scientific reason was seen as supernatural and divine while technical activity was seen as natural and lower class, the Greeks had no reason to combine the two. Thus a characteristic feature of classical Greek science and technology was their separation from each other. This separation of technē and logos also held true for the Romans, but for altogether inverted reasons: Romans were preoccupied with maintaining military and economic power, and so technology interested them more than divine science, which was disappointingly useless. Science is powerless, and thus generally uninteresting to a Roman.

Politically speaking, Roman society was considerably different from ancient Greek societies. Many of the Greek city-states were democracies where political power was achieved by convincing public speaking; rhetoric and logic where highly valued civic skills, and as such, scientific reasoning was similarly prestigious. Rome, however, was not a democracy, but (at first) a republic run by the wealthiest citizens who gained political office by "public demonstrations of power" both militarily and economically.[37] In fact, Rome was a highly militarized society whose economy was dependent on conquest—i.e., stealing the wealth of neighboring countries like Carthage or Gaul. Political power and military power were typically vested in the same fabulously rich person (e.g., Julius Caesar) and, as such, the "Romans were at heart a people interested in practical knowledge.... Making nature do your bidding was more essential than right reasoning."[38] You didn't need to be a philosopher if you could kill anyone you disagreed with.

Unsurprisingly, the Roman military juggernaut eventually conquered the culturally Greek successor states that had formed after the early death of Alexander the Great. In this respect "the conquering Romans found themselves in a somewhat ambiguous position. There could be no doubt of the superiority of Greek thought in almost every intellectual area.... Yet the Romans were conquerors and the Greeks their vassals and slaves. The Romans might well ask themselves of what use scientific learning and speculation was in a world dominated by military power."[39] The famous Greek scientist Archimedes was a case in point. Archimedes lived in Syracuse, a Greek city on the island of Sicily. Uncharacteristically, he is said (according to Plutarch) to have turned his scientific knowledge to practical application, devising many unique technologies to fight the Romans who were besieging his city. Sadly, his inventions were unable to prevent the Romans from breaching the walls and gates. With the Romans running rampant within, Archimedes went back to working out geometry puzzles by drawing them in the sand. When a Roman soldier came to tell him to surrender, Archimedes told the soldier to wait until he had finished his math. So the Roman soldier killed him. Legendary or not, this story illustrates the ideal Greek scientist as aloof from the practical concerns of the world, while his death at Roman hands illustrates why the Romans—who were all about efficiency and expediency—saw scientific expertise as generally powerless and useless.

The Romans did not wantonly destroy the intellectual heritage of the Greek world, however. Following their military conquest model of economics (i.e., taking other countries' money), they put captured Greek science to their own practical uses wherever they could: studying "Greek philosophy was seen as a good method to discipline the mind just as the legionnaires disciplined the body; both prepared the elite of Rome for their role as masters of the world."[40] Most scientific writings during the Roman period were collections of existing research which they thought had technological applications, rather than new scientific work: "the emphasis was placed on the wholly practical aspects of scientific enquiry. Given this technological orientation, the most important Roman scientific writings tended to be works of synthesis (from previous authors) and arrangement, rather than original investigations."[41] Even the famous armament of the Roman legionary was 'borrowed' from the cultures they conquered: the mail shirt (*lorica hamata*) and Montefortino helmet (with iconic neck and cheek flaps) were 'adopted' from the Celts to the north of Rome; the short, high-carbon steel stabbing sword (*gladius hispaniensis*) was taken from Spain; and the large oblong body shield (*scutum*) was probably adopted from the Italian hill tribes after the round hoplite-style shield they copied from the Greeks or Etruscans failed to suffice. In such a context of tool collecting, scientific advancements were few and far between; Ptolemy's astronomy and Galen's medicine stand out as rather lonely exceptions.

As with the Greeks, the two activities of making (technē) and knowing (logos) remained largely separate for the Romans—the only difference being a shift in evaluation. For both cultures, science was purely theoretical and thus pointless: but that pointlessness was a *good* thing to the Greeks (because it signified liberty and divinity), but a *bad* thing to the Romans (because it signified military defeat and economic stagnation). On the other hand, technological power signified undignified slavery to the Greeks but military, economic, and political prestige to the Romans. In either cultural context, there would have been no unity of the craftsman and the theorist; there would have been no technoscience. Thus, when Hero of Alexandria built a steam engine in approximately 62 CE, it was viewed as a mere curiosity or toy. Combining the theory of a steam engine with a practical application—say, to make steam trains and a railway system—was an *idea* which would have occurred to neither the Greeks nor the Romans. Even if they had had the requisite materials available

A modern reenactor depicting the body armor, helmet, sword, and shield adopted by the Romans.

for a railway network (e.g., advancements in metallurgy and mining, etc.), it would have taken a different way of culturally conditioned thinking to combine technē and logos such that creating steam trains and a railway network would have actually been seen as desirable. To see how such an idea of combining knowledge and power could arise, we will need to examine the European Middle Ages.

Notes

1 Martin Fichman, *Science, Technology, and Society: A Historical Perspective* (Kendall/Hunt, 1993), 8.

2 George Grant, *Technology and Justice* (University of Notre Dame Press, 1986), 12.

3 For public examples of scientific squeamishness about philosophy, simply search the web for what Lawrence Krauss or Neil deGrasse Tyson have to say about it.

4 David H. Lund, *Making Sense of It All: An Introduction to Philosophical Inquiry*, 2nd ed. (Prentice Hall, 2003), 2.

5 Thales wasn't altogether naturalistic, however. One of his other fragmentary aphorisms holds that *all things are full of gods.*

6 W.K.C. Guthrie, *The Greek Philosophers: From Thales to Aristotle* (Harper & Row, 1950), 24.

7 Aristotle, *Politics*, 1259a9, quoted in Philip Wheelwright, ed., *The Presocratics* (Prentice Hall, 1997), 45–46.

8 Plato, *Republic*, 514a–521c.

9 Aristotle's student was Alexander the Great, a Macedonian who conquered the Persian empire and spread Greek culture across much of the Mediterranean world. This marked the end of classical Greece and the start of the transitional Hellenistic period, where Greek culture was synthesised with Indigenous traditions of the Near East. The Romans emulated Alexander and conquered the Mediterranean world, although under their rule the character of scientific research changed markedly (as we shall see later in this chapter).

10 Aristotle, *Metaphysics*, 981b10.

11 Aristotle, *Politics*, 1.v.5.

12 Andrew Ede and Lesley B. Cormack, *A History of Science in Society: From Philosophy to Utility*, 3rd ed. (University of Toronto Press, 2017), 22.

13 Sergio Sismondo, *An Introduction to Science and Technology Studies* (Wiley-Blackwell), chapter 15 (1st ed., 2004) or chapter 14 (2nd ed., 2010).

14 Aristotle, *Metaphysics*, 981b1.

15 Aristotle, *Metaphysics*, 981b5.

16 Aristotle, *Metaphysics*, 981a25.

17 Aristotle, *Metaphysics*, 981b25–982a1.

18 Aristotle, *Metaphysics*, 982a20.

19 Aristotle, *Metaphysics*, 982a5.

20 Aristotle, *Metaphysics*, 982a10.

21 Aristotle, *Metaphysics*, 982a25.

22 Aristotle, *Metaphysics*, 982b25.

23 Aristotle, *Metaphysics*, 983a5.

24 Aristotle, *Metaphysics*, 981b15.

25 Aristotle, *Metaphysics*, 981b15–20.
26 Aristotle, *Metaphysics*, 981b20.
27 Ede and Cormack, *History of Science*, 6.
28 Aristotle, *Metaphysics*, 982a15.
29 Aristotle, *Metaphysics*, 982b25.
30 Robert C. Scharff and Val Dusek, *Philosophy of Technology: The Technological Condition—An Anthology* (Blackwell, 2003), 5.
31 Scharff and Dusek, *Philosophy of Technology*, 5.
32 Martin Heidegger, "The Question Concerning Technology," in *Basic Writings*, ed. David Farrell Krell (HarperSanFrancisco, 1993), 314.
33 Heidegger, "Question Concerning Technology," 313–14.
34 Heidegger, "Question Concerning Technology," 314.
35 Heidegger, "Question Concerning Technology," 315.
36 Heidegger, "Question Concerning Technology," 315.
37 Ede and Cormack, *History of Science*, 30.
38 Ede and Cormack, *History of Science*, 29, 30.
39 Fichman, *Science, Technology, and Society*, 11.
40 Ede and Cormack, *History of Science*, 29.
41 Fichman, *Science, Technology, and Society*, 12.

Chapter Eight

The Medieval West

LATIN EUROPE

ONE OF THE CHARACTERISTICS OF MODERN SCIENCE AND TECHNOLOGY is their functional equivalence as technoscience. By contrast, one of the characteristics of science and technology in classical antiquity was their mutual separation from each other. Certain historical, intellectual, and social factors had to come into play before modern, Western technoscience could become anything approaching a unified endeavor. These various factors are remarkably complex, but they all come to a head at the end of the European medieval period. Speaking possessively as a member of the modern Western intellectual tradition himself, the celebrated historian of technology Lynn White, Jr. boldly argues that "both our technological and our scientific movements got their start, acquired their character, and achieved world dominance in the Middle Ages...."[1] As such, if we as readers are supposed to understand modern Western technoscience as a social phenomenon, it behooves us to examine "fundamental medieval assumptions and developments" with respect to both natural philosophy and technology.[2]

> **Medieval Period:** the period of European history between the fall of the Western Roman Empire and the so-called Renaissance or early modern period; the word *medieval* means "middle age," from the Latin words *medium* and *aevum*.

The notion of "the Middle Ages" (or the "medieval" period, taken from *medium* and *aevum*, the Latin for "middle" and "age"), however, is somewhat problematic. What is this age in the middle of, anyway? The answer is that the very notion of a "medieval" period is already a part of the usual story. The people of the European middle ages—colloquially known to us as Vikings, princesses, knights in shining armor, Robin Hood, dirty peasants, etc.—certainly wouldn't have referred to themselves as living in-between two epochs, because to do so, they would have needed paranormal knowledge of the future. Looking back, however, we are in a position to say (if so inclined) that the middle ages were in-between two remarkable epochs: the classical era of Greek and Roman civilization, on the one hand, and *our* present civilization—modernity—on the other hand. If we like to see ourselves as the pinnacle of a long trajectory of "progressing continuity,"[3] then the middle ages are an unbecoming hiccup or bump in the road between the wonders of antiquity and the glories of the here and now. (Whether we *ought* to think of things in this way, of course, is what is under question in this book.) Therefore, the very idea of a middle age is uncomplimentary to that age, viewing it as a step backwards in the otherwise upward march of progress. This implicit denigration is even more so the case when speaking of the early middle ages, for this period (roughly from 500 to 1000 CE) is colloquially known as 'the Dark Ages.'

The Dark Ages and Western Monasticism

In 286 CE, the Emperor Diocletian split the Roman empire into two administrative units, the eastern half having its capital at Constantinople (now Istanbul) and later becoming known as the Byzantine empire. The capital of the western empire, the city of Rome itself, was sacked by Germanic invaders in 410 and 455 CE, with the conventionally recognized fall of the Western Roman empire coming in 476 CE when Odoacer declared himself king of Italy. If the Romans themselves had never been particularly interested in producing a context within which scientific research could flourish (see Chapter Seven), the new Germanic rulers of Italy, France, Britain, Spain, and North Africa were even less interested. Unsurprisingly, therefore, "much of Greek scientific thought was lost in the almost universal disruption of learning which followed the final collapse of the [Western] Roman Empire."[4] If the inhabitants of Western Europe at the time could not have said they were in 'the middle' of two great civilizations, there is some justification for suggesting that those inhabitants would have seen their times as 'dark,' at least with respect to education and research, if not also economy, military capacity, inventiveness, and agricultural production. There was an almost post-apocalyptic awareness[5] of the loss of classical civilization:

> The early Middle Ages were tinged with a certain pessimism.... Throughout Europe there were, literally, concrete examples that the past was better than the present, as the remains of the power of Rome dotted the landscape. Ruins of aqueducts, roads, and coliseums were a continual reminder of lost power and lost knowledge.[6]

If anything, then, the general mood would have been the opposite of progressive; rather, everything seemed to be in decline, 'progressively' getting *worse*.

St. Benedict of Nursia (480–547 CE).

But as we know, the world didn't actually come to an end, and so neither did scientific learning or technological innovation. In fact, even if you think the world is coming to an end, one reasonable response to that sense of loss is *conservativism*, the attempt to save or *conserve* from further decay what little yet remains. Additionally, one can also attempt to rescue what has been lost from total oblivion, and perhaps even bring it back. This mission of protection and rescue was one of the functions of medieval monasteries, the flourishing of which began with St. Benedict of Nursia's (c. 480–547 CE) rulebook for monastic life.

Technically, Western monasticism sets apart a group of Christian men (i.e., monks) or women (i.e., nuns) for religious contemplation and prayer. But there were practical implications for this form of religious life as well. Ideally, monasteries were self-sufficient communities, and so had to supply their own food. However, these institutions were usually founded on cheap land donated by the local baron or lord, and as such the land was not particularly fertile. Monasteries had to be quick to adopt innovative agricultural practices to increase productivity, or careful to retain advanced farming techniques from the Roman period that might have otherwise been forgotten. Monasteries also had a highly organized labor pool at a time when that was very rare. As a result, monasteries became remarkably productive, pioneering land reclamation below sea level in the Low Countries, and even to this day producing world famous cheeses and beers. Despite vows of poverty, monasteries grew increasingly wealthy.

The preservation of written texts, meanwhile, went hand-in-hand with contemplation and prayer. Without scriptures, for example, there's very little to inform your prayer or meditations. Moreover, the British monk Alcuin became the minister of education for Charlemagne's Frankish empire, and under his direction monasteries and cathedrals were required to establish schools so that (especially) priests could learn how to read. This increase in literacy was to serve an empire-building purpose: without writing, long distance communication breaks down, which makes holding together far-flung territories difficult. So not only were monasteries a crucial source of literacy and clerical training, they also had an interest in preserving much of the written material that survived the fall of the Western Roman empire. In the absence of the printing press, monastic *scriptoriums* painstakingly copied ancient and contemporary manuscripts to both save the old ones from decay and to make those texts more widely accessible. Not only were these books another source of wealth for monasteries (hand-copied books were terribly expensive and prestigious to own), but the process of copying encouraged a number of innovative information technologies as well. These included the book (*codex*) instead of the scroll, spaces between words, punctuation (like the comma), and lower-case letters (the Romans, by contrast, wrote in all caps). Therefore, monks were not only agricultural conservators and innovators, but informational conservators and innovators as well.

Monasticism: a system of religious communities of men or women who devote their lives to contemplation and prayer.

Broadly speaking, monasticism was part of a trend in the Early Middle Ages to try to bring back the glorious knowledge of the classical world. Charlemagne wanted to bring about a new 'Holy Roman Empire' (a name that came to be used for the German-speaking territories of Europe, until Napoleon's victory at Austerlitz in 1806), and monasteries became the repository of the knowledge—both textual/theoretical and practical—that had been saved from the cataclysm of the Germanic invasions. But as we'll see below, even though the Christians of the early middle ages weren't *only* trying to copy what the Greeks and Romans had done, their attempts at conservation and innovation were being surpassed in the Islamic world.

The Islamic Golden Age

Although the Western Roman empire fell to Germanic tribes about 1,500 years ago, the Greek-speaking Eastern Roman or 'Byzantine' empire did not. While the Byzantine Romans "enjoyed an amazing prosperity and vigor for over a thousand years,"[7] they were, like the classical Romans before them, not particularly renowned for scientific or technological innovation. White remarks that "the Greek East ... seems to have produced no marked technological innovation after the late 7th century, when Greek fire was invented."[8] The Byzantine empire frequently engaged in warfare against a new power ascendant in the Mediterranean, the Arabic-speaking Islamic states that arose after the death of the prophet Muhammad (c. 570–632 CE). By the end of the Umayyad Caliphate in 750 CE, the Islamic world stretched from modern-day Pakistan across North Africa and into Europe to include modern-day Spain and Portugal. This meant that Islam came to control many of the intellectual centers of the Mediterranean world, including Alexandria, giving their scholars access to crucially important academic material. Unlike the Germanic rulers of the old territories of the Western Roman empire, "the Islamic invaders were prepared to accept much of the cultures of the peoples whom they conquered."[9] Arabic-speaking culture and empire fostered a scientific and technological vibrancy that outshone the early medieval West.

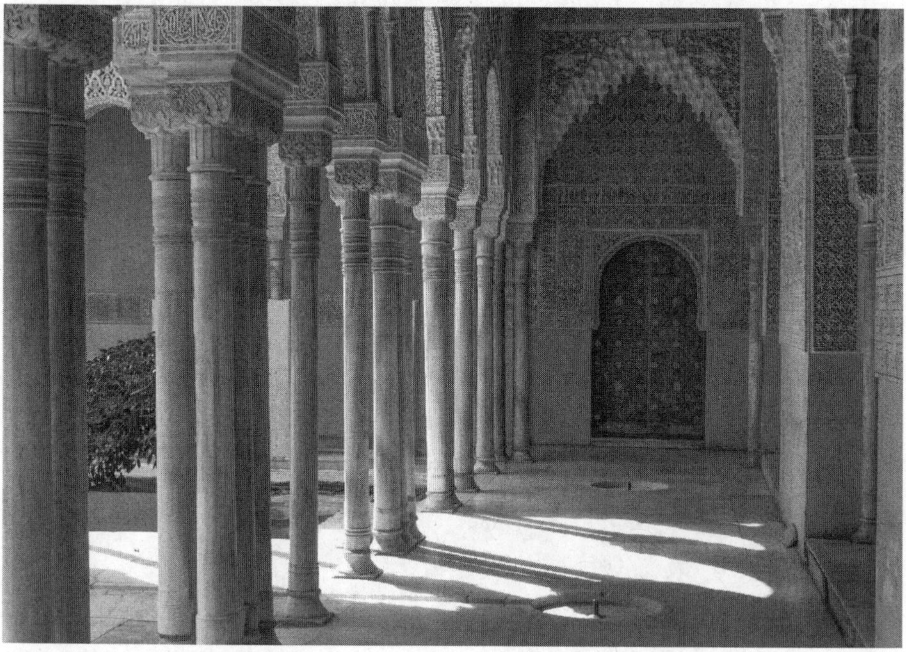

The Court of the Lions in the Alhambra, Granada, Spain.

Arab scholars had access to many more works of Greco-Roman science than did Western European monks, as well as Persian, Indian, and even Chinese material via the famed Silk Road. The Abbasid caliphs funded a massive translation operation, and by the year 1000 CE, almost every surviving work of Greek science had been translated into Arabic. But this undertaking was not simply a passive process of replication and assimilation. "The absorption of Greek science and philosophy into Islamic culture facilitated the development of an indigenous system of science"[10] because it was difficult not to approach pagan texts with a critical eye when reading them from a religiously and culturally external perspective. Arab scientists thus carried on a dialogue with authoritative Greco-Roman texts, sometimes correcting them "when later observations proved the earlier ones false,"[11] and also constructing original treatises of their own.

Islamic scientific expertise was perhaps most pronounced in the field of medicine. Arab physicians became more skilled than the classical Greeks and Romans, and especially the monastic doctors of medieval Europe. Medicine, moreover, is a practical and technical field that is nonetheless explicitly informed by theory, and its importance in the Arabic-speaking world may be one reason why Islamic scholars had a more hands-on approach to science than did the Greeks and Romans. Mathematics was another field in which the Arabic-speaking world excelled, so much so that a number of scientific terms in modern English have Arabic etymological roots: e.g., algebra, algorithm, alkali, elixir, zenith, alembic, and alchemy. Even the decimal system of numerical notation is known today as "Arabic numbers," for although the Arabs themselves learned it from the Hindus of India, Western Europeans learned of it via the Islamic world.

From the perspective of Western Europeans at the time, therefore, the ascendancy of Islamic culture may have been yet another reason to refer to their own age as 'dark'—the great lights of science, technology, and industry appeared to be in the hands of non-Christian 'infidels.' From the perspective of the Islamic golden age, on the other hand, "the appearance of the barbarous and ill-educated knights of Western Europe" during the Crusades "seemed of little threat to the power of the Islamic world."[12] However, the Crusades arose at an inopportune time for Islamic science, as the Seljuk Turks were also beginning to cross the frontier of the Islamic world. In the resulting social context of "political turmoil that disrupted all aspects of society and the swings from a high level of religious tolerance to strict fundamentalism that occurred almost overnight when leadership changed hands, making it politically dangerous to engage in work that suddenly might be deemed unacceptable,"[13] the Islamic golden age faltered and eventually ended.[14] The prolonged contact between crusading Western Europeans and both Islamic and Byzantine cultures, however, fostered a new intellectual and technological vigor in the West which would inadvertently establish the conditions necessary for what we now call modern Western technoscience.

Ibn Sina or Avicenna (c. 980–1037), one of the greatest natural philosophers of the Islamic tradition.

Scholasticism and the Medieval Synthesis

When the Romans conquered the Greek-speaking world, they felt the need to incorporate that intellectual heritage into their own particular cultural style. So too did the Muslims when they conquered the southern half of the Greco-Roman world. It should not be surprising, then, that Latin Christians also critically appropriated this Greco-Roman-Islamic intellectual heritage, especially after the capture of Toledo and its many libraries in 1085 CE. As already mentioned, early medieval scholars had already been preserving and copying classical manuscripts and, like Islamic scholars, had been critically interacting with Greco-Roman thought as they incorporated it into a Christian context. Many Christian philosophers favored Platonism (see Chapter Seven) because its otherworldly emphasis on pure form, and its distrust of the body, were seen as meshing well with both monastic asceticism (self-denial) and Christian doctrines of (otherworldly) salvation and heaven.

St. Thomas Aquinas (1225–74).

As Western Christians reconquered Spain from the Muslims, however, more and more works of Aristotle became accessible to Latin-speakers, and his comparatively this-worldly focus (see Chapter Seven) increasingly came into vogue. Some scholars saw Aristotle's philosophy as a threat to Christianity, while others (most famously St. Thomas Aquinas) sought to show how Aristotle could be critically appropriated by Christianity. Indeed, the religion itself had its own this-worldly elements, especially the central figure of Christ who was held to be both fully divine and fully human. In this way, Christianity could be seen as affirming the goodness of the material world and connecting the supernatural and natural realms in a way that was rather foreign to either Plato or Aristotle.

The "schoolmen" (i.e., scholars) of medieval universities thus spent several centuries attempting to fine-tune a unity between science and religion, a delicate synthesis of Greek, Roman, Jewish, Islamic, and Christian intellectual resources. This "scholastic" project was carried out in universities, unique institutions of higher learning that, while sanctioned by official Church authorities, were not completely controlled by them:

> [T]he universities stood in a complex relationship to the larger structure of the Catholic Church. They were seldom under the complete control of any one bishop, and thus they ... allowed the debate about the primacy of faith or reason to be played out within their walls and cities. While several scholars were imprisoned for their impious views, the fact that these debates could take place at all speaks to the power and independence of these institutions.[15]

The vigorous intellectual environment of the medieval university is well illustrated by the format of debate or *disputatio*. A question or thesis statement was proposed, and students were to defend positions on either side of the issue, before the master stepped in to resolve the debate. In essence, modern graduate and doctoral degrees follow the same format:

> This skill extends to the modern day with the dissertation and defence system used to obtain a PhD, a doctorate in philosophy [*Philosophiae Doctor*]. The system supposes that the thesis is an argument made by a student who publicly defends it against questions posed by scholars knowledgeable in the field; our continued use of the method created by medieval scholars indicates how robust an educational system it is.[16]

Medieval universities provided an international network of scholars all using a common technical language (Latin) and engaging in a form of education premised upon rational skepticism. Add to this Aristotle's this-worldly emphasis on the five senses, and the medieval university and scholasticism provided several of the necessary conditions for the flourishing of science as we understand it today.

The synthesis between Christianity and Greco-Roman and Islamic science achieved by the scholastics was known as the *scala natura*, the Ladder (or Hierarchy or Great Chain) of Being. Within this theoretical model of all things, every possible kind of being had a place in the system, from sheer nothingness or void at the very bottom to Absolute Being or God at the very top. Science, technology, religion, society, and every creature under the sun (even angels and monsters) were all a part of a seamless and ideally stable unity:

> Aristotle's hierarchical cosmos—from formless matter through the "great chain of being" to the "intelligences" that moved the planets and, finally, to the "Unmoved Mover" (upon which all form and motion in the universe depended)—was analogous to the medieval Christian cosmos. This latter was predicated on a hierarchy stretching upward from serf, bourgeois, noble, king, angel and, finally, to Christ.... The views of the Greeks had been incorporated along with Christian doctrine in a vast synthesis, backed by the powerful authority of Church and State.[17]

The synthesis connected what we now call the natural and supernatural levels of reality. Similarly, what we now call science and technology were connected and each had their proper place in society, which was hierarchically ordered. While all the parts depended on each other—e.g., the wealthy depended on the working class for food, the working class depended on the fighting class for protection, and science depended on nature as the basis of knowledge—it was generally assumed that it was better to be wealthy, male, and human than poor, female, or an animal. While the Great Chain of Being incorporated into one unified system many of the things that the Greeks and Romans kept separate, the system still perpetuated an hierarchical evaluation of those things, a value-assessment that would have been very familiar to the Greeks and Romans.

The Great Chain of Being.

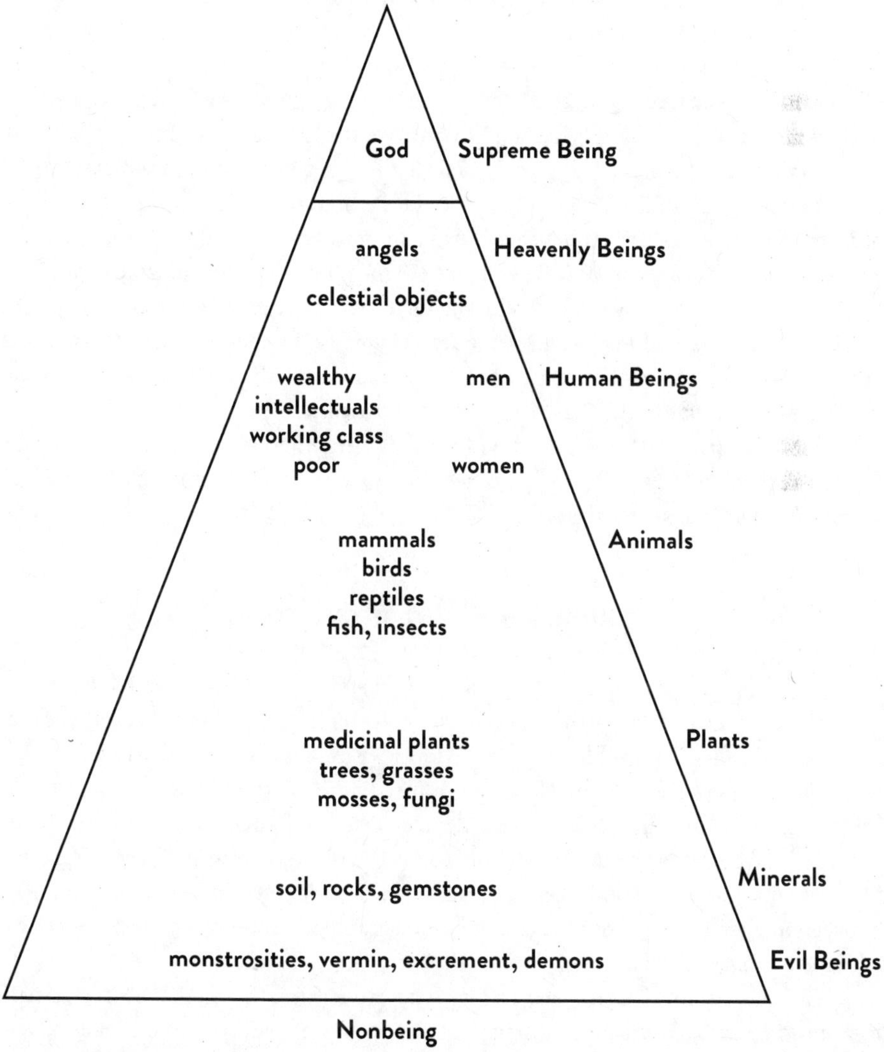

On the one hand, this hierarchical vision of the cosmos can be celebrated as a "[spiritually diverse] ecosystem, with its precise niches and unifying flow of energy."[18] As one contemporary theologian remarks:

> When a medieval man went out on a starry night and looked up at the heavens, he saw, in one sense, just what you and I see in modern times: innumerable dots of light on a black background. But when he came to explain to himself *what it was* that he saw—that is, when he tried to *understand* what he was looking at, he came up with something very different from our understanding.... To him, ... the stars and planets moved, not in empty space, but in a vast envelope which he called "mind" or "wisdom"; and they moved, not in obedience to mute physical laws, but by *desire* for the highest good.... The stars in the sky and the blood in his veins were both participants in a vast, harmonious, and, most important of all, loving universe.[19]

On the other hand, it can be argued that the medieval synthesis was "a view whose hold needed breaking."[20] It might be difficult to discover the laws of the circulation of blood, or Kepler's three planetary laws, if you believe that blood or planets are being moved by love or wisdom. Indeed, the static nature of the hierarchy of being could be a limit to improvement and discovery, just as it was a limit to social mobility: "to criticize any single aspect of the medieval legacy of science meant criticizing the whole, carefully balanced, Christian framework."[21] Medieval universities were full of criticism and debate, of course, only you had to be careful about how you went about criticizing things, lest you be seen as criticizing the whole synthesis. In any case, the medieval synthesis did break down, and out of the debates at the end of the middle ages and throughout the (so-called) Renaissance the modern synthesis of science and technology began to take form. To understand this development, we need to look more carefully at the tensions within the medieval synthesis itself.

Medieval Rationalism

My rough characterization of the intellectual tensions at the end of the middle ages will use decidedly modern terms: rationalism and empiricism. We must recognize that no medieval thinker would have used the term 'rationalist' or 'empiricist' to describe themselves; these terms, technically, apply only to schools of thought that arise well into the modern period, with philosophers like Descartes and Locke. However, insofar as the 'science wars' of the late twentieth and early twenty-first centuries "have strong resonances with traditional philosophical issues,"[22] we can identify helpful commonalities between intellectual trends in the late middle ages and controversies about science and technology that permeate the entire modern age, including our own time.

With this in mind, then, we can begin by saying that the medieval "scholastics were rationalists at heart in that they argued reason was required to understand the universe...."[23] In itself, that's not a particularly controversial claim. Surely you'll need reason to understand anything! But in more detail, rationalists (be they medieval or modern) emphasize the intellectual, mental, or cognitive aspects of human life over against the more physical, bodily, or empirical aspects. As we saw above, both these intellectual and physical aspects of life are integrated into the Great Chain of Being, but the hierarchical placement of the mind above the body is strongly suggestive of the Greek valorization of the rational over the natural (see Chapter Seven). Likewise, the medieval rationalists—albeit to varying degrees—could be fairly skeptical of the lower, more physical orders of reality. St. Bonaventure, a Franciscan monk and broadly Platonic philosopher, was concerned that the human mind could be "distracted by cares" (i.e., mundane or worldly priorities), "obscured by phantasms" (i.e., images of physical objects), and "allured by concupiscence" (i.e., sexual desire or temptation). For the mind to be set upon the path to God, therefore, it must somehow *go beyond* "this sensible world."[24] As we'll see below, it might be a bit odd to call a Franciscan monk a 'rationalist,' but St. Bonaventure certainly exhibits a Platonic skepticism towards the physical that is characteristic of rationalism.

The method of reasoning preferred by medieval rationalists was what we now call deductive logic, syllogistic reasoning from general statements or universal axioms to conclusions about particular things. For example, if we know—as a universal axiom—that all human beings are mortal, and if we also know—about a particular person—that Vashti is a human being, then it is 100 percent certain—by force of deductive logic—that Vashti will eventually die. Moreover, we could say that we know this truth about Vashti *in advance* of Vashti actually dying. We do not need to wait around to see if Vashti is mortal; rather we know, independent of any corroborating experience of Vashti's own death having occurred, that she is in fact mortal. In this sense, logical deduction can give us *a priori* knowledge about particulars, knowledge that we can have about particular things that is in principle available to us by force of reasoning, *logically prior* to having a physical experience of the conclusion being true. Thus, although medieval scholastic philosophers "were not opposed to observation and relied heavily on the works of Aristotle, an astute observer of nature, *a priori* reasoning was held to be the primary road to new knowledge."[25]

Deductive Reasoning: the logical process of deriving 100 percent certain conclusions from a formally valid set of premises; often involves applying a general rule to a specific instance.

***A Priori* Knowledge:** knowing a thing in advance of having an experience of that thing happening (e.g., "nothing is inevitable except death and taxes").

In this way, rationalists value universal axioms as highly as Aristotle did, for the man who "knows all things ... has in the highest degree universal knowledge; for he knows in a sense all subordinate objects."[26] The knowledge of subordinate objects (especially particular things) by the deductive method does not lend itself particularly well to the technical control of those things, however. You aren't likely to invent new technologies because you know with 100 percent certainty that Vashti will die. Intellectual mastery of things, therefore, is valued more highly by rationalists than technological mastery.

Likewise, universal truths about nature are more highly valued than particular physical things themselves. These universals themselves are usually the formal or final causes of things (see Chapter Seven), rather than the efficient or material causes of things. As such, the medieval rationalists sought to discover the "essence" of something rather than the process by which something could have been made. This is, roughly, what they meant by 'scientia,' a method of demonstration rather than manufacture. In this respect too, then, the scholastic synthesis echoed the characteristically Greek emphasis on what we would now call pure or theoretical science over against applied science or technology.

Universal axioms, however, are not usually discovered by reason alone. Rather, human reason must *abstract* generalizations or universal axioms away from the experience of particular things. This involves a second method of reasoning called *induction*, the extraction of statistically relevant generalizations from a limited number of particular experiences of things. Returning to our earlier example about mortality, we do not come to a universal understanding that "all humans must die" simply by thinking. Rather, we (or others before us) have numerous individualized experiences of other humans inevitably dying, and then we use reason to extrapolate from this limited number of examples to what we think to be (at least) a highly probable law: all humans will die. Induction, then, is the logical process by which reason identifies patterns in certain sets of physical things and abstracts those patterns into a lawlike generalization or universal axiom, or even a claim about the essence of something. This method of knowing cannot occur in principle without experience of the physical world, and as such it knows things *a posteriori*, i.e., after the fact. We first experience many things, and then we realize the common features behind those many things. On the way up to universal axioms, our method is inductive generalization and our knowledge of essences or axioms is a posteriori; but on the way back down to claims about other particular things, our method is logical deduction and our applied knowledge of particulars can be (relatively) a priori. In either case, the emphasis is always placed on universal knowledge: the generalization, the form, the essence, or the axiom.

Inductive Reasoning: the logical process of deriving more or less probable conclusions from a statistically relevant set of premises; often involves deriving a general rule from a limited sample.

***A Posteriori* Knowledge:** knowing a thing after having an experience of that thing happening (e.g., "seeing is believing").

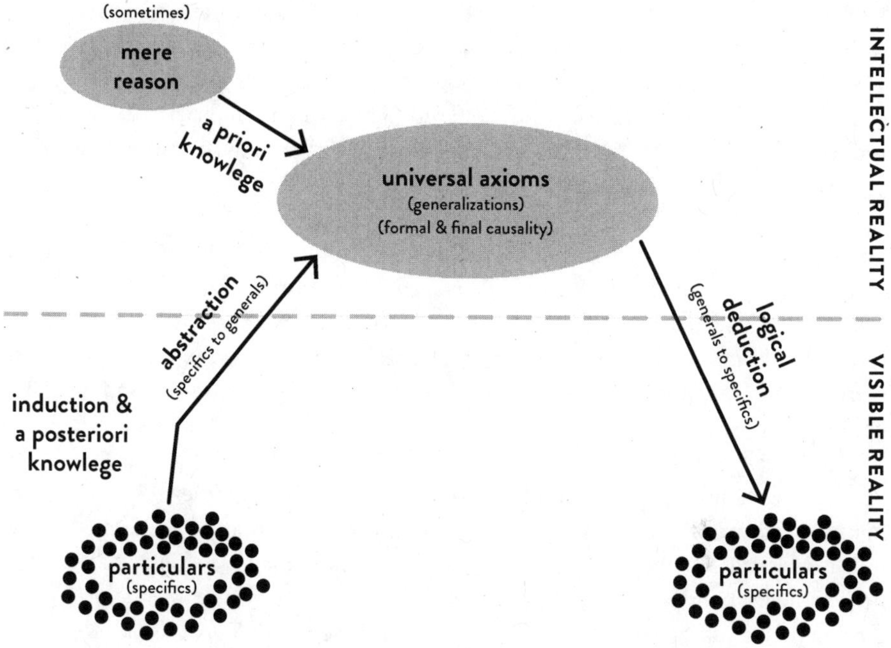

Rationalism, deduction vs. induction, and *a priori* vs. *a posteriori* knowledge.

At least with reference to science, then, St. Thomas Aquinas says that "a thing makes itself known in the soul by its exterior appearance, since our cognition takes its beginning from sense, whose direct object is sensible qualities."[27] In this respect, all knowledge is a posteriori, for even our relatively a priori knowledge of other particulars is deduced from generalizations that are themselves known via the a posteriori experience of sensible objects. This would exclude Aquinas from the early modern definition of 'rationalism,' for in that later context, thinkers such as Descartes or Leibniz argued that at least some (but certainly not all) axioms could be known completely a priori, such as the law of non-contradiction or the rules of arithmetic. Regardless, what characterizes medieval rationalism for our purposes are its values: reason is more important than nature, universals are more interesting than particulars, deduction is more reliable than induction, and knowing the essence of things is more meaningful than manipulating or controlling particular things. As we have seen, in these respects the predominantly rationalist medieval synthesis mirrors the broadly Greek reasons for keeping science and technology separate from each other.

Medieval Empiricism

While the medieval synthesis was the product of largely rationalist medieval scholars, the eventual breakdown of that synthesis can be traced to comparatively *empiricist* medieval scholars. 'Empirical' means that which pertains to the five senses, and so an empiricist will emphasize and prioritize the role sensation plays in human knowledge more than a rationalist would. As with rationalism, however, empiricism is also an explicitly modern category; medieval scholars didn't refer to themselves as empiricists either. In the modern context, empiricists defined themselves against rationalists by denying that a priori knowledge of essences or axioms was possible. *All* knowledge of universals had to be a posteriori for modern empiricists, whereas modern rationalists did think it was possible that at least *some* knowledge of universals could be a priori. By that modern definition, Aquinas and Aristotle would qualify as empiricists, because each claimed that all knowledge begins with sense experience. But our task in this chapter is not to apply modern definitions of rationalism and empiricism to medieval scholars, but rather to identify commonalities or resonances between these modern categories and medieval emphases or values with respect to the status of scientific knowledge and its relation to technological pursuits.

Thus far I have argued that the rationalist emphases in medieval scholasticism hierarchically prioritized science above technology (and wisdom and virtue above both, but why would we moderns talk about that?). In what remains, I will try to illustrate how empiricist emphases in later medieval scholarship subverted this hierarchy and created the conditions in which technology and science might begin to meet on equal terms. Philosopher George Boas argues that "when one compares science as it was before the fourteenth century and that which it became after that date, one sees that only a strong emotional propulsion would have produced that change of interest. That propulsion," he claims, "came from the Franciscans."[28] In other words, the impetus towards empiricism and its subversion of the scholastic hierarchy of Being had its decisive moment in a novel interpretation of Christianity, one attributed to St. Francis of Assisi (1181/2–1226 CE).

St. Francis is venerated as the patron saint of animals and the environment; he was known for preaching sermons to birds and wolves. He also embraced poverty and preached sermons in the vernacular dialects of the common folk, rather than using Latin, the universal language of scholarship. He wrote "the Canticle of the Sun," a song in the Umbrian dialect of Italian, wherein he praises God "through all your creatures," including everything from his "Brother Sun" in the sky to his "Sister Bodily Death" down at the bottom.[29] In effect, "Francis tried to depose man from his monarchy over creation and set up a democracy of all God's creatures."[30] He was such a radical that, as White remarks, the "prime miracle of St. Francis is that he did not end [up burnt] at the stake, as many of his left-wing followers did."[31]

Francis saw in the person of Christ something which upset the medieval status quo: the unification of divinity with lowly physicality. As noted earlier, Christians believed that Christ was both fully God and fully human. Moreover, Christ in the person of Jesus was a Palestinian Jew trained as a carpenter. Given the hierarchies of the classical and medieval worlds, a working-class non-Roman would not have been the obvious choice for the King of Kings or the divine ruler of the universe. Yet this was the central claim of Christianity, and the Franciscans embraced its revolutionary implications:

St. Francis of Assisi (c. 1181–1274).

> God joined with man formed on the sixth day, the eternal joined with temporal man, born in the fullness of time of a Virgin—the most simple joined with the most composite, the most actual with the most passive and mortal, the most perfect and immense with the little, the most highly unified and all-inclusive with the composite individual distinct from all else, namely, Jesus Christ.[32]

By equating the poor and the physical with the divine in the figure of Christ, the hierarchical separation of the rich from the poor and the divine from the physical was called into question. If Jesus Christ is the only Son of God, thought Francis, then why should wealthy military rulers be any more socially important than lowly peasants? Why should preaching be delivered in Latin rather than the language of the common people? Why should the abstractions of reason be more important than practical knowledge about plants or animals? Why should aristocratic science be kept separate from, and more highly valued than, working class technology? Franciscanism, at least, encouraged the unification of what the classical and medieval worlds had kept separate, even the subordination of what had been more highly valued to that which had been seen as more lowly. Shockingly, it made sense to a Franciscan to combine what the Greeks had kept so separate: slavery and divinity.

Franciscanism encouraged, therefore, a particularly empiricist emphasis on the 'lower' things in the medieval synthesis that rationalism found less interesting, namely sensible things and our sensible knowledge of those things. For medieval empiricists, physicality is not something suspicious which abstraction allows the human mind to intellectually escape from. Rather, the whole point of intellectual activity would be to learn more about particular things, and in the process learn more about the mind of God which is 'clothed' by the physical world. Induction could be seen as *more important* than deduction. In fact, learning about God by studying the physical world was known as 'natural theology,' and was practiced by both medieval rationalists and empiricists. However, the empirical emphasis on physical things meant that

Natural Theology: learning about the nature of God by studying the nature of physical things.

> in the Latin west by the early 13th century natural theology was following a very different bent. It was ceasing to be the decoding of the physical symbols of God's communication with man and was becoming the effort to understand God's mind by discovering how his creation operates. The rainbow was no longer simply a symbol of hope first sent to Noah after the Deluge: Robert Grosseteste, Friar Roger Bacon, and Theodoric of Freiberg produced startlingly sophisticated work on the optics of the rainbow, but they did it as a venture in religious understanding.[33]

Rather than using induction to discover universal forms or axioms, here it could be turned towards the discovery of physical operations or processes by which certain physical things brought about other physical effects. Thus did the medieval empiricists begin to seek knowledge of the material and efficient causes of things rather than focus on the formal or final causes as the medieval rationalists did. Truth took on a more embodied, incarnate, dirty, and working-class character than was typical of "aristocratic, speculative, intellectual" scholasticism.[34] This empirical impulse went on to incorporate both experimentation and technology within the scientific enterprise.

Natural Magic

Much of Chapter Four of this book was spent dismantling the common notion that magic is a false and faulty attempt at performing practical or technological tasks. We saw that myths and ritual behaviors were not about the operations of the physical world per se, but rather about the symbolic meaning of various features of human experience. By the end of Chapter Five, however, we raised the possibility that agricultural social contexts may have provided some impetus for viewing religion or magic as a way of enhancing practical control over a recalcitrant natural world. Then at the end of Chapter Six, we saw that the technology of writing facilitated the view that the natural world could be intellectually, if not practically, mastered by human reason. But in the late medieval context, the distinction between intellectual (scientific) and practical (technological) mastery of the world became increasingly unstable. As we've seen above, a theological understanding of God's mind could be achieved (in part) by understanding how physical processes operate. Understanding magic as a method for manipulating physical things would certainly be a live possibility in the middle ages, if not earlier.

Contrary to common stereotypes of the middle ages as the golden age of witch burning, magic was an accepted part of regular medieval life. (The famous witch hunts of Europe and North America actually occurred well after the end of the medieval era.) Whether magic was socially acceptable or not depended on whether it was seen to derive from demonic/heretical sources or not. All magic (good or bad) was understood as harnessing 'occult' or unknown forces, unlike 'mundane' processes (like baking bread) which were clearly knowable to regular people. But simply because an explanation for an event might be 'occult' or deeply mysterious did not necessarily mean that the occult force itself was a devil or some other threat to Christianity. Therefore, it was possible to practice magic in a way consistent with Christianity.

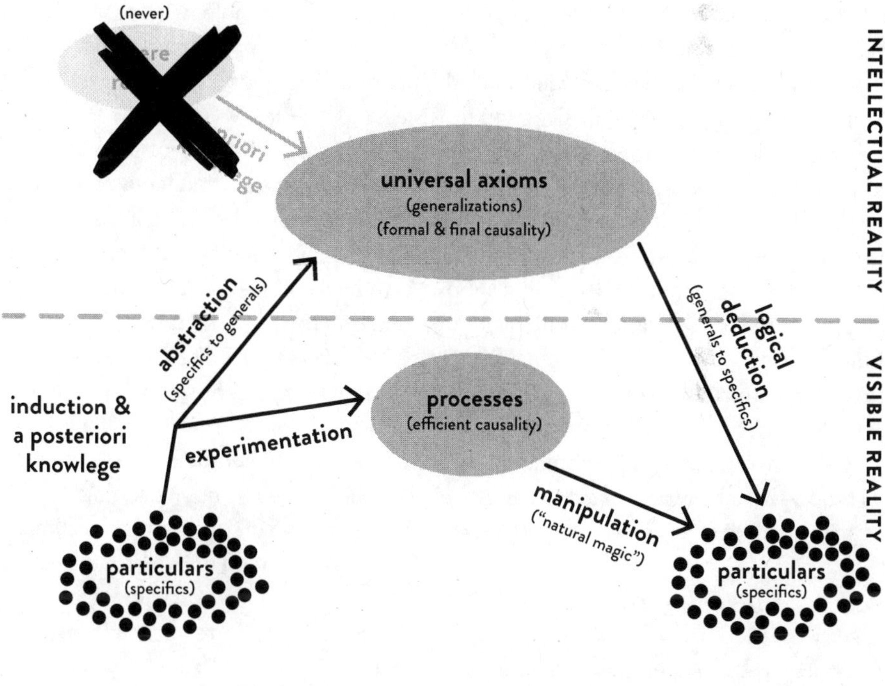

Empiricism, experimentation, and natural magic.

The idea of *natural magic* was the least likely to be confused with demonology or heresy because the occult forces involved were themselves natural processes, albeit ones very difficult to discern through the usual methods. The *Book of Secrets*, possibly written by the Dominican monk St. Albert the Great, is a manual of medieval magic, but it "carefully avoids the issue of witchcraft, supernatural powers of either good or ill, or calling on the powers of supernatural beings."[35] It attempts to describe processes by which one could control "the unknown through naming and describing it."[36] Amazing physical effects could be obtained (according to the book) if one learned the secret processes by which natural changes occurred. Moreover, the discovery of these processes could be accelerated by *experimentari*, a Latin word recognizable to English readers as *experiments*. Roger Bacon, a Franciscan natural philosopher mentioned earlier, "speculated about the design of underwater and flying vehicles, and supported the idea of experiment as a method of discovering things about nature."[37] Umberto Eco, the Italian scholar of medieval philosophy, even suggested that Franciscan natural magic—aka experimental science and technical innovation—could be employed to overcome the decline felt after the fall of the Roman Empire, using science and technology to hopefully rebuild what had been lost and even improve upon the wisdom of the ancients.[38] Roger Bacon certainly advocated for the inclusion of experimental science in the training of priests, who would learn about God by studying and manipulating natural processes (in addition to studying religious texts). Therefore, in so far as suspicious religious elements (such as demons) were not involved in natural magic, the legitimation of what the medieval synthesis saw as "lower-class, empirical, action-oriented" technology led to a vision of science that was simultaneously practical.[39]

However, St. Francis' affirmation of the physical, the poor, and the lowly can be seen as having a paradoxical effect. On the one hand, Francis valued nature as something good in itself and worthy of study in its own right: "His view of nature and of man rested on a unique sort of pan-psychism of all things animate and inanimate, designed for the glorification of their transcendent Creator, who, in the ultimate gesture of cosmic humility, assumed flesh, lay helpless in a manger, and hung dying on a scaffold."[40] Human knowledge was supposed to be humble—loving and respecting all aspects of God's creation—rather than prioritizing the rich, the powerful, and the intellect over against everything below them on the Great Chain of Being. This is why medieval empiricists would have found the inductive study of natural processes to be both legitimate and important.

On the other hand, by legitimating the inductive study of natural processes, human technical pursuits were elevated to the level of a science. The whole point of doing *theoria*, from this perspective, is to make better *technē*. As a result, medieval empiricism allowed for the combination of *intellectual* mastery (of a recalcitrant physical world) with *practical* mastery (of a recalcitrant physical world). The agrarian activity of "coercing the world around them—plowing, harvesting, chopping trees, butchering pigs" became not just the activity of peasants or slaves,[41] but the object of scientific study and religious enlightenment. Thus, when in the book of Genesis, Adam gives names to all the animals God created, it was understood to establish human "dominance over them. God planned all of [creation] explicitly for man's benefit and rule: no item in the

Anthropocentrism: the belief that nothing in the universe matters more than human beings (i.e., *anthropos*, Greek for "human"), and that all things should be made to serve human purposes.

physical creation had any purposes save to serve man's purposes."[42] White argues that medieval Western Christianity placed humanity in a hierarchically superior position over against nature. Therefore, the elevation of human technical reason to a natural theology and philosophy didn't liberate nature from the oppression of the plough so much as make the agricultural and literate mastery of nature one of humanity's highest callings.

It is unlikely that St. Francis would have been enthusiastic about such a project, but it is perhaps the most familiar vision of technoscience yet presented in this book. The usual story sees modern Western technoscience's project of mastering the physical world as basic to human nature, but now we're seeing it as a product of some very particular medieval contingencies: "viewed historically, modern science is an extrapolation of natural theology, and ... modern technology is at least partly to be explained as an Occidental [i.e., Western], voluntarist realization of the Christian dogma of man's transcendence of, and rightful mastery over, nature."[43] The will to mastery that Grant saw in modern Western technology (Chapter Three) finds its first expression in the late middle ages, and that didn't have to happen (even though it did). Thus, when White says that "modern Western science was cast in a matrix of Christian theology,"[44] he's not saying it to be critical of medieval Latin-speaking Christians. He's saying this so that we think critically about science and technology in our own time. In essence, his point is that *modern Western technoscience* "is the most anthropocentric religion the world has seen."[45] Contrary to the usual story, Western science and technology are religious, and of a type that sees no value in the material world unless we can transform it into something valuable for ourselves.

Nominalism

The empiricist preferences for discovering the processes by which things are made (i.e., efficient causality)—and thus elevating technical operations to the level of a science—reached a crucially important juncture in the person of William of Ockham (c. 1280–c. 1349), another Franciscan monk. Ockham's memory is preserved to this day in modern science's use of Ockham's Razor, also known as the principle of parsimony. This principle holds that when a scientist is faced with two equally powerful explanations for the same set of phenomena, the *simpler* of the two explanations should be preferred as more likely to be true. With respect to the medieval synthesis, however, this principle had a radical implication. Both rationalistic and empirically minded medieval scholars held to a two-fold division of reality. At the upper end of the synthesis was the metaphysical realm of formal and final causes, ideals, universals, or essences. Below it lay the physical realm of material and efficient causes, particular and individual things, and the processes by which these things can be made or manipulated. We have seen how the medieval

Ockham's Razor: given two or more equally powerful explanations of a set of phenomena, always choose the simpler explanation (also known as the principle of parsimony).

rationalists placed more emphasis on the 'ideal' dimensions of reality, while the medieval empiricists placed more emphasis on the 'visible' dimensions of reality. In general, however, both perspectives were happy to admit to the proposition that there are multiple layers to reality, including that which lies beyond what can be seen.

To modern ears, the idea that there are at least two different levels of reality—one we can see with our eyes, the other only accessible through reason—sounds rather odd. Even if we can't see atoms or subatomic particles, we tend not to think that physics gives us rational insight into a world distinct from the one we live in; physics is supposed to give us a unified theory of everything, not split it into two fundamentally different realms. Indeed, part of the usual story is that there is *only one* level of reality, namely the empirical or physical. This world, the one made out of matter, is the only 'real' world, or so we are expected to believe. We can thank William of Ockham for that general idea. If we were to apply Ockham's Razor to a two-fold division of reality, we would have to ask whether the postulation of ideals, universals, essences, or formal and final causes provided any explanatory power that an experimental understanding of processes or material and efficient causes does not. If the answer is no, then we should reject the notion of an 'ideal' level of reality where universal essences exist, and affirm only the realm of physical particulars as real. If so, there are no essences to things; there is no truth 'out there' in some vaguely abstract realm of pure form. There is only the here and now.

This single-level picture of reality is called *nominalism*: it literally means *name-ism*, from the French *nom* and the Latin *nomen*, both meaning 'name.' Nominalism is the belief that there is only one kind of reality: particular things. There is no higher level of reality where universals or essences exist beyond the physical instantiation of things. If things do not have essences, then all that happens when human scientists group things into categories like 'zebra' or 'horse' is that a *name* is applied to some things, and another name is applied to some other things. But the names themselves do not identify something that is real; there is no real difference in kind between a horse and a zebra. Rather, the application of names is simply a convenient process of classification by which human beings sort things into useful categories.

Nominalism: there is no order in the universe save that which we impose upon it.

Nominalism has some vexing implications for science. If it's true, then *universal laws of nature do not actually exist*. Science, insofar as it says anything categorical or universalizing about material objects, is not identifying any truths about reality, but simply *projecting* a falsely imagined order *onto* reality for the sake of human ease and use: *names*. Science as a lawlike explanation of material phenomena (see Chapter Two) cannot possibly be true, because the truth is not 'out there.' The truth is simply that there are individual things, and anything we might want to say about them is just that: stuff we say about them.

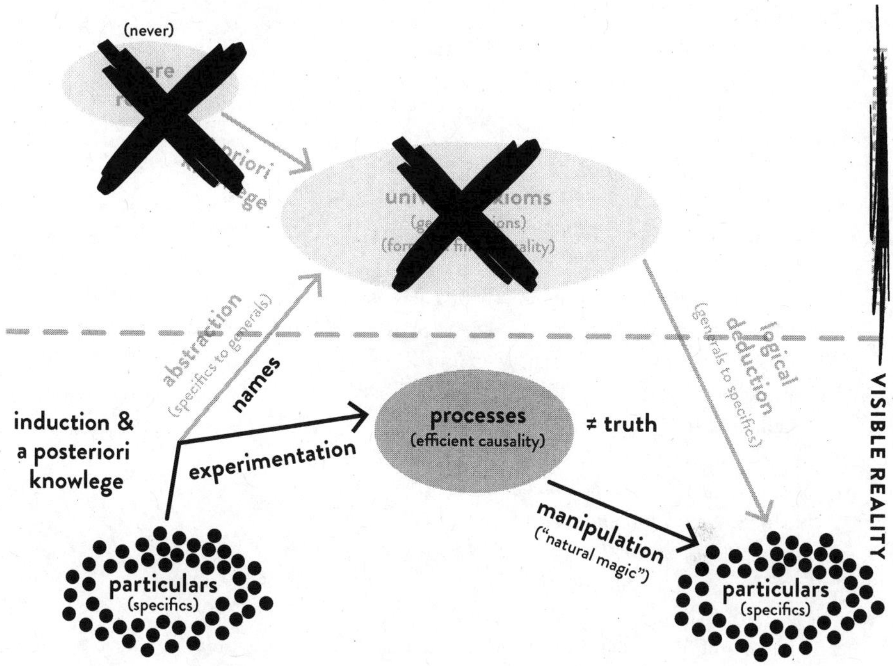

Nominalism, the denial of universals, and truth.

Nominalism sounded the death knell for the medieval synthesis (albeit not singlehandedly), and it can be seen as an outgrowth of the empiricist emphasis on knowledge of process rather than forms. If we are really interested in discovering material and efficient causes for the making or manipulation of things, we are not discovering what things really are. Rather, we're discovering what we can do with those things. We discover how to impose force onto things, but we do not discover the truth about things, any more than the statement "your mother was a hamster" is true because a muscle-bound bully threatens to beat you up if you disagree. So just as the knowledge of processes allows us to harness the quasi-magical power to create or control things technologically, the scientific 'knowledge' of universal laws of nature (etc.) are not truths but the result of a process of *forcing* 'convenient and helpful' names onto things. There is no order in the universe which science discovers; science is rather the imposition of order *onto* the universe. In this way, *theoria* (science) is interchangeable with *technē* (manipulation). Both of them amount to nothing more than forcible sorting.

Nominalism, therefore, may be an unexpected and unwelcome implication of the late medieval unification of science and technology. In fact, the notion that techno-science is merely brute force is one of the controversies about science that we will examine in Chapter Fifteen. But that is an issue that concerns the modern world, to which we now turn.

Notes

1 Lynn White, Jr., "The Historical Roots of Our Ecologic Crisis," *Science* 155, no. 3767 (1967): 1204–05.
2 White, "Historical Roots," 1205.
3 George Grant, *Technology and Justice* (University of Notre Dame Press, 1986), 11.
4 Martin Fichman, *Science, Technology, and Society: A Historical Perspective* (Kendall/Hunt, 1993), 17.
5 Technically, we should say "pre-apocalyptic," because many people at the time thought the end of the world was around the corner—i.e., they didn't think it had quite happened yet.
6 Andrew Ede and Lesley B. Cormack, *A History of Science in Society: From Philosophy to Utility*, 3rd ed. (University of Toronto Press, 2017), 69.
7 Fichman, *Science, Technology, and Society*, 18.
8 White, "Historical Roots," 1206.
9 Fichman, *Science, Technology, and Society*, 16.
10 Fichman, *Science, Technology, and Society*, 16.
11 Fichman, *Science, Technology, and Society*, 16.
12 Ede and Cormack, *History of Science*, 65.
13 Ede and Cormack, *History of Science*, 64.
14 Cliff Bekar and Richard G. Lipsey, "Science, Institutions, and the Industrial Revolution," *The Journal of European Economic History* 33, no. 3 (2004): 747–49.
15 Ede and Cormack, *History of Science*, 77; cf. Bekar and Lipsey, "Science," 719.
16 Ede and Cormack, *History of Science*, 84.
17 Fichman, *Science, Technology, and Society*, 22–23.
18 Paul Shepard, *Nature and Madness* (University of Georgia Press, 1982), 82.
19 Robert Farrar Capon, *The Third Peacock: The Goodness of God and the Badness of the World* (Image/Doubleday, 1972), 62.
20 Capon, *Third Peacock*, 63.
21 Fichman, *Science, Technology, and Society*, 23.
22 Ian Hacking, *The Social Construction of What?* (Harvard University Press, 1999), 82.
23 Ede and Cormack, *History of Science*, 84.
24 Bonaventura, *Itinerarium Mentis in Deum*, 4.1.
25 Bekar and Lipsey, "Science," 714.
26 Aristotle, *Metaphysics*, 982a20.
27 Aquinas, *Questiones Disputatae de Veritate*, Q.1, A.10, reply, trans. Robert W. Mulligan (Henry Regnery, 1952).
28 George Boas, "Introduction," in St. Bonaventura, *The Mind's Road to God*, trans. George Boas (Liberal Arts Press, 1953), xx.
29 Francis of Assisi, "The Canticle of the Sun," trans. Bill Barrett, accessed 10 September 2008, http://www.webster.edu/~barretb/canticle.htm.
30 White, "Historical Roots," 1206.
31 White, "Historical Roots," 1206.
32 Bonaventura, *Mind's Road to God*, 6.5.

33 White, "Historical Roots," 1206. Note that while Theodoric of Freiburg was a Dominican monk, Roger Bacon was a Franciscan monk and Robert Grosseteste was a highly sympathetic teacher of Franciscan monks, although he was not one himself.
34 White, "Historical Roots," 1204.
35 Ede and Cormack, *History of Science*, 80.
36 Ede and Cormack, *History of Science*, 80.
37 Ede and Cormack, *History of Science*, 81–82.
38 Umberto Eco, *The Name of the Rose*, trans. William Weaver (Harcourt Brace Jovanovich, 1983), 63.
39 White, "Historical Roots," 1204.
40 White, "Historical Roots," 1207.
41 White, "Historical Roots," 1205.
42 White, "Historical Roots," 1205.
43 White, "Historical Roots," 1206.
44 White, "Historical Roots," 1206.
45 White, "Historical Roots," 1205.

IV

Early Modernity

Chapter Nine

The Baconian Project

ONE OF THE PURPOSES OF THIS BOOK HAS BEEN TO PREPARE YOU, THE reader, to understand modern Western technoscience as a socially contingent reality with a unique and defining intellectual history. We have now reached the point where we can finally speak about modern Western technoscience directly. We have already seen that most human cultures over the course of our species' history were uninterested in what we now call 'scientific' explorations of reality, and that certain social changes had to occur before such explanations could become meaningful. We also saw that ancient scientific explanations were rarely if ever turned towards the purposes of developing or improving techniques and technics, while certain contingent factors in the European middle ages—including novel interpretations of Christianity—allowed the notion of a 'science of techniques' to gain a foothold. And so, as we turn to the modern world, we also turn towards the notion of a techno-logos, a logic or science or theory that is simultaneously practical and tool-oriented.

At this stage, however, the examination in this book will also move towards a thematic rather than chronological organizational structure. To be sure, there are numerous historical periods within modernity, and one could focus STS investigations on (for example) the Scientific Revolution before moving onto the Industrial Revolution. But the broader interdisciplinary project we are pursuing here seeks to identify the 'flavor' of Western technoscience—the sociological, philosophical, and historical dimensions of science and technology—or, in Francis Bacon's words, the "signs which are taken from the origin and birth-place of the received philosophy...."[1] In this part of the book, therefore, we will examine several characteristic aspects (or 'signs') of modern Western culture that left their mark on technoscience (if not also vice-versa), as well as some instructive countercurrents against it. We will nevertheless separate this thematic exploration into three very basic chronological parts of three chapters each: early, classical, and late modernity. One of the odd things about modernity, after all, is that it is surprisingly old.

Perhaps anything that is 'modern' *shouldn't* be 500 years old, but modernity is anyway. The word *modern* means "the right now," and it was first used in the 1500s to refer to the 'now' that existed then. It's ironic that modernity is so old, because modernity's very nature makes it uninterested in old things. You might be uninterested in a 500-year-old 'now,' but that's probably because you're a modern reader. Modernity is, by definition, uninterested in its own history. In so far as the usual story would have us believe that we can learn nothing important about ourselves, our world, or science and technology by studying how we acquired these things over time, the usual story is modern. Modernity is, in this sense, the counsel to forget those very things which STS wants to bring to light. In that sense, STS is not modern, but the usual story is. To do STS, therefore, requires us to think critically about modernity, because without that, we won't be thinking critically about science or technology either.

It is a matter of historical fact that the shift from the middle ages to modernity wasn't a clean break, but was a complex and contested process. As we saw in Chapter Eight, the usual story positions the middle ages conveniently in-between the (supposed) Golden Age of classical antiquity and the (supposed) Golden Age of the modern here and now. The very idea of the Renaissance reinforces this picture; it is the idea of a "re-birth" (*re-naissance* in French) of all the glories of Greece and Rome that were (supposedly) lost or suppressed in the middle ages. The usual story would like us to think, therefore, that the Renaissance was the start of the big turnaround which resulted in modern technoscience, especially in so far as scientists (supposedly) started thinking for themselves and performing experiments rather than just copying and repeating what ancient authorities said.

The Renaissance: the (supposed) era after the middle ages where Western culture had a "re-birth" of the golden age of classical antiquity, restoring what was (supposedly) lost or suppressed in the medieval era.

But this isn't actually the case. The Renaissance was hardly different from the middle ages with respect to science. In both periods, "the starting point of all scientific research was the firm conviction that everything to be known had already been discovered and recorded in antiquity."[2] Even if you don't think this is the case now, any good scholar will nevertheless survey the existing literature to see if their idea truly is original, or if it's already been done, before embarking on a new project. There's no point in reinventing the wheel, after all. So the scholars of the Renaissance did not criticize medieval scientists for

> basing their science on writings from ancient times, because they themselves did the very same thing, but they rather criticised scholastic science for employing the wrong methods for commenting on those writings.... It was only in the

seventeenth century, when the discoveries of Galileo Galilei and Isaac Newton showed that no one could deny anymore that the classical authors were wrong about a number of points, that doubt about the normativity of classical antiquity became widespread.[3]

Therefore, science didn't become magically modern as soon as Ockham's Razor called universal essences into question. Rather, early modernity was characterized by a drawn out struggle for a new way of understanding scientific knowledge and the production of technologies, a vision with its roots in the middle ages but not one automatically accepted by society at large. Modern Western technoscience was first and foremost a controversial *intellectual proposal*, and its most prominent proponent was the arch-empiricist Francis Lord Bacon, with a little help from the arch-rationalist René Descartes.[4]

Knowledge as Power

Francis Bacon (1561–1626) was an aristocrat and a courtier to Elizabeth the First, Queen of England. He later became Lord Chancellor to King James I, and in that position he would have "oversee[n] the use of torture in an age when evidence obtained by torture was considered reliable."[5] We will see below why this sad fact is important for understanding the scientific method he proposed! Because Bacon knew (like a good STS scholar) that the advancement of science was a social activity, he also knew that society itself had to change in order to achieve what he saw as scientific progress. Specifically, science and society had to break free from what he saw as the chains of the middle ages. Bacon thus wrote what he hoped were highly motivational documents that would inspire the sort of change that would lead to a revolution in science. (His writings were so inspirational, in fact, that his followers—most notably, Robert Boyle—founded the Royal Society to fulfill his dreams, and it remains one of the most prestigious scientific organizations in the world today.) The primary goal Bacon set forth as the value to motivate this change was *power*. If today we take for granted the notion that 'knowledge is power,' historically we have him to thank for the prominence of that idea.

Francis Bacon (1561–1626), engraving by James Posselwhite (1823).

The notion that knowledge is power is directly related to the unification of science and technology. If science is understood as a form of knowledge, and technology as the power to manipulate physical objects, then a unified technoscience will see knowledge as being interchangeable with the power to manipulate physical objects. This is what Bacon means when he says "Truth therefore and utility are here the very same things."[6] Truth is scientific knowledge, and utility is the practical use-value of that knowledge. Something which is useless, or which lacks the power to change the world, is not true. Recall the nominalist position from Chapter Eight: science, in that view, is *forcing* the world to conform to our categorizations, in much the same way that the statement "your mother was a hamster" is true when a muscle-bound bully threatens to beat you up if you disagree. It doesn't matter what 'the truth' is (in some abstract sense of essence) if you have *the power* to make the universe do what you want it to. This is why Richard Dawkins will say, in our century, that science is true because it works. Science is (supposedly) true *because* science produces successful technology. To a Baconian, weird philosophical questions about the epistemological status of science don't really matter because science gives us power, and power is the same thing (or just as good) as truth. Science is therefore understood to be the same thing as technology.

Equating knowledge with power is why the notion of a scientific theory's 'explanatory power' sounds entirely reasonable to our ears; our conception of science is already Baconian, thanks to our upbringing within the usual story. This is also why the phrase 'experimental science' sounds redundant to us; what other kind of (authentic) science could there possibly be, we'd like to ask, if it wasn't experimental? If you remember from Chapter Seven, the classical Greeks and Romans didn't think that experimentation would lead to truths about nature because experimentation technologically changes the nature of the thing under investigation. An experiment is an artificial arrangement involving technics and techniques, applying human power to things rather than discovering what things are naturally, or really like in themselves (we saw in Chapter Eight, however, that Roger Bacon—no relation to Francis Bacon—was an exception to this general rule). But if you think, as (Francis) Bacon did, that science and technology are essentially the same thing, then there's really no reason to worry about using experiments in science. So what grounds do we have for thinking that "knowledge and human power meet in one," then?[7]

The Cartesian Grounding of Experimental Science

To achieve his revolution in the sciences, Bacon had to give a convincing account of why powerful experimentation technology was synonymous with scientific truth. Across the English Channel in France was a philosopher laying the intellectual basis for such a transformation: René Descartes (1596–1650). Descartes had been educated in the scholastic tradition, and then traveled the world as a soldier in his search for truth. He came to the conclusion that the scholastic method of education (especially the *disputatio* or dissertation; see Chapter Eight) was inadequate because it did not lead to *certainty*. Because debate was at the heart of the scholastic tradition, everything it claimed as true was in fact debatable, and everything debatable is by definition uncertain!

Bacon had a similar complaint: "the wisdom of the Greeks was professorial and much given to disputations; a kind of wisdom most adverse to the inquisition of truth."[8] Descartes wanted certainty and Bacon wanted power, but in either case it appeared to them that the Greco-Roman-Islamic-Christian tradition only gave them more words. Rather than giving absolutely certain knowledge or the power to make the world behave the way you wanted it to, premodern Western science was simply a lot of old-fashioned hot air.

Portrait of René Descartes (1596–1650), engraving by William Holl the Younger (1833).

One of the characteristic flavors of modernity, therefore, is the *repudiation of tradition*: it doesn't matter what old, dead babblers have said—we might as well get rid of all that blather and start all over again from scratch! In the words of Descartes (ironically, an old dead man), "thus I realised that once in my life I had to raze everything to the ground and begin again from the original foundations, if I wanted to establish anything firm or lasting in the sciences."[9] The other old dead man, Bacon, concurred: "It is idle to expect any great advancements in science from the superinducing and engrafting of new things upon old. We must begin anew from the very foundations, unless we would revolve forever in a circle with mean and contemptible progress."[10] If we're going to actually get technoscientific progress, according to Descartes and Bacon, then we need to stop listening to old, dead men (i.e., all the ancient and medieval scientists who thought experimentation wasn't a good method of knowing) and go back to the absolute beginning, whatever that is.

If you're at all familiar with the usual story, you will assume that the 'absolute beginning' of science are *facts*. Truly modern scientists don't give two cents for what Aristotle had to say about horse's teeth; real scientists will simply look a horse right in the mouth and see those teeth for themselves. That's why the motto of the Royal Society is *nullius in verba*, Latin for "take nobody's word for it." Science is supposed to be true because it's based on cold, hard facts, not on any traditional or literary authority. However, if knowledge is synonymous with human power, then facts are secondary at best. If nominalism gets rid of universal essences (see Chapter Eight), then all we've got left are individual things and *whatever we say they are*. Facts don't have to factor into that equation. Recall again that Aristotle didn't think experiments would help science because technological manipulations of things would alter them so much that their formal essence would not be known (see Chapter Seven); but if "forms are figments of the human mind," as Bacon says,[11] then we don't have to worry that our experimental power over things won't reveal essences to us—essences don't exist in the first place. He's a nominalist, an empiricist so extreme that he doesn't believe there are any essences that could even be known *a posteriori*. Even if facts are supposed to be universal truths about the world, they do not exist prior to human categorization. For modern science, therefore, the absolute beginning that all of technoscience's certainty and power rests on is not facts but *human cognition*.

This is Descartes' most well-known philosophical contribution: all knowledge, all certainty, all science ultimately starts in the human mind, not in the physical world. The existence of everything is dubitable, says Descartes, including the existence of physical facts! For example, maybe you're just a ghost having a dream about physical facts (you'll never know for sure that you're not). There is, however, one undoubtable thing: *you*, as a (ghostly?) thing which thinks, are the thing doing the doubting. Human mental existence is *absolutely certain* because it is the condition for the possibility of doubting anything else. That's why Descartes famously said "I think, therefore I am" (*cogito ergo sum* in Latin; "je pense, donc je suis" in French), and it is the absolute beginning. He thought he could know this *a priori*. Everything else, including modern science, only gets established on the pre-existing certainty of human thought. This is why Descartes was a rationalist par excellence.

Of course, both Descartes and Bacon believed that science should be *about* empirical facts, but those facts can only be known as such *after* their existence has been proven on the basis of the human mind. In this sense, then, all things that exist have to be conceived of as *products of the human mind*. To the modern mind, the only things which can be known by humans with certainty are the things which humans have made. This is why experiments could lead to knowledge: experiments are themselves products of human making (technical arrangements) and thus are fully transparent to human knowing:

> The use of the experiment for the purpose of knowledge was already the consequence of the conviction that one can know only what he has made himself.... The experiment repeats the natural process as though man himself were about to make nature's objects.[12]

The Cartesian/Baconian argument goes something like this, then: 1) to make something is to know it inside and out; 2) technologies are products of human making; 3) experiments are a technological operation; 4) therefore, experiments (and only experiments) give us true knowledge because they allow us to know something inside and out. Anything else would lie outside human power.

Almost 200 years later, the philosopher Immanuel Kant praised the experimental methods of scientists like Galileo and Boyle for understanding that "reason has insight only into that which she herself produces on her own plan, and that she must move forward with the principles of her judgements according to a fixed law and compel nature to answer her questions, but not let herself be led by nature.... Reason ... must approach nature ... as an appointed judge who compels the witnesses to answer the question which he himself proposes."[13] Human reason (science) does not follow the direction of nature's leading but rather forces nature to answer those questions which human beings put to it. That is why it can be certain instead of doubtful: human reason (science) is calling the shots, taking everything apart and putting it back together again experimentally. Because of Descartes' (rationalist) anthropocentric postulation of the human mind as the fundamental basis of scientific certainty, Bacon's (empiricist) project of making experimental science equivalent to technological power was able to get off the ground.

A Technoscientific Method

Bacon was well aware that a few exceptional medieval monks had engaged in experimental science, but he viewed their successes in this area as random accidents, because they had no theory or logic to guide and structure their experimentation. They were, in a manner of speaking, pure empiricists—just waiting around to see what would happen—when they would have done better (in Bacon's opinion) by mixing their empiricism with a rationalism that would lead to "methods for discovery or plans for new operations."[14] Bacon's scientific revolution would require a new method that exponentially increased the pace of scientific discovery and technological innovation. This is what the Latin title of his book, *Novum Organum*, literally means: a new logic, a knowledge tool, a theory of experimentation, or a 'scientific method.' As it turned out, the particulars of Bacon's experimental method were not as influential as those which came afterward (and, as it also turns out, there is no one scientific method anyway),[15] but what matters for our purposes here is that the *very idea* of a method for experimental science is techno-scientific, implicitly combining the logical with material application.

Up to this point, we might be forgiven if we've imagined Bacon as a secular scientist, the epitome of what the usual story imagines to be the objective perspective of modern science, in contrast with the deeply religious natural philosophers of the middle ages and earlier. But Bacon was not an atheist (nor was he a scientist), and yet he cast his project for natural philosophy in profoundly religious terms. In fact, he viewed his experimental method as an extension of his Christian religion. He saw God as the Creator of the physical universe, and science was an extension of natural theology (see Chapter Eight). Science was decoding God's mind by unravelling the processes by which God created things, thus discovering the rational laws that God used to make the universe. Presumably God doesn't do things randomly but follows a rational plan, and technoscience should as well. Properly organized experiments—which treat all things as products of the human mind—are therefore a reflection of God's mind, which helps to take away some of the sting of being a nominalist. Nominalism might be making up truths by forcing the world to conform to the human mind, but that's no worse than what God did when he made (up) the universe. On the one hand, then, when Baconian technoscience follows the proper method, it will yield discoveries and inventions that "are as it were new creations, and imitations of God's works...."[16]

On the other hand, the nominalist position holds that there are no essences in God's mind, so all that remains are the hidden *processes* within things that God used to create them (assuming that—contrary to nominalism—these processes are not themselves inventions of the human mind). All that experimentation could hope to reveal are the processes by which something is made or built. A Baconian experiment's purpose is to re-create the causal process by which God originally made something, as if the human being was the creator of the thing instead of God. Once a human being is able to replicate that causal process, the efficient cause—the cause of the maker (see Chapter Seven)—is revealed to the human mind. The knowledge of efficient causality gives the technoscientist the power to re-produce objects at will. Unlike the formal and final causality denied by nominalism, the knowledge of efficient causality (the only notion of 'cause' retained by modern technoscience) gives us the power to harness and manipulate things, the same power that God had when he created the world. An experimental method which reveals the processes behind things fits seamlessly into Bacon's will to power: experimental technoscience is supposed to give human beings the revolutionary knowledge they need to perfectly master physical things with divine force.[17] It gives us knowledge as power.

In sum, Bacon thinks humans can gain this powerful knowledge (this techno-logos) only if they follow a rationally organized scientific procedure. Technological progress will be far too slow and unpredictable if discoveries only appear by accident or by trial and error. The only certain way to make 'divine' inventions is to turn science into a kind of technology, science as a form of experimental manipulation, a reversal of the Greek priority of science over technology. Bacon calls this a "true induction," as any good empiricist would.[18] At the same time, he knows that induction must be systematically organized by reason in order to *reliably* produce 'useful' (and therefore 'true') results. In this way, early modern science's project of unifying *technē* with *logos* combines them both while putting knowledge at the service of power over the physical world. This dominance of technology over science distinguishes Western technoscience from every other form of scientific investigation or technical operation known to history.

Signs from the Birthplace of Modern Science

Modern Western technoscience has a particular flavor, then, much of which comes from an old, dead man who was living decidedly in his 'here and now' almost 500 years ago. That Baconian flavor is not the delicious flavor of bacon; it is the flavor of power. Put another way, the Baconian project of modern science (which, in our era, is functionally equivalent with modern technology) embodies a particular set of *values*, or things that are set forth as goods or motivational goals. If we are going to be able to think for ourselves about modern technoscience, we need to know what those values are, rather than blindly accepting them without even realizing that's what we just did. Let us look a bit more closely, then, at the value-signals which appear at the birth of modern Western technoscience.

Recall that Bacon proposed his 'new logic' in motivational terms, hoping to inspire social change that would result in a scientific revolution. Also recall that he saw his scientific method in religious terms (as did virtually all scientists of the early modern period). To think, therefore, that modern science is value-neutral or free from social entanglements (including religion) is factually inaccurate. It's remarkable, in fact, how much supposedly secular science is inspired by religious themes (see Chapter Two). For example, Bacon actually thought that modern science and technology would essentially re-establish paradise.[19] According to his understanding of the Bible, Adam and Eve lived in the Garden of Eden at (almost) the very beginning of time, and like good Baconians, they knew everything that God knew. That is, Bacon thought the first human beings lived in a paradise where they knew all the scientific truths of the universe and had the unlimited power that came with it.

Adam and Eve being tempted by Satan in the Garden of Eden, etching by Rembrandt (1638).

Sadly, however, Adam and Eve sinned against God and were cast out of paradise, forgetting all of their science and losing all of their technology in the process. The way to get it back was, apparently, by colonizing the (so-called) New World. Conveniently, just as Europeans were claiming dominion over the Americas, the Baconians of the Royal Society (especially Boyle) were claiming that technoscience gave them divinely ordained dominion over the entire natural world.[20] It's irrelevant whether Bacon and his inspired followers viewed the Garden of Eden as historical fact or as a psychologically satisfying myth (see Chapter Four). For Bacon, the purpose of modern technoscience is to regain all that was lost in a (literal or mythological) primordial fall from grace. Science will give 'us'—eventually—divine knowledge that will enable 'us' to rebuild paradise—in the colonies. If sin is ignorance, then salvation is the opposite of ignorance: technoscientific knowledge. Insofar as the usual story (often via science fiction) suggests that science and technology will lead to a utopia where replicators make food, clothing, and shelter free for all and where anyone can fly to new worlds beyond the stars, it is re-enacting an early modern interpretation of a very old religious and cultural story, and one tainted by morally and theologically problematic involvements in empire, conquest, and settler-colonialism. That's where some of the values expressed in techno-utopianism come from.

Of course it's difficult to establish a paradise on Earth, and Bacon has a number of explanations for why technoscience's colonial project hits so many roadblocks. If perchance your science textbooks ever made mention of Bacon by name, it's probably with respect to his 'Idols,' those things which the usual story says interfere with technoscientific progress. According to Bacon, there are four *false gods* (idols) that scientists have to overcome if they're to attain truly divine knowledge and power: illusions of the five senses (the idol of the tribe), personal biases (the idol of the cave), linguistic confusion (the idol of the marketplace), and dogmatism (the idol of the theater). Unlike these false gods which give us all kinds of wrong ideas, Baconian science provides us with "the Ideas of the divine" (rather than the human) mind. Scientific truths are "the true creator's own stamp upon creation, impressed and defined in matter by true and exquisite lines."[21] It's as if Bacon loses his nerve here and forgets his nominalism, speaking like a rationalist who can access the universal essences in God's mind: "For I am building in the human understanding a true model of the world, such as it is in fact, not such as a man's own reason would have it to be...."[22] Never mind his momentary lapse of theoretical consistency; it's important to note that if false gods are interfering with technoscience, then the true god is technoscience itself. Science and technology, for Bacon, are the "true religion"[23] (along with Christianity, of course) whereas everything else is heresy. Science and technology are therefore religious weapons used in the conquest of everything, including other people.

The usual story would like us to forget the oddly religious and shamefully imperialistic origins of modern science, but Bacon just won't let it go. If divine technoscience is the way to overcome false gods and bring back the Garden of Eden, then giving humanity divine power is the path to salvation. Modern technoscience is literally the solution to all our problems! Bacon says that "the true and lawful goal of the sciences is none other than this: that human life be endowed with new discoveries and powers."[24] These new powers are supposed "to relieve and benefit the condition of man," which is why ancient inventors were "rewarded ... with divine honours and sacred rites...."[25] But modern scientists and engineers are even more godlike, he thinks:

> Again, let a man only consider what a difference there is between the life of men in the most civilized province of Europe, and in the wildest and most barbarous districts of New India; he will feel it be great enough to justify the saying that "man is a god to man," not only in regard of aid and benefit, but also by a comparison of condition. And this difference comes not from the soil, not from climate, not from race but from the arts.[26]

Bacon is saying that early modern Europeans were like gods in comparison to the Indigenous inhabitants of North and South America, not because of anything genetically intrinsic to Europeans per se, but because of European *technē*. His quasi-religious view of science is technologically racist, justifying European superiority over non-European cultures because of the latter's supposed lack of technological advancement. It is a matter of historical fact that European imperialists lacked the technologies for survival in the New World until they learned them from the Indigenous inhabitants (see Chapter Four). It is also a matter of historical fact that the European colonization of the Americas had (and still has) a sordid record of committing atrocities and cultural genocide. But given the values that Bacon rallies in support of his project, it should not be surprising that Western technoscience finds itself implicated in colonization and related horrors. For Bacon, the whole point of science is power, and the whole point of power is conquest.

So the next time Western science and technology is said to be the silver bullet which will singlehandedly solve all the problems of another (so-called) underdeveloped nation, not only will you know where that idea came from, you will also see it as anything but neutral or objective: Western science and technology is instead "best understood as co-constituted with colonialism" and militarism.[27]

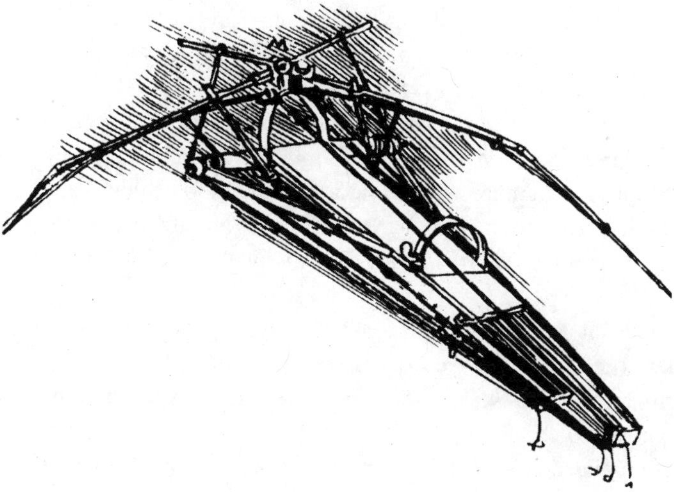

Ornithopter (fuselage) designed by Leonardo da Vinci (15th–16th century).

We have seen above that the mastery of things is the goal of Baconian techno-science. If knowledge is power, then might makes right and justice is trial by combat. In an intentionally ironic statement, however, Bacon says that "Nature to be commanded must be obeyed."[28] This means, on the one hand, that the scientist and technician must obey (submit to) the laws of nature if they're going to make anything that works. So it sounds like the technoscientist isn't conquering anything, but being passive and neutral! But on the other hand, once we know how nature operates (by submitting to it), we can turn that knowledge to our own purposes and make things happen that nature on its own would never let happen (e.g., flying machines). That way, humans can conquer (subdue) nature by using her own rules subversively. Therefore, scientific objectivity and technological neutrality actually disguise a will to power over other people and things:

> [C]olonial practices of "just observing" and "just reporting what was seen" were often framed as having no consequences at all for what was subsequently done by others (such as militaries and corporations) with those observations and reports of them. Yet in fact colonial scientists were always also commenting on the value of local plants and indigenous practices for Europeans, and on how "nature" (including the indigenes) could be improved, as they collected samples and renamed indigenous plants and animals. Scientific exploration inherently makes use of "imperial eyes."[29]

Thus by obeying nature—or just observing it—does modern technoscience conquer it and those people it sees as obstacles.

Conquest, in fact, is the highest calling of the human spirit, according to Bacon:

> Further, it will not be amiss to distinguish the three kinds and as it were grades of ambition in mankind. The first is of those who desire to extend their own power in their native country, which kind is vulgar and degenerate. The second is of those who labor to extend the power of their country and its dominion among men. This certainly has more dignity, though not less covetousness. But if a man endeavor to establish and extend the power and dominion of the human race itself over the universe, his ambition (if ambition it be called) is without doubt both a more wholesome and a more noble thing than the other two. Now the empire of man over things depends wholly on the arts and sciences. For we cannot command nature except by obeying her.[30]

Bacon is saying that it's perfectly fine for someone to try to gain power over their country's political system (he was a big fan of Machiavelli), but it's far better to try and conquer other countries and colonize new lands. But the best thing of all—and this is supposed to motivate people to get on board with his modern technoscientific revolution—is cosmic megalomania, making (some) human beings the imperial masters of the entire universe. Terraforming Mars is only the beginning, for a Baconian. Sound familiar? (We will return to the postcolonial critique of Western technoscience later in Chapter Ten.)

Feminist Critiques of Technoscience

Immediately above I said "(some) human beings" in reference to being the masters of the universe, but perhaps I should have said "He-Man." You may have already noticed above how a Baconian refers to nature with feminine pronouns, whereas God and human conquerors receive masculine pronouns. For Bacon, sexism is the final value infused into Western technoscience to motivate its adoption. One of the failings of Greek science, according to Bacon, was that it amounted to "the talk of idle old men to ignorant youths." Bacon complained that the Greeks "were always boys," for "they have that which is characteristic of boys; they are prompt to prattle, but cannot generate; for their wisdom abounds in words but is barren of works."[31] Here the lack of new technological works in ancient and medieval science is explicitly compared with children incapable of *generating*, which means a lack of sexual maturity and the inability to reproduce (*genesis* means to give birth to new life). Bacon is comparing the technoscientific conquest of nature (and other countries along the way) with the ability to impregnate women. In this context, when he talks about how technoscience should "penetrate into the inner and further recesses of nature,"[32] it starts to sound a lot like rape.

Bacon himself didn't think this was a problem, of course, as the sort of "lurid sexual imagery"[33] he used was supposed to motivate people (i.e., men) to adopt the modern approach to science. Many feminist scholars have noted the presence of this toxic form of masculinity in modern science. The philosopher Mary Midgley says it better than I can:

> Bacon had dismissed the Aristotelians as people who had stood impotent before Nature, destined to "never lay hold of her and capture her." Aristotle (said Bacon), being a mere contemplative, had "left Nature herself untouched and inviolate." By contrast, Bacon called upon the "true sons of knowledge" to "penetrate further" and to "overcome Nature in action," so that "passing by the outer courts of nature, which many have trodden, we may find a way at length into her inner chambers." Mankind would then be able, not just to "exert a gentle guidance over Nature's course," but to "conquer and subdue nature, to shake her to her foundations" and to "discover the secrets still locked in Nature's bosom." Men (Bacon added) ought to make peace among themselves so as to turn "with united forces against the Nature of Things, to storm and occupy her castles and strongholds." By these means scientists would bring about the "truly masculine birth of time" by which they would subdue "Nature with all her children, to bind her to your service and make her your slave."[34]

I would hope that sexual assault is something that would no longer motivate people to get involved in technoscience, but it was distressingly effective in the social context of the early modern world. Rather than being seen as a nurturing mother, Nature was anthropomorphized as a seductive female who was sexually unavailable to the masculine technoscientist. He would therefore have to use his power to find, chase, catch, fight, and cage her. Once she was trapped, Nature could be

> wooed, harried, vexed, tormented, unveiled, unrobed, and "put to the question" (i.e., interrogated under torture), forced to confess "all that lay in her most intimate recesses," her "beautiful bosom" must be laid bare, she must be held down and finally "penetrated," "pierced" and "vanquished" (words that constantly recur).[35]

The early modern Baconian technoscientist is a precursor of the twenty-first-century incel.

The historian of science Carolyn Merchant points out that this viciously sexist language against feminine nature was an attempt to morally justify the increasing domination of the natural world by early modern mercantile and commercial powers. The Earth was now supposed to be considered "a wicked stepmother" whose physical body was hiding and protecting all the ores and other raw materials desired by new industries.[36] At the same time, women in Europe were losing their economic power as early capitalism squeezed women out of the trades and crafts and medical science replaced midwifery.[37] 'Proper womanhood' now meant "confinement to the newly invented sphere of private domesticity, [in] contrast not only with the public sphere of men but also with the perceived degraded conditions of colonized women—indigenes, slaves, peasants."[38]

This economic context transformed the role of women into either "a psychic resource for their husbands" or into witches,[39] symbols of power rooted in nature and not controlled by men (the vast majority of witch hunts did not occur in the middle ages but rather the early modern period, especially the seventeenth century).[40] In the case of both women and nature, then, it was supposed to be appropriate to "min[e] the female flesh for pleasure" or to "penetrat[e] nature and shap[e] her on the anvil, ... allowing all manner of assault."[41] Thus would Bacon see no problem with using torture as a model for the experimental scientific method.

Even if this is only how modern technoscience (knowledge as power) understood itself *symbolically*, it is morally problematic in the extreme. If we were to throw Bacon's words back at him, we might say that "the signs which are taken from the origin and birth-place of the received philosophy are not good,"[42] only now we're referring to modern technoscience. But as feminist scholars point out, these signals within modern technoscience are not just symbolic: they have real effects on the lives of real people. Out of this comes the feminist critique of technoscience: "scientific knowledge does not equally benefit everyone; it produces even more unequal social relations ... thus emphasizing the need for new ways of producing science."[43]

Unsurprisingly, the formation of modern technoscience over two centuries in the cultural crucible of rampant sexism and abuse resulted in a number of deeply ingrained barriers to participation in science and technology by women. Despite decades of social progress, women are still underrepresented in many scientific fields and especially among the scientific elite who function as gatekeepers into the discipline.[44] There are structural obstacles to female participation in science, for example the demanding tenure process at research universities which "usually comes at the time when women are most likely to be interested in having children and spending time with their young families."[45] Some feminist scholars argue that with the removal of such barriers, female participation will increase and eventually equalize with male participation. But others point out that this assumes that science itself is gender-neutral, as if it lacked any of the value-laden historical baggage it receives from its origins and cultural contexts.

Feminist Critiques of Technoscience: using feminist theory to identify and redress female exclusion in science or technology, and to propose improvements to the technoscientific status quo.

This led some feminist scholars to suggest that there can be a female approach to science that would create materially different results, a female science that offers an alternative to the science of the usual story. Other feminist research problematizes this approach, pointing out that there is no single 'female perspective' on anything, let alone science and technology. This leads to questions about whether there is an 'essence' to being a woman at all, itself an issue directly related to the controversy over nominalism (see Chapters Eight and Fifteen). The feminist critique of science (and to a lesser extent, technology)[46] has been ongoing since the 1970s, and gender is still an active and open-ended area of STS research. But given the social issues that are at stake, it is a crucially important and timely one. Even when we discover that the idea of modern technoscience is roughly 500 years old, it raises moral questions that we have to face in our contemporary here-and-now. STS will no longer allow us to take the usual story about modern science and technology for granted.

Notes

1 Bacon, *Novum Organum*, §71.
2 Peter Raedts, "Wat is middeleeuws?," in *Cultuurgeschiedenis van de middeleeuwen: Beeldvorming en perspectieven*, ed. Rob Meens and Carine van Rhijn (Wbooks/Open Universiteit, 2015), 24; my translation.
3 Raedts, "Wat is middeleeuws?," 25.
4 It took many centuries for Bacon's proposal to actually take effect in science and technology, hardly showing up in practice until about 1850 (albeit a bit earlier in the chemical industries). See Lynn White, Jr., "The Historical Roots of Our Ecologic Crisis," *Science* 155, no. 3767 (1967): 1203.
5 Andrew Ede and Lesley B. Cormack, *A History of Science in Society: From Philosophy to Utility*, 4th ed. (University of Toronto Press, 2022), 145.
6 Bacon, *Novum Organum*, §124.
7 Bacon, *Novum Organum*, §3.
8 Bacon, *Novum Organum*, §71.
9 Descartes, *Meditation One*, AT 17.
10 Bacon, *Novum Organum*, §31.
11 Bacon, *Novum Organum*, §51.
12 Hannah Arendt, "The 'Vita Activa' and the Modern Age," in *Philosophy of Technology—The Technological Condition: An Anthology*, ed. Robert C. Scharff and Val Dusek (Blackwell, 2003), 360.
13 Immanuel Kant, Preface to the Second Edition of the *Critique of Pure Reason*, in Lewis White Beck, ed., *Kant Selections* (Prentice Hall, 1988), 97.
14 Bacon, *Novum Organun*, §8.
15 Wenda K. Bauchspies, Jennifer Croissant, and Sal Restivo, *Science, Technology, and Society: A Sociological Approach* (Blackwell, 2006), 21.
16 Bacon, *Novum Organum*, §129.
17 Bacon, *Novum Organum*, §13, §29.
18 Bacon, *Novum Organum*, §14.
19 For a book-length study of Bacon's technoscientific theology, see Steven Matthews, *Theology and Science in the Thought of Francis Bacon* (Ashgate, 2008).
20 Sarah Irving-Stonebraker, "From Eden to Savagery and Civilization: British Colonialism and Humanity in the Development of Natural History, c. 1600–1840," *History of the Human Sciences* 32, no. 4 (2019): 64–68.
21 Bacon, *Novum Organum*, §124.
22 Bacon, *Novum Organum*, §124.
23 Bacon, *Novum Organum*, §129.
24 Bacon, *Novum Organum*, §81.
25 Bacon, *Novum Organum*, §73.

26 Bacon, *Novum Organum*, §129.
27 Banu Subramaniam et al., "Feminism, Postcolonialism, Technoscience," in *The Handbook of Science and Technology Studies*, 4th ed., ed. Ulrike Felt et al. (MIT Press, 2016), 408; cf. 418.
28 Bacon, *Novum Organum*, §3.
29 Subramaniam et al., "Feminism, Postcolonialism, Technoscience," 418.
30 Bacon, *Novum Organum*, §129.
31 Bacon, *Novum Organum*, §71.
32 Bacon, *Novum Organum*, §18.
33 Dee Carter, "Unholy Alliances: Religion, Science, and Environment," *Zygon* 36, no. 2 (2001): 362.
34 Mary Midgley, *Science as Salvation: A Modern Myth and Its Meaning* (Routledge, 1992), 77.
35 Midgley, *Science as Salvation*, 77.
36 Carolyn Merchant, "Mining the Earth's Womb," in *Machina Ex Dea: Feminist Perspectives on Technology*, ed. Joan Rothschild (Pergamon, 1983), 109; cf. 107.
37 Carolyn Merchant, *The Death of Nature: Women, Ecology, and the Scientific Revolution*, 40th anniversary ed. (HarperOne, 2020), 151–55.
38 Subramaniam et al., "Feminism, Postcolonialism, Technoscience," 416.
39 Merchant, *Death of Nature*, 155.
40 Peter T. Leeson and Jacob W. Russ, "Witch Trials," *The Economic Journal* 128, no. 6 (2018): 2066–105.
41 Merchant, "Mining the Earth's Womb," 112, 113, 111.
42 Bacon, *Novum Organum*, §71.
43 Subramaniam et al., "Feminism, Postcolonialism, Technoscience," 408.
44 Sergio Sismondo, *An Introduction to Science and Technology Studies*, 2nd ed. (Wiley-Blackwell, 2010), 40.
45 Sismondo, *Introduction to Science*, 43.
46 See, for example, Judy Wajcman, *Feminism Confronts Technology* (Pennsylvania State University Press, 1991), and Anabel Quan-Haase, *Technology and Society: Social Networks, Power, and Inequality*, 2nd ed. (Oxford University Press, 2016).

Chapter Ten

The Doctrine of Progress

THE USUAL STORY IS QUINTESSENTIALLY MODERN. THE STORY MODERNITY tells itself about its own science and technology has a number of elements, and the Baconian will to power is one of the most prominent. Another prominent characteristic of the usual story is *progress*, so much so that it may be its single most defining feature. Modern science progresses, and so does modern technology, right? But what is progress? It's nicely encapsulated in the optimistic spirit of Robert Cabana, director of the Kennedy Space Center, who said this back when NASA was moving towards using commercially built spacecraft to send astronauts into orbit: "This is a truly exciting time for human space flight in our nation. And believe me, it's only going to get better as we charge off into the future."[1] Believing that things only get better in the future is the doctrine of progress. The central idea of modern culture and its reason for being is progressive optimism about science, technology, economics, social behavior etc.

This idea of progress was already built into Bacon's intellectual project. Recall that he was worried that progress would *not* occur unless science and industry were revolutionized along the lines that he proposed. He thought that his scientific revolution would only happen if certain societal conditions were in place. But we discovered that at least some of these societal conditions were morally problematic, like colonialism and sexism. Perhaps we would now like to say that it would have been better had colonialism and sexism *not* been built into Baconian technoscience. But if they hadn't been, then modern Western science and technology might have turned out much differently than they actually did. If Bacon had died prematurely, or had not been born into the nobility, or had made different career choices, technoscience could have taken at least a somewhat different path.

But what if we *didn't* think that those social and personal factors could have been different? What if we thought that Bacon was *destined* to become the inspiration for the power-seeking vision of Western technoscience? Or, if not him, then *someone* was destined to do this, because the human spirit is simply the kind of thing that will eventually use science and technology to actualize a supposedly innate desire for power? That is to say, what if we think it's impossible for technoscience to have taken a different path than the one it actually did? If this is what we think, then modern Western technoscience could not have turned out much differently than it actually did. A Baconian technoscience (even if it had a different name) was bound to happen. We would say that human nature is not just *homo sapiens* (the wise hominid), but also *homo faber* (the manufacturing hominid), because knowing *always was* making and knowledge *always was* power. We would think that technological power is our species' "dominant and interminable goal."[2] If we were to think these things, we would be thinking in terms of the doctrine of progress. We would be thinking that there's only one way for modern technoscience to develop; it had to develop the way that it did, and that's why it did. That's what the modern belief in progress says about science and technology, and that's the usual story.

A Short History of Technoscientific Progress

According to the usual story, there never was any other way for science or technology to be except the way that it happens to be right now. If we did things differently, we wouldn't have *any* (good) science and technology. Science and technology *had* to 'evolve' this way, for there is only one way for them to be, and that is the objectively true way. Modern science and technology are therefore supposed to be the products of a long process of progress, and—according to this view—science and technology are going to continue progressing indefinitely into the future. This is supposed to be something we 'already know,' something we can take for granted and ignore. In the modern Western context within which I write this book, the usual progressive story "tends to arise unreflectively."[3] Progress is our automatic explanation for almost anything good that happens, certainly the go-to explanation for any recent scientific discovery or technological innovation. It's also the automatic solution to anything bad that might happen: technoscientific advancement will supposedly eliminate any problem we may encounter in the future. Understanding science and technology as part of a larger progressive narrative is second nature to globalized Western minds, operating "like all ideology, largely 'behind our backs,' hardly entering consciousness...."[4] If we are going to think critically about science and technology in our society, then we are going to have to become conscious of the doctrine of progress.

If progress sees nothing optional about modern technoscience, then at the basic level modern technoscience is simply what human beings have always done or wanted to do. The only difference is the superficial fact that 'we' happen to be better at it than other people (in the past or elsewhere) were. This view is what George Grant (in Chapter Three) called "progressing continuity." But it is difficult to square the progressive view of technoscience[5] with the historical facts of ancient, medieval, and non-Western cultures. As we have already noted, progress is in fact a decidedly modern idea. It only starts to take hold of the mind of some people in seventeenth-century Europe, which is rather geographically limited and comparatively recent. Outside this social context, "there was, even in the best and most vigorous times, no proclaimed *idea* of a future of *constant progress* in the arts."[6]

Certain aspects of the modern idea of progress can be found in earlier cultures, but with important differences. The ancient Mesopotamians and classical Greeks believed in humanity's "gradual rise in the past from a savage to a more civilised state,"[7] but this agrarian notion of the emergence of order from wild chaos (see Chapter Five) had its limits. For the Greeks, to use *technē* to radically liberate humanity from its natural limitations was to challenge the gods, an act of hubris that deserved divine punishment, as seen in the myth of Daedalus and Icarus. Moreover, the idea of a Golden Age, which we also get from the Greeks, was actually anti-progressive. A Golden Age was a time in the mythical past when human tools were made out of gold. The Iron Age of Greece and Rome with its less-than-golden tools was seen as a technological *decline* by comparison, not a step forward! Christianity, however, transferred "the golden age from the past to the future," and thereby "substituted an optimistic for a disillusioned view of human destiny."[8] But this Christian golden age of the future was largely about an otherworldly, spiritual salvation (see Chapter Eight), not a thisworldly vision which would incorporate science and technology into the 'salvation' of the material world (see Chapter Nine). When Christian salvation became secularized during the European Enlightenment, however, perpetual progress became a "cardinal doctrine of the educated classes"[9] and a basic tenet of modernity. Scholars therefore consider progress to be "the dominant faith of modern man"[10] and "modernity's key foundational narrative."[11] Its cultural dominance is why modern Westerners generally assume that science and technology are automatically going to get even better in the future than they are right now.

Contemporary examples of seeing technoscience progressively abound. The on-again, off-again slogan of the Audi car company is *Vorsprung durch Technik*, German for "springing forward (progressing) through technology." In an offhand remark, an Israeli futurist historian says that he doesn't "try to stop the march of technological progress" because he "won't be left with any choice anyway."[12] A group of 218 scientists published a warning in the *Wall Street Journal* that called environmentalism "an irrational ideology which is opposed to scientific and industrial progress and impedes economic and social development."[13] These examples assume that technoscientific progress is real, and that technoscientific products (like an Audi) are proof of progress' reality.

Modern science and technology are tied so tightly to the idea of progress, however, that it's difficult to say whether 'progress' is supposed to make technoscience happen, or if technoscience is supposed to make 'progress' happen. Science is said to be both the "prime mover" of progress *and* the vehicle through which progress itself happens, generating "new discoveries and applications."[14] Similarly, technology is said "to be a guarantor of continued progress"[15] while progress is said to be "an inherent drive" of technology itself.[16] So technoscience gives us progress, but progress gives us technoscience?

Clearly, modern Westerners are convinced that technoscientific progress is real without having a particularly clear understanding of what that means.

The word *progress* literally means a "through-motion." A medieval 'progress' was the tour a monarch would make through their royal territories. Progress, therefore, makes modern technoscience fundamentally restless. Technoscience is always in motion. Neither science nor technology can stand still or rest satisfied with their achievements; that would not be modern. Modern technoscience is "an enterprise and a process" whereas non-modern sciences, technics, and techniques, by contrast, were possessions or states.[17] Technoscience never stops moving because it's always progressing, which might be why we don't have a clear understanding of what it means. A concept of science and technology that is always moving is difficult to pin down:

> The common man, before whose eyes the marvels of science and invention were constantly displayed, noted the unprecedented increase in wealth, the growth of cities, the new and improved methods of transportation and communication, the greater security from disease and death and all the conveniences of domestic life unknown to previous generations, and accepted the doctrine of progress without question: the world was obviously better than it had been, obviously would be better than it was. The precise objective toward which the world was progressing remained, however, for the common man and for the intellectual, somewhat vague.[18]

Therefore, by the nineteenth century, the "laws of progress became self-evident" even if no one was really sure what they were.[19] In fact, the positioning of the vague goals of progress and technoscience in the fluid future meant that it was inherently difficult to think critically about them. If the idea of progress is always floating around just past the edges of your mind's eye, you'll be tempted to think it's there, almost within reach, and less likely to think that the entire notion is just stringing you along. Our job in this chapter is to do a better job of examining it than that.

Limitless Progress

By locating the goals of technoscientific progress in the out-of-reach future, progress itself remains indefinite and amorphous. We do not have to ask if scientific discovery is aiming at a body of knowledge that, once it has been discovered, will make further scientific research unnecessary. We know both that it is possible for science to eventually know everything, and that scientific discoveries will never end. But how can you know everything *and* never stop learning new things? For the time being, we're supposed to carry on as we have been doing and not be bothered by such pesky issues. The future-orientation of progress is therefore a way of getting us to avoid asking critical questions about science in the present.

The same thing is true of technology. There's supposedly no point in asking whether we'll ever reach a point of "technological saturation" where we find an equilibrium between our needs or wants and the technological means to satisfy them.[20] Whatever the answer to that question is, in the here and now we clearly don't have enough technological power to satisfy all our desires. Thus, technological progress must continue on unabated, or at least that's what we're supposed to think. The vagueness of technoscientific progress is thus a way of reinforcing belief in technoscientific progress itself, even though we're *still* not sure what that means!

Progress, therefore, is supposed to be fundamentally without limit. The question of any limit or end to progress is forestalled into the indefinite future, and so for all intents and purposes, progress will "go on forever and forever in the same direction...."[21] Progress is the idea that "in the post-medieval world, the options for [hu]mankind are literally infinite,"[22] an idea characteristic of the early modern period and the usual story we continue to operate with today: "In both its capitalist and communist versions, the great promise of modernity was progress without limit and without end."[23] Like good moderns, we think that "endless progress in [scientific] knowledge" is possible, and that technology makes "indefinite augmentation of wealth and living standards" possible.[24] It would be just as foolish to say, today, that no more technoscientific progress is necessary or possible, as it was shortsighted for Charles Perrault to say over 300 years ago that the year 1697 was "the very summit of perfection."[25] Rather than reaching an endpoint, technoscientific progress will "prepare humanity to live on other planets or beneath the Earth's oceans,"[26] and after that, who knows? The common refrain among many physicists and engineers that humanity must learn interstellar travel, extraterrestrial terraforming, and space colonization is a direct application of the doctrine of progress. Nothing, not even the universe, can limit our ability to go everywhere and do anything (just as Bacon had dreamt). The usual story operates on Buzz Lightyear's ironic motto *to infinity and beyond*, even though the sentiment is comically and literally impossible to actualize. We don't realize that this belief in reaching infinity is "wholly unprecedented" and only a few centuries old.[27]

Technoscientific progress means that there is no end in sight to our Baconian mastery of nature, and that there *can* be no end in sight either. If progress is limitless, *you shouldn't try to stop progress*. Modern science and technology must not be limited. We will discover whatever scientists are going to discover, regardless of any doubts unimaginative naysayers might have. We will develop whatever technologies are coming next, regardless of what concerns critics might have. Nobody should stand in the way of progress. Besides, who would dare stop technoscientific advance? Our science gives us the knowledge of the gods (see Chapter Seven), and our technology gives us the power of gods (see Chapter Nine). Anthropocentric humanity is the greatest thing the universe will ever see (see Chapter Eight), and by occupying the unlimited realm of infinity and eternity, we replace the gods. *We are* 'the God-particle' coming to know itself as such, and external limits to the progress of our material power are blasphemous. The usual story is beholden to nothing, and everyone is supposed to believe in it.

Inevitable Progress

If progress is limitless, it cannot be stopped. If it cannot be stopped, progress is inevitable. "Technology," as philosopher Hans Jonas says, "is destiny."[28] Progress is "generally understood as a portrayal of past events that *leads inexorably* to a future state,"[29] a future state that either happens to be the 'now' of the present moment or an extrapolation of present trends into an all-too-familiar utopia of the future. It's not only that we ought not to limit or stop progress; there is literally no possible way to stop progress! Charles Bury claimed that "civilization has moved, is moving, and will move in a desirable direction," while Charles Van Doren claimed that progress necessarily affirms "some 'irreversible' and thus inevitable 'patterns of change.'"[30] Historian Sidney Pollard simply defines the idea of progress as "the assumption that a pattern of change exists in the history of mankind [that] consists of irreversible changes in one direction only, and that this direction is towards improvement."[31] Progress is understood to be a linear (if not parabolic) and "steady ... line [of] almost uniform improvement throughout history...."[32]

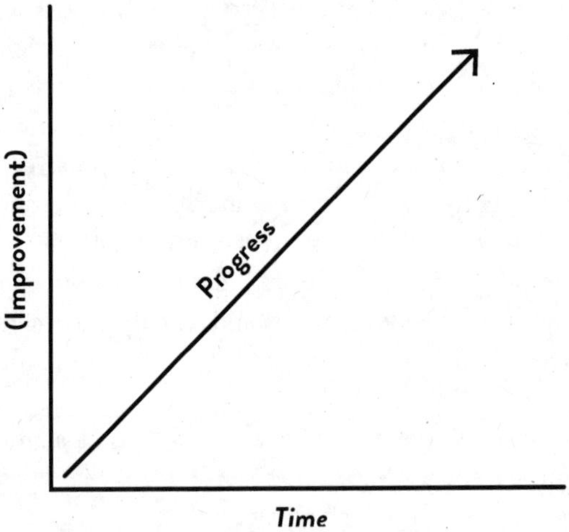

Progress as improvement over time.

Science and technology are also a part of the inevitable aspect of progress, but in more of an off-center way. Of course, it would go without saying that scientific and technological development are supposed to proceed on their predetermined trajectories towards absolute truth and power, but technoscience—including, for the first time, *social science*—plays an important instrumental role for the advance of progress in general. In the eighteenth century, intellectuals believed that progress could be achieved by directing human reason—including technoscience, of course—to the problems of human society. By the nineteenth century, progress was thought to happen regardless of whether human reason was in charge of human life:

> After the middle of the [nineteenth] century natural science invested the doctrine of progress with a more materialistic implication. Progress was still regarded as the result of a force external to man; but the force was to be found not above but inherent in the phenomenal world.... Guided by these preconceptions thinkers abandoned the effort to hasten progress by describing utopias and turned to the search for the inevitable law by which progress had been and would be achieved.[33]

There was supposed to be some "hidden pattern" in human history, science, and technology,[34] and it was the job of especially the 'social scientist' to figure that pattern out. That hidden pattern might even be thought to be human nature itself, as if humans were 'naturally' (and not socially) driven by "some inherent desire to pursue progress...."[35] Regardless, progress was understood as a force or law of nature, and like good Baconians, scientists would gain control over that law by 'obeying' what they thought were its basic principles. From Auguste Comte (who coined the term 'sociology') to Adam Smith (whose theory of the 'invisible hand' fortuitously transformed individual greed into collective economic benefit) to G.W.F. Hegel (who defined progress in terms of the *zeitgeist*, the 'spirit of the time') to Karl Marx (who described the economic laws that would inevitably overthrow capitalism and usher in a communist utopia), science could *harness* the forces behind progress and implement techniques for more efficient and relentless improvement.

Because of its inevitability, progress could be understood to be just a matter of time. Newer things were automatically more progressive than older things. And because progress always means improvement, newer things were automatically *better* than old ones: "to be old-fashioned or to be 'out of date' was to lack value."[36] Lewis Mumford defines the doctrine of progress as the assumption "that a later point in time *necessarily* carries a greater accumulation of values...."[37] This denigration of the old went hand-in-hand with the modern repudiation of tradition (see Chapter Nine): there couldn't be any point in studying (for example) the intellectual history of science and technology, because current science and technology are automatically better than anything which came before. If someone were to ask why a social, scientific, or technological change was occurring, the progressive answer could simply be "because it's today, the year [insert date]." Thus the poet Tennyson compared progress to an unstoppable locomotive speeding down a track through time.[38] "Triumphant in its war against all forms of traditionalism," technoscientific progress would "roll hopefully onward, a juggernaut that crushes those that fall beneath its wheels."[39] The upward trajectory of progress throughout time would leave everything else behind in the dust where they belong.

Intellectual Progress

At this stage, however, we should ask ourselves 'progress *in what*?' What is progressing, exactly? If we want to say scientific discovery or technological innovation is progressing, these qualities (i.e., discovery and innovation) are still quite abstract and vague. The answer, in part, is that progress is supposed to be an increase in *intelligence*—a measurable quality, we'd like to think. The quintessentially modern view of progress is "that evil and misery are the result not of innate failings, but rather of ignorance, prejudice, and poverty...." Therefore, as human intelligence inevitably increases over time, "the problems of humanity can [and will] be solved by greater use of reason, better education, and increased material prosperity."[40] If technoscience equates knowledge with power, then the more powerful our technologies are, the more knowledgeable we are. Wealthy, powerful, and technologically advanced people are automatically more rational, scientific, and correct than impoverished, weak, and 'backward' people—such is the nature of progress. Leverage over the material world makes us smarter; this is why we have 'smart classrooms' with 'smart boards,' 'smart phones,' or 'smart bombs.' Those things are *smarter* because they're technologically better (whatever that is). Therefore, if Western technoscience is unlimited and unstoppable intelligence, then progress becomes *the standard for rational thought*. Technoscientific progress equals rational progress, and anything other than technoscientific progress is regressive nonsense, literally irrational.

The third characteristic of the modern doctrine of progress, therefore, is that humanity is "climbing steadily out of the mire of superstition, ignorance, savagery, into a world that [will] become ever more polished, humane, and rational...."[41] Progress sees technoscience as literally rationality itself, in contrast to myth, superstition, and religion (see Chapter Six). Anything other than modern science and technology is stupidity. For example, the famous French critic Voltaire "held that the social effects of religion were often pernicious and an obstacle to progress" (even though he was not an atheist himself).[42] Unlike the medieval synthesis which tried to unify science and religion (see Chapter Eight), modern technoscience saw itself as progressively *displacing* religion: the more science and technology progressed, the more rational and the less religious people would become (see Chapter Eighteen for an extended discussion of the relationship between science and religion in STS).

The funny thing is that by getting rid of religion, technoscientific progress *replaces* it. Advocates of progress recognized that "as religious orthodoxies ... declined in importance, ... a belief in progress and a commitment to aid it ... provided one important source of meaning in life."[43] Technoscientific progress thus functions as a myth, giving meaning and significance to modern life (see Chapter Four). This ought to be unsurprising, for as we noted earlier in this chapter, the doctrine of progress is a secularization of the Christian vision of salvation. Lynn White, Jr. even called progress a "Judeo-Christian heresy."[44] The modern idea of progress is well-suited to the category of myth and religion because it takes science and technology and ties them to a transcendent system that is "more enduring and significant than [we are]."[45] Thus can advocates of progress claim that technoscience is "celestial in its theme, human in its means, but literally superhuman in its goal."[46] If the scientists writing in the *Wall Street Journal* can explicitly call environmentalism a set of anti-progressive "false gods,"[47] then the usual story can also explicitly say that progress is "at once a science, and a religion...."[48] Because technoscientific progress alone is truly rational, it is also the only true religion (whereas all the other non-scientific religions are irrational).

Moral Progress

Underlying these three characteristics of the doctrine of progress (i.e., limitlessness, inevitability, and rationality) is an atypically scientific characteristic: *value*. The usual story presents both science and technology as objective and neutral, suggesting that technoscience is supposed to stand aloof from issues of goodness or badness. Values and evaluations are assumed to be merely 'subjective opinions' or 'emotional responses' in comparison to science's unbiased rational appreciation of facts. But progress isn't neutral at all. We've already seen that progress is all about improvement or betterment. According to the doctrine, any change over time is an increase in value, which is why progressives assume that being 'old-fashioned' or 'out of date' is *bad*. Progress supposedly increases intelligence, which is of course *good*; nobody is supposed to think that higher levels of rationality are bad! Therefore, value is built right into the notion of progress, and as such is built right into the notion of modern technoscience, even though modern technoscience is supposed to be independent of value. How are we to make sense of this inconsistency?

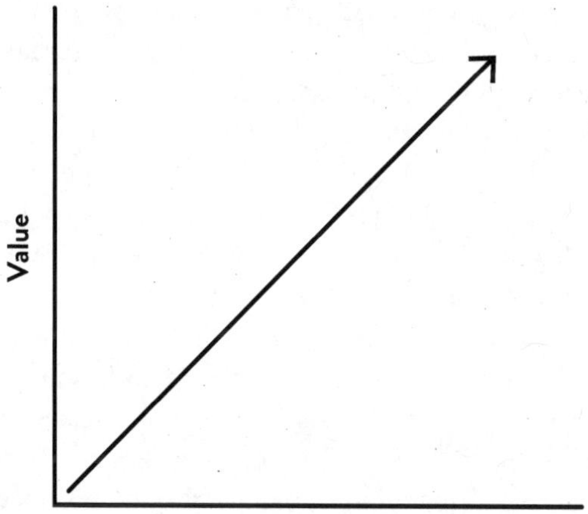

Progress as value increase over time.

Many, though not all, early modern thinkers equated what they saw as improvements in rationality with improvements in morality. If stupid choices are bad choices, then good choices will increase as stupidity decreases. Education, therefore, was expected to lead to both intellectual and moral progress. Intelligent and educated people were assumed to be automatically better people, and human reason could be applied to improve human behavior, just as scientific reason had been applied to improve technologies:

> reason could be used to perfect the science of government, and to engineer both social and moral progress. [This] could lead to a golden age of greater happiness and the abolition of war.... The rule of reason would ... rid the world of tyranny and superstition, increase equality, improve health, prosperity and happiness, and even lead people to exercise voluntary birth control "rather than foolishly to encumber the world with useless and wretched beings."[49]

This, then, was the impetus behind the new and modern idea of the social sciences: turning scientific experimental methods towards human society (rather than just nonhuman nature) would achieve moral improvement just as natural science had created improved technologies. The great hope of the American and French revolutions was that with the overthrow of monarchies, the last vestiges of medieval oppression would be replaced with scientific reason, technological efficiency, economic growth, and moral progress.

In the context of modern Europe and North America, therefore, it was hardly possible to separate science or technology from a grander narrative of moral progress towards the perfect state of human society. As science progresses in terms of discovering new truths, "scientific progress is taken to imply progress for society at large, as scientific technologies relieve the burden of daily living."[50] Replacing Aristotle with modern technoscience was supposed to go hand-in-hand with replacing outdated social structures and behaviors with rationally advanced and 'progressive' institutions and behaviors. In that kind of intellectual environment, it wouldn't make sense to imagine that a new scientific discovery could be bad. Technological progress was also liberating, as it produced many labor-saving devices. 'Advances' in technology were by definition 'good,' otherwise they wouldn't be 'advancements' at all. Progress sees technology as "being of predominant worth," the "transcendent end" or goal of progress and human society itself (i.e., telos or final cause; see Chapter Seven).[51] The implication is that more power is always better than less power, just because it is. This is, of course, not a particularly enlightening explanation for *why* it's always better.

Therefore, the doctrine of progress assumes that when reason, science, and technology increase, so does value. Western technoscience simply makes everything better. If we think (as does so much of Silicon Valley) that the internet is "a force for good, supporting inclusion and democracy,"[52] our thinking is firmly embedded in the progressive narrative. If we think that giving "One Laptop Per Child" (OLPC) will automatically empower children in sub-Saharan Africa, we think that progress guarantees that advanced technology will always be valuable in any and all social contexts, and we also assume that Africans lack the goods that Western technoscience provides.[53] If, on the other hand, we subject technoscientific progress to questioning, asking if it

Intrinsic Value: the value internal to (or inherent in) a thing. An intrinsically valuable thing is pursued for its own sake, not for the sake of something other than itself.

really is expanding or enriching human (or nonhuman) life, that would commit "the ultimate heresy."[54] Such questioning would be utterly meaningless and irrational. You simply cannot question the goodness of progress. Technoscience is therefore the standard of value too, the standard by which goodness or badness is measured: "Life was judged by the extent to which it ministered to progress, progress was not judged by the extent to which it ministered to life."[55] It wouldn't be possible to evaluate scientific or technological progress in terms of badness or goodness, because scientific and technological progress *is* the standard by which goodness or badness are measured! Technoscientific progress is intrinsically valuable: not valuable in relation to an external standard, but valuable in relation to itself, for itself.[56]

This is how modern Western technoscience can claim to stand above questions of value as the objectively true though neutral pinnacle of human knowing, while simultaneously claiming to be exceedingly valuable, intrinsically good, and automatically creating better people. As the standard of value itself, it is simultaneously above and beyond all the petty things which are measured against it, while also being the highest good. This is also how technoscientific progress can be endless and limitless: it is its own goal, pursued for its own sake, and so never reaches an end other than the moving target of itself. In the words of Lewis Mumford,

> progress was motion towards infinity, motion without completion or end, motion for motion's sake. One could not have too much progress; it could not come too rapidly; it could not spread too widely; and it could not destroy the "unprogressive" elements in society too swiftly and ruthlessly: for progress was good in itself independent of direction or end.[57]

Technoscientific progress is understood as limitless because the goal it is aiming at is itself. So long as it's progressing, it is reaching its goal. Its endpoint is endless, and its value is self-justifying. It is literally meaningless (according to the usual story) to question the value or objectivity of modern Western technoscience. Progress makes STS theoretically impossible.

In the early modern period, then, progress was thought to be self-evident.[58] It was simply guaranteed that things in the past were worse than they were in the present, and that things in the future would be better than things in the present. History *had to* reveal that living conditions in the middle ages were worse than in industrialized Europe, that cultural norms in ancient Rome were more oppressive than they were in Victorian England, that Bronze Age warfare was more bloody and vicious than in the twentieth century, and that hunter-gatherers were more impoverished and hungry than the urban homeless of today. Whether these claims were factually accurate "did not even occur to the exponents of Progress as possibilities for investigation. [That was] ruled out automatically by the theory itself."[59] Progress thus asserts itself with such force that it puts itself beyond assessment by either facts or norms.[60] Progress and technoscience are untouchable, the previously mentioned juggernaut which "moves on relentlessly...."[61] We have no choice but to sit back and enjoy the ride (if we can).

Material Progress

Yet for all this talk of value and intelligence, the idea of progress is still explicitly materialistic. When it really comes down to it, the modern notion of progress isn't about the onward march of 'spirit' but rather improvements that we unreflectively call 'quality of life.' Quality of life is supposed to mean a physical quantity of value that can be measured, like money or leisure time or number of automobiles or televisions per household. "*Material* progress" (however that's defined) is a "mainstay of the modern belief in human progress."[62] Because there cannot be any legitimately scientific or technological measurement of abstract or quasi-spiritual qualities like love or virtue or the meaning of life, progress cannot be about those things. Rather, progress is about getting your head out of the clouds and back down to the physical things on planet Earth.

Ultimately, then, progress cannot be about human moral progress! On the one hand,

> the truly modern conception of progress [focuses on] not human perfectibility but the combination of continued progress of reason in science and technology, and a concomitant surge in economic growth and material prosperity.[63]

Yet in the same breath, material progress cannot stop itself from claiming to deliver lofty value-positive outcomes for subjective human experience:

> These two versions of progress—technological advance and economic growth—seemed to promise to all [hu]mankind a sure material route to progress in happiness and social welfare.[64]

It should not be surprising, then, that the doctrine of progress cannot keep moral improvement distinct from material improvement. "Improvement" is unavoidably a term about value, and it's notoriously difficult (if not impossible) to separate values from moral judgments. Moreover, if (as we saw earlier in this chapter) progress sees morality as a result of education and intelligence, and if knowledge and power are the same thing, then technological power or material prosperity will be indistinguishable from moral progress. Therefore, the usual story assumes that we don't have to "give much thought to moral progress," because it "assume[s] that it goes hand in hand with the material."[65] We think we know, thanks to modern economic science, that we can achieve "unbounded progress in welfare and an indefinite augmentation of wealth—a morally desirable solution for society—without the need for the moral perfection of the citizens of that society."[66] If we will all get rich eventually, won't that make us all good? "We take for granted that progress now means ... merely an increase in the technological means of living a comfortable, middle-class life."[67] I could continue by giving a list of examples of amazing gadgets that we're all supposed to believe make our lives infinitely better than those of our ancestors, but what's the point of that? The list would be horribly out of date in a few years' time, and you would just laugh at it. So let's just pretend that

I'm writing in your present, and that we're all suitably impressed by the current tools we have at our disposal for manipulating material particles and energy in the best ways ever. They're the best, they make us happy, and everything's great for everyone, right?

Manipulating material particles and energy in the best way ever is exactly what technoscientific progress is supposed to mean. It doesn't matter if those materials are petroleum hydrocarbons, money in the bank, or chemicals in your brain; the "implied ambition of science is to be able to explain all the phenomena of the physical universe, and to reduce those residual parts of the universe and of human experience that purportedly lie outside the realm of science to a form of physical activity and nothing else."[68] The progress of modern science includes the goal of explaining materially and quantifying *everything in the universe*, yourself included. We may not be there yet, but on the assumption that progress is real, we'll get there eventually and the results will be great. Therefore, the usual story which has been in play "since the seventeenth century," conceives of science and technology as "a materialistic ... and reductive philosophy that implies a mechanical and deterministic view of human behavior and of social organisation."[69] The mastery of nature, human or nonhuman, is the only thing we have to get better at, and we will get better at it, because that's just how humans are (so we moderns think). The usual story tells us—actual historical and anthropological facts be damned—that human beings have always been doing this, trying to use science and technology to 'make the world a better place.' It's as if 'we' always were modern:

> The capital fact of man, as distinguished from the lower animals and from plants is that he does not have to take the world as he finds it, that he does not merely adapt himself to his environment, but that he himself is a creator of his world. If our ancestors had taken and left the world as they found it, we should be little more than erected monkeys today.[70]

According to modernity's doctrine of progress, if it wasn't for technoscience's manipulation of the material world, humans would be nothing but a bunch of naked apes. And that would suck, because without the increasingly powerful manipulation of physical particles, the world would be an incredibly shitty place. This profoundly modern idea, however, has remarkably racist implications.

The State of Nature

Even though the goals of progress are always off in the distance of the indeterminate future, progress still defines itself in opposition to something very specific that it doesn't want to be: *nature*. 'Living naturally' is the great enemy of progressive and modern technoscience. 'Living naturally' isn't even supposed to be meaningful (hence the scare quotes); progress means there's only one way to live decently, and that's the way of the future, the way of the stars where Buzz Lightyear goes, not Earthly nature! Progress is not about living in 'backwards' conditions, any more than it is about following traditional

An Inuit hunter in traditional clothing.

ways of life. Progress is about overthrowing tradition and overcoming physical limitations (see Chapter Nine), otherwise we'd be no better than animals (which, of course, would be utterly unacceptable). Progress, therefore, defines itself via the negation of what it (falsely) considers to be the natural: the oppressive, barbaric, savage, animalistic, and uncivilized 'State of Nature.' That is to say, the doctrine of progress reflects the intellectual and social framework of an agriculturally conditioned society (see Chapter Five).

Early modern theorists set forth the hypothesis of the State of Nature (to be understood either literally or metaphorically) to both explain and justify their doctrine of progress. The State of Nature was supposed to be a period of time when human beings lived in what early modern people considered a 'natural' situation. They thought living naturally meant people were completely free to do whatever they wanted, just like they thought wild animals were (the word "wild" literally means *free*, or to be able to follow your own *will*). In the State of Nature, humans were thought to be under no legal or moral constraints because it was assumed that there would have been no social organization in a natural situation. The 'natural' state of humans was thought to be completely individualistic and anti-social, as if the opposite of modern Western culture was absolutely no culture at all.

The idea of the State of Nature was not formulated on the basis of any anthropological research. In fact, it literally never existed. However, European contact with African and American cultures during the ages of exploration and colonization was understood to confirm what the Europeans already 'knew' about themselves: European superiority over against other humans who had not 'progressed' as far as Europeans thought Europeans had. John Stuart Mill, the great British moral philosopher, used the example of "brutal and degraded" Patagonians (South American hunter-gatherers) and "Esquimaux" (the Inuit hunter-gatherers of North America) to show why he didn't think God had created the natural world as a perfectly good place.[71] The idea was that if God had actually created nature as a good place, then the people who lived most in tune with nature (i.e., the Patagonians and Inuit) would have been as advanced and refined as nineteenth-century Europeans. The fact that these people were not the same as nineteenth-century Europeans was 'proof' that the natural world was a terrible place and that God needed humans (i.e., European men) to use technoscientific progress to help him make it better.

To live in the State of Nature, therefore, was entirely undesirable. In *Leviathan* (1651), Thomas Hobbes famously stipulated that life in the State of Nature was "solitary, poor, nasty, brutish, and short." Humans lived there in perpetual terror of the dangers supposedly lurking around every corner. Red in tooth and claw, nature was continually trying to kill people. On Mill's estimation, nature was (and still is, in its untamed form) violent, bloodthirsty, cannibalistic, painful, disorganized, unintelligent, sexist, and fascist (racism didn't make his list, perhaps because he didn't see a problem with it). Humans therefore have a moral obligation to *not* live in harmony with nature, lest they emulate all those natural vices. The people who do live in the State of Nature (or close to it) are virtually subhuman because they resemble the awfulness of nature in the raw, and thus Europeans thought themselves to be justified in enslaving those people and taking their land for colonies, all in the name of progress. At this juncture we really should take an explicit look at the role of colonialism and racism in modern Western technoscience.

Colonialism, Racism, and STS

The postcolonial critique of Western technoscience is that there is no modern Western science and technology without empire. Europeans *invented* the idea of modernity to help them make sense of their discovery of what was, for them, the 'New World' Modernity is a center–periphery system, with self-proclaimed modern people at the center and nonmodern everybody else out beyond the edges of respectability. Empires get formed by the core homeland conquering the peripheries, and empires are justified by ideologies of the superiority of the center. The Doctrine of Progress (after all this analysis, it probably deserves capital letters at this point) away from the State of Nature is one such ideology:

> The ideology of race emerged during the heyday of European expansion around the world, as colonial empires attempted to justify the socioeconomic privilege of white colonists, the plundering of 'their' colonies, and the subjugation of indigenous groups and involuntary migrants in far-flung colonies. This race ideology was based on the belief that white Europeans were culturally, mentally, and physically more evolved than and thus superior to colonized peoples.[72]

This belief in superior European racial evolution walked in lockstep with Western science and technology. Science and technology are both "the motors of modernity and its progress" and the supposed proof of the superiority of the center.[73] Scholars of colonialism and race, in addition to those of gender (Chapter Nine) and indigeneity (Chapter Four), point out that the *benefits* of modern Western technoscience tend to accrue to people associated with the center, while the *costs* tend to be borne by people relegated to the peripheries. Even though the prefix "post" in postcolonialism suggests that colonialism is over, scholars in these fields recognizes that "Colonial assumptions are by no means features of the past."[74] This can be seen in the persistence of the narrative of Progress and the belief that race is a biological, not social category.

Postcolonial Critique of Technoscience: rather than being neutral and objective, there is no modern Western technoscience without empire. It has been and still is complicit in imperialism, colonialism, racism, and slavery.

Empires require knowledge about their subject peoples so that they can be more effectively governed.[75] Early modern Europeans 'knew' they were better people than their colonized subjects because they themselves had 'progressed' materially: their science was (supposedly) universally true, and their technology was (supposedly) advanced, not backwards or primitive. This belief was held in spite of the fact that colonial powers had to *learn* how to survive in their 'New World' from the local knowledge of the Indigenous inhabitants (see Chapter Four), and utilize Indigenous technologies to do so (e.g., consider the role of the canoe in the fur trade). The Doctrine of Progress nevertheless told Europeans that "'mankind' had progressed ... 'from Infancy to manhood,'"[76] at least as far as they were concerned, thanks to an inevitable, universal "process of development from nature to civilization."[77] To govern their colonial subjects, they just had to identify what stage of development their populations were in to determine their strength, economic potential, and susceptibility to disease. This was one of the things that the supposed *science of race* was for.

Unsurprisingly, modern Europeans identified the colonized peoples of Africa, the Americas, and the British Isles (i.e., Ireland and Scotland) as savages, the earliest and most primitive stage of progressive development away from the State of Nature.[78] Biological taxonomy, the emerging practice of categorizing life forms in terms of their genus, species, etc. was therefore applied to differences *within* the human species (not just between species) and weaponized as hierarchy.[79] Racializing these supposed hierarchies transformed savagery and civilization from cultural or economic categories into scientific, biological, or physical ones. Doing so meant that a vast number of human beings were considered closer to animals than to civilized people, providing a (supposedly) scientific justification for colonization and slavery.

European involvement in the slave trade was already ongoing, but through those commercial channels the first orangutans and chimpanzees were 'discovered' by Western scientists.[80] In addition to being unsure if the two types of ape were actually a single species, the question arose whether these animals were human or whether (some) humans were more like animals. The great apes came to be (at that time) considered human, but that commonality was *why* humans whose supposed racial characteristics were animalian enough were deemed suitable for slavery (see Aristotle in Chapter Seven). Apes weren't suitable as slaves because they couldn't talk and were lazy, whereas racialized savages were suitable as slaves because they could be taught to talk and forced to work.[81]

Moreover, early modern scientists thought that the physical characteristics of humans could be modified by their life conditions: climate, labor, economics, diet, etc. For example, the imagined harshness of living a life of foraging (see Chapter Five) was believed to create weakness and other undesirable physical traits.[82] The level of a person's technological progress therefore affected their race! Conversely, this allowed the Western imperialists to think that "savage lands and peoples" could be civilized—brought further out of the State of Nature—by "bringing Christianity, capitalism, agriculture, nuclear family, and so-called reasoned inquiry" to change savage ways of life for the better.[83] Thinking agriculturally, the Doctrine of Progress saw science and technology as the means to tame and domesticate the wild other, in this case, other people.

But if (social) science can discover the universal laws of human development and hurry them up with the correct technologies or policies, this can happen both in conquered territories and the core homeland. Well into the twentieth century, modern Europeans constructed a science of eugenics (literally "good birth") to improve racial gene pools by controlling who was allowed to reproduce (much like breeding livestock). According to one eugenicist, "forces have been at work determining for progress" the racial purity of Caucasian people,[84] but unless science can take control of this process, such progress will only happen by chance, and slowly. Interpreting biological evolution in a progressive, teleological, and racist way, eugenicists thought that evolution wouldn't reach its ultimate goal (whatever that was) quickly enough without the guiding hand of technoscience. Eugenics was the inspiration behind the Nazi German policy of "racial hygiene" as well as policies and laws in both the United States of America and Canada that included the forced sterilization of people deemed defective.

Eugenics: a social movement and a science of the nineteenth and twentieth centuries which tried to improve the human race by controlling human reproduction.

The eugenics law in my home province was repealed three years before I was born, but the harms of it and of many other forms of colonial and racist technoscience continue, for example in modern medicine:

> Racist theories motivated unethical experiments on Black Americans. Physicians have used race-specific standards to interpret pulmonary function tests. Psychologists used race-specific norms to interpret neuropsychological tests of retired football players who had had concussions. They have exhibited bias in the assessment and management of pain. Some widely used technologies underperform in Black and Brown people: pulse oximeters,

> for instance, can overestimate blood oxygen levels in patients with darker complexions and delay identification of patients in need of treatment.... race based beliefs and practices, especially the use of racial categories, remain widespread [in medicine].[85]

Importantly, however, "socially defined racial and ethnic groups ... are not biological or attributive categories...."[86] Some medical outcomes may be racially coded, but these results have more to do with family history, environmental conditions, or social inequalities than "superficial traits" like skin color.[87] Race is therefore more about shared social experience than the "genetic homogeneity"[88] that was supposed to naturalize the lower socio-economic status of people at the imperial peripheries, as if the people there deserved it.[89] Even so, the usual story's Doctrine of Progress away from the State of Nature—including its imperialism, colonialism, racism, and slavery—is all too often "embedded in medical practices and institutions." It is therefore important to "reset the defaults in medicine" in this respect,[90] as well as Western technoscience more generally.

So when boosters celebrate "the fact that technological innovation allows higher incomes without backbreaking labour, as well as longer life expectancy and freedom from many debilitating diseases,"[91] we need to recognize that these sentiments replicate the colonialist and racist usual story of Progress away from the State of Nature. It assumes that societies other than mainstream Western society are automatically characterized by backbreaking labor, shorter life expectancy, and debilitating disease. It may not occur to us that most epidemic diseases (like influenza) only crossed the species barrier into humans *after* agriculturalists started living in close proximity to domesticated animals, that life expectancy actually *dropped* after the invention of agriculture, and that hunter-gatherers work *fewer* hours per week (and medieval peasants worked fewer days per year) than Westerners currently do. The Doctrine of Progress doesn't want to be evaluated in light of actual historical, anthropological, or archaeological facts, however. Rather than seeing the rise of modern Western technoscience as a socially contingent phenomenon, the usual story wants us to see progress as ironically natural: it is supposed to be natural selection "that alone has raised us from the beast and the worm,"[92] and so it's only natural that 'we humans' should desire to use science and technology to elevate and extract ourselves out of the State of Nature. It's supposed to be natural for (Western?) humans to eventually develop the (Western!) science and technology they need to leave nature behind.

Early modern thinkers had a variety of theories for how 'people' (i.e., Western cultures) supposedly liberated themselves from the clutches of nature. The political philosopher John Locke's theory was that early humans formed a 'social contract' whereby they agreed to certain rules that would structure a society and allow them to build civilization together, rather than be completely individualistic and free. The key notion here is that Progress is supposed to occur when humans submit themselves to 'reason' rather than to the laws of nature. 'Reason' for Locke was predominantly political and economic, but it could just as well have been scientific. Submission to rational limits was also technological, as progressing away from nature required hard work. Guided by scientific reason, human beings supposedly add 'labor value' to 'raw nature' and technologically make it worth something (finally!).

The late modern scientists writing in the *Wall Street Journal* say nothing less, three centuries later: "humanity has always progressed by increasingly harnessing Nature to its needs and not the reverse.... [The instruments of] Science, Technology, and Industry ... are indispensable tools of a future shaped by Humanity, by itself and for itself."[93] Thus are 'we' supposed to escape the "animal-like existence" of the State of Nature with a strong work ethic and technoscientific rationality.[94] Modern Western technoscience will fix all that is disgusting about nature while elevating our species away from it and towards the unbelievable utopia awaiting us in the future. The myths of Progress and the State of Nature propose that by sacrificing nature and human freedom to modern science and technology, society will achieve great goods beyond all imagining. There you have it: this is the story modernity likes to tell itself about itself. Progress away from the State of Nature is the usual story, and it is deeply flawed.

Notes

1 Robert Cabana, *Global National*, newscast, 3 August 2018.

2 Hans Jonas, "Toward a Philosophy of Technology," *The Hastings Center Report* 9, no. 1 (1979): 38.

3 John C. Caiazza, *The Crisis of Progress: Science, Society, and Values* (Transaction, 2016), 7.

4 Mick Smith, "The State of Nature: The Political Philosophy of Primitivism and the Culture of Contamination," *Environmental Values* 11 (2002): 410.

5 The progressive view of technoscience is often called the "Whig history of science" in STS, "stadial history," or (as its eighteenth-century advocates called it) "conjectural history." It was especially associated with the Scottish Enlightenment, including Adam Smith and David Hume (see Chapter Fourteen).

6 Jonas, "Philosophy of Technology," 35; original emphasis.

7 Richard Bronk, *Progress and the Invisible Hand: The Philosophy and Economics of Human Advance* (Little Brown, 1998), 28.

8 Carl Becker, "Progress," in *Encyclopaedia of the Social Sciences* (Macmillan, 1934), 496.

9 Lewis Mumford, *Technics and Civilization* (Harcourt Brace, 1934), 182.

10 Bronk, *Progress*, 7.

11 Smith, "State of Nature," 408.

12 Yuval Noah Harari, interviewed by Nate Hopper, *TIME*, 27 February/6 March 2017, 116.

13 "Beware of False Gods in Rio," *Wall Street Journal*, 1 June 1992, A12.

14 Caiazza, *Crisis of Progress*, 8, 89.

15 Rick Szostak, *Restoring Human Progress* (Cranmore, 2012), 256.

16 Jonas, "Philosophy of Technology," 35.

17 Jonas, "Philosophy of Technology," 34.

18 Becker, "Progress," 498.

19 Mumford, *Technics and Civilization*, 182.

20 Jonas, "Philosophy of Technology," 34.

21 Mumford, *Technics and Civilization*, 182.
22 Caiazza, *Crisis of Progress*, 26.
23 Ronald Wright, *A Short History of Progress* (Anansi, 2004), 6.
24 Bronk, *Progress*, 47, 7.
25 Quoted in Szostak, *Restoring Human Progress*, 96 note 54.
26 Szostak, *Restoring Human Progress*, 79.
27 Jonas, "Philosophy of Technology," 37.
28 Jonas, "Philosophy of Technology," 35.
29 Caiazza, *Crisis of Progress*, 7.
30 Quoted in Szostak, *Restoring Human Progress*, 72.
31 Quoted in Wright, *Short History*, 3.
32 Mumford, *Technics and Civilization*, 182.
33 Becker, "Progress," 497–98.
34 Szostak, *Restoring Human Progress*, 73.
35 Szostak, *Restoring Human Progress*, 73.
36 Mumford, *Technics and Civilization*, 184.
37 Mumford, *Technics and Civilization*, 184; original emphasis.
38 "Fool, again the dream, the fancy! but I *know* my words are wild, / But I count the gray barbarian lower than the Christian child. / …Mated with a squalid savage—what to me were sun or clime? / I the heir of all the ages, in the foremost files of time— / ….Not in vain the distance beacons. Forward, forward let us range, / Let the great world spin for ever down the ringing grooves of change. / Thro' the shadow of the globe we seep into the younger day; Better fifty years of Europe than a cycle of Cathay" (Alfred Lord Tennyson, *Locksley Hall*, 1842, https://www.poetryfoundation.org/poems/45362/locksley-hall).
39 Smith, "State of Nature," 410.
40 Bronk, *Progress*, 2.
41 Mumford, *Technics and Civilization*, 182.
42 Bronk, *Progress*, 47.
43 Szostak, *Restoring Human Progress*, 85.
44 Lynn White, Jr., "The Historical Roots of Our Ecologic Crisis," *Science* 155, no. 3767 (1967): 1205.
45 Becker, "Progress," 495.
46 Caleb Williams Saleeby, *Parenthood and Race Culture: An Outline of Eugenics* (Moffat Yard, 1916), 18.
47 "Beware of False Gods in Rio," *Wall Street Journal*, 1 June 1992, A12.
48 Saleeby, *Race Culture*, ix.
49 Bronk, *Progress*, 49, concluding with a quotation from Nicolas Marquis de Condorcet.
50 Caiazza, *Crisis of Progress*, 25.
51 Jonas, "Philosophy of Technology," 38.
52 Lawrence Landweber, quoted in Matt Taibbi, "The Facebook Menace," *Rolling Stone*, 19 April–3 May 2018, 45.
53 For more on the One Laptop Per Child program, see Anabel Quan-Haase, *Technology and Society: Social Networks, Power, and Inequality*, 2nd ed. (Oxford University Press, 2016), 161–62.
54 Mumford, *Technics and Civilization*, 185.
55 Mumford, *Technics and Civilization*, 185.

56 Contrast this notion of intrinsic value with that of instrumental value from Chapter Three.
57 Mumford, *Technics and Civilization*, 184.
58 Mumford, *Technics and Civilization*, 182.
59 Mumford, *Technics and Civilization*, 183.
60 Jonas, "Philosophy of Technology," 41.
61 Jonas, "Philosophy of Technology," 35.
62 Bronk, *Progress*, 7; emphasis mine.
63 Bronk, *Progress*, 51.
64 Bronk, *Progress*, 51.
65 Wright, *Short History*, 4.
66 Bronk, *Progress*, 51.
67 Caiazza, *Crisis of Progress*, 89.
68 Caiazza, *Crisis of Progress*, 21.
69 Caiazza, *Crisis of Progress*, 5.
70 Saleeby, *Race Culture*, 1.
71 John Stuart Mill, "Nature," in *Three Essays on Religion*, 2nd ed. (Longmans, Green, Reader and Dyer, 1874), 41.
72 Ramya M. Rajagopalan et al., "Race and Science in the Twenty-First Century," in *The Handbook of Science and Technology Studies*, 4th ed., ed. Ulrike Felt et al. (MIT Press, 2016), 350.
73 Banu Subramaniam et al., "Feminism, Postcolonialism, Technoscience," in *The Handbook of Science and Technology Studies*, 416.
74 Subramaniam et al., "Feminism, Postcolonialism, Technoscience," 414.
75 Sarah Irving-Stonebraker, "From Eden to Savagery and Civilization: British Colonialism and Humanity in the Development of Natural History, c. 1600–1840," *History of the Human Sciences* 32, no. 4 (2019): 64, 69, 70–71, 76.
76 Bruce Buchan and Linda Anderson Burnett, "Knowing Savagery: Australia and the Anatomy of Race," *History of the Human Sciences* 32, no. 4 (2019): 122; quoting (in part) Reverend John Walker (1731–1803).
77 Irving-Stonebraker, "Savagery and Civilization," 64; cf. 71, 74.
78 Silvia Sebastiani, "A 'Monster with Human Visage': The Orangutan, Savagery, and the Borders of Humanity in the Global Enlightenment," *History of the Human Sciences* 32, no. 4 (2019): 81.
79 Buchan and Burnett, "Knowing Savagery," 117–18.
80 Sebastiani, "'Monster with Human Visage,'" 82.
81 Sebastiani, "'Monster with Human Visage,'" 81, 84–85.
82 Buchan and Burnett, "Knowing Savagery," 120.
83 Subramaniam et al., "Feminism, Postcolonialism, Technoscience," 419.
84 Saleeby, *Race Culture*, 14.
85 Itali Bavli and David S. Jones, "Race Correction and the X-Ray Machine: The Controversy over Increased Radiation Doses for Black Americans in 1968," *New England Journal of Medicine* 387, no. 10 (2022): 947.
86 Lundy Braun et al., "Racial Categories in Medical Practice: How Useful Are They?" *PLOS Medicine* 4, no. 9 (2007): 1423.
87 Bavli and Jones, "Race Correction," 950.
88 Braun et al., "Racial Categories," 1424.

89 Rajagopalan et al., "Race and Science," 351.
90 Bavli and Jones, "Race Correction," 950, 951.
91 Szostak, *Restoring Human Progress*, 256.
92 Saleeby, *Race Culture*, 16.
93 "Beware of False Gods in Rio," A12.
94 Smith, "State of Nature," 408.

Chapter Eleven

The Mechanistic Worldview

AS WE HAVE SEEN, MODERN WESTERN TECHNOSCIENCE COMBINES THE practical hands-on power to manipulate objects with the theoretical, scientific explanation of the nature of things. This notion that science and technology are basically the same thing yields two of the three major characteristics examined in this section of the book: the Baconian will to power and the progressive belief in the inevitable and unlimited increase of that power. The unification of science with technology also yields a third major characteristic: the mechanical worldview. This aspect was noted back in Chapter Two, when we were coming up with tentative definitions of science:

> All of modern biology and, indeed, all of modern science takes as its informing metaphor the clock mechanism described by René Descartes in Part V of his *Discourses*.... Modern science sees the world, both living and dead, as a large and complicated system of gears and levers.[1]

Viewing the world as if it's a giant machine fits right into the doctrines of Baconianism and Progress, because machines are *powerful*, and supposed to get *more* powerful. The world, however, isn't obviously like a machine; at the very least, its living bits appear to be more organic than mechanical. Machines aren't alive, because they're just a complex pile of parts and particles. This is odd, because piles of dead things don't sound very powerful. Moreover, living things are active, pursuing various goals and generally busying themselves with the business of living. Gears and levers, meanwhile, don't do anything by themselves. They might amplify or transfer power, but in themselves, they're as static as a lump of pure matter. The machine view of the world goes so far as to conceive of the entire universe as if it were a clock, using a tool as the model for scientific explanation, and yet—in the end—it's not clear where the power which runs the technoscientific cosmos-machine comes from. Who or what winds the clock? In this chapter we will explore the irony of a dead machine worldview which cannot produce powerful results without reintroducing mystical forces into its understanding of things.

Perhaps it's strange to think of modern Western technoscience as an interpretive schema or worldview, but that's what the usual story is: a set of prereflective assumptions that tells us what we 'already know' about anything we could possibly encounter. A worldview is a way of seeing possible experience, or a theoretical framework for understanding all of reality. The usual story is a worldview because it provides the framework for what a technoscientific explanation of anything has to be: knowledge has to be power, progress has to be unstoppable, and everything has to be mechanical. Mechanistic explanations of fundamentally inert material particles (supposedly) comprise the nature of all truth.

Sir Isaac Newton (1643–1727), portrait by Godfrey Kneller (1689).

As a result, technoscience in the modern Western world takes on the characteristics of myth (see Chapter Four) and even legend, like the apple which (didn't actually) hit Sir Isaac Newton (1643–1727) on the head. True, Newton's theory of universal gravitation was inspired after he saw an apple fall to the ground, but what's remarkable about this anecdote is that it's as much a part of the modern Western cultural vocabulary as the fruit (not necessarily an apple) that Eve gave to Adam in the garden of Eden. Newton is the super-scientist of the Western tradition, at least until we get to Einstein. He didn't just contribute to the mechanistic framework of modern Western technoscience; he *saved* it! His role in modernity is so legendary that his scientific framework

> formed the picture of the world in which everyone now alive was brought up. For it is safe to say that relativity and quantum physics have not yet been taken for granted as are Newton's notions of time, space, place, motion, force, and mass.... [T]o have been brought up in the Newtonian world certainly does shape [our] consciousness, as it does to have been brought up an American rather than a Frenchman, [or] a Christian rather than a Moslem. It is an element of culture, and to exist in a culture with no notion whence it came is to invite

> the anthropologist's inquiry rather than to live as an educated [person], aware and in that measure free.[2]

STS is an attempt to live as educated persons with respect to science and technology, to be aware of where technoscience came from, and therefore free with respect to how we'll respond to it. So we need to understand that even though quantum physics is probably true, everyone around here lives their day-to-day lives in Newton's machinelike universe of matter and force, as if it were so obviously true that nobody needed to notice it. Our default worldview is that everything boils down to particles and power, following the same mechanical laws that govern how machines work. But STS does not simply take this for granted. The mechanistic worldview may be a major part of Western culture's operating system, but rather than simply being trapped within that status quo we can and should be able to consider options for thinking and living differently, should that be necessary.

The Nature of Machines

To understand the notion of a mechanistic worldview, first of all, we'll need to find out what a machine is. As you may recall from earlier chapters in this book, words and concepts that are taken for granted today do not always mean what we thought, once we've taken a closer look at them. This remains the case with machines. The word "machine" has etymological roots in the Greek word *mechanē*, which simply means a contrivance, an invention, or an artifact. For example, a chariot or a siege tower would have been considered a war machine by the Greeks. Linguists speculate that the Greek word has its own roots in the much older (and therefore unwritten) Proto-Indo-European root **magh*, which encompassed the semantic range of "to be able" or "to have power." In fact, the modern German verb *machen* (which would almost sound like "machine" if you were to read it like an English word) means both to make or to do, while the German noun *macht* means power or force. These two aspects are reflected in the English word "might," which can mean both *power* (being "mighty") or the ability to do something ("I might do that"). The concept of a "machine," therefore, implies a *device that is able to apply power* to something. It has the power to be powerful. If the nature of the universe is mechanical, and the nature of machines is power, then the fundamental nature of the universe is Baconian. That's what happens when you combine your definition of science with the definition of technology!

Automatic: to be self-acting or self-operating.

Nevertheless, there is nothing in the basic meaning of "machine" that implies automation or self-regulation. Machines do not necessarily have a 'life of their own' (at least, they're not supposed to; see Chapter Seventeen). This is why the idea of machine thinking in artificial intelligence research is so revolutionary: machines existed for several thousand years but nobody thought they were *alive*. To be *auto-matic*, however, did imply life. *Autos* is the Greek word for "self," while *matos* is the Greek word for thinking or animation (i.e., being filled with a soul). So a *self-acting machine* would be considered to be "automatic," but that is a very unique sort of machine. It may be obvious to point this out, but the whole point of trying to create machine learning and artificial intelligence is that (normally) machines are *not* intelligent or capable of learning, because *machines are fundamentally dead and inanimate.*

Death, then, is the first characteristic of the modern mechanistic view of the world: all things in the universe move, but all things are nevertheless fundamentally lifeless and static. Machines are generally composed of inert parts[3]—e.g., metal gears, silicon chips, rubber belts—and so when those types of components form your picture of the basic structure of reality, you are supposed to see everything as mindless and bland. No wonder machines can't think! They're nothing but *mere matter*—and the Western tradition has traditionally assumed that mind and matter are completely different things.

If the mindless material particles that make up the mechanical universe are fundamentally inert, though, how do they even *move*? As non-automated, they cannot move themselves; they must be acted upon by something external to themselves. Machines are arrangements of intrinsically inert parts that move because of the way they're arranged, because of an organizational structure imposed on them from the outside. Machines are fundamentally *systems of governance* that orchestrate the relationships between otherwise inert parts so that they can work together to produce a desired effect. A machine is therefore "a combination of resistant parts, each specialised in function, operating under human control, to transmit motion and to perform work…."[4] It doesn't matter if the machine is thousands of human laborers following orders given by a bureaucracy, or the powertrain of a Ford F-teenthousand pickup truck;[5] without a system of governance, machines cannot function and effectively revert back to an inert lump of matter.

Mechanism: a contrivance of parts which, in virtue of their arrangement, possesses the ability to produce an effect.

But even though externally governed machines move, they can't think; and if they can't think, they can't feel, right? René Descartes argued that animals were machines, and as such animals could neither think nor feel.[6] This made experimenting on animals rather easy, at least in terms of avoiding moral qualms about causing suffering: if you can't cause a clock to suffer, why worry about experimenting on a live cat? Similarly, Lewis Mumford points out that "it was the king alone who had the godlike power of turning live men into dead mechanical objects."[7] This is why, if you're a part of the modern economic system, your boss may not care much about your feelings. It's nothing personal, it's just the mechanism of business. Of course, if you own or are in charge of a machine, you'll take care of it to make sure that it works, but you don't have to *really* care about hurting its feelings. Cogs are replaceable, after all. So material things—especially mindless, unfeeling, dead, and mechanical things—have no value in and of themselves. Machines are tools, and the value of a tool (as is so often assumed) is found entirely in what you (or someone else) can use them for (see Chapter Sixteen). This is why 'use value' is the same idea as *instrumental value*: an instrument is a tool, and a tool is (supposedly) only good if you can use it to get what you want. This is why calling another person "a tool" is an insult!

Inasmuch as early modern technoscience viewed the universe as a giant machine comprised of smaller machines and innumerable material particles, the universe and everything in it had to be fundamentally inert and dead, mindless and unthinking, unfeeling and insensate, and instrumentally valuable only. Even if plants and animals *look* alive, that's only because their complicated systems of mechanical governance cause them to grow and move. But what about individual people? Are we machines too? Descartes thought our bodies were mechanical, which is why he called his own body a "corpse,"[8] but our ability to think and feel implied to him that there was a special non-mechanical aspect to individual human beings: the ghost in the machine.[9] This non-mechanical, non-material, and thinking-feeling substance inside human beings was what Descartes thought made us uniquely important compared to merely instrumentally valuable animals, but it didn't take early modern natural philosophers long to reject Descartes' ghostly mental substance as unmechanical and unscientific. After all, how is a ghost supposed to push a machine around, rather than float right through it? That is why today, when a computer scientist says that your brain is a "three pound computer made out of meat,"[10] nobody bats an eyelash. We treat ourselves as if we are machines too, thanks to the usual story we don't even notice is there.

> **Instrumental Value:** the value a thing has for getting something else that is valuable (other than itself); "use value."

The application of these broad mechanical characteristics to literally everything in the universe was simply a matter of course for the trailblazing early modern technoscientists. What really got them excited was the *operations* or *functions* of mechanism, and thereby seeing everything in the universe in terms of *work*. Machines are things that work, both in the sense that they should function or operate, and in the sense that they perform important and powerful tasks. The 'workings' of any natural phenomena were taken to be the same, in principle, as the 'workings' of a machine; these are the *processes* by which a thing in nature does or makes whatever it does or makes. If a scientist can understand or know what those 'natural' manufacturing processes are, then they can be *harnessed* technologically to reproduce those effects at will. Mechanistic science is therefore begging to turn into industrial technology. Nature is just a factory waiting to be built (or enslaved, which is much the same thing). Either way, it's time to get productive!

A *productive* science is precisely what Bacon wanted. An actual machine, therefore, copies nature's manufacturing processes, but does not have to copy nature's forms, essences, or appearances (see Chapter Eight). A factory does not have to look like a forest, or like anything, really. It just has to do its job. A horseless carriage (which is an old-fashioned name for an automobile) does not have to look like it is being pulled by horses so long as it has *horsepower*. That's why, when you ask someone with a fast car, "how many horses have you got under the hood?" they can't pop the hood and show you a bunch of horses. Rather, they'll show you a mechanism that can produce or make the power of many hundreds of horses (see Chapter Eight). Therefore, machines *work* because 1) we know the efficient causes of things (which is the only sense of causation that modern technoscience recognizes), 2) they reproduce natural processes by manufacturing the effect, not the form, and 3) they produce power.

The "New Philosophy"

This mechanistic framework was the "new philosophy" (aka science) that replaced Aristotelianism in seventeenth-century Europe, about a century before the first Industrial Revolution. The new mechanical philosophy was considered an improvement over scholastic science because it produced explanations that were *helpful*. Helpful in what sense, though? Mechanistic explanations, unlike those of Greco-Roman and medieval natural philosophy, were supposed to "make clear how the phenomenon to be explained is produced."[11] If the whole purpose of modern science is to *produce* new technological powers, mechanical explanations would be 'helpful' for such a project. On the other hand, if a scientific explanation didn't allow you to (re)produce anything, then that explanation would be literally useless and unproductive. After all, machines are just complicated tools; and if a tool doesn't have a use, it isn't really a tool at all.

This emphasis on usefulness was a radical departure from the systems of value that had governed premodern science; scientific explanation now had to expose the causal mechanisms inside everything, because science's new value was to be found in its Baconian ability to allow humans to invent new tools (especially machines). Science itself had become a tool for creating new tools. So the next time someone asks you what you are going to *do* with your education (see Chapter One), remember that they are asking you a quintessentially modern question. They are assuming that your *education is a tool* that has to be 'useful' if it is to have any value, and that your *job* is to get a job where *you* can be 'useful and productive' in contributing to, well, whatever it is you're supposed to be useful and productive for. If you're a machine, remember that your value is instrumental, dependent on the goals of the system of governance put in place over you. Science—and you—only 'work' if you both 'help' whatever social project is in charge.

In overview, the new (and 'useful') mechanical philosophy of nature operated on the belief that

> all natural phenomena, no matter how complex, all the sensible and insensible properties and behaviours of bodies, can be causally explained in terms of the arrangement and motion (or rest) of minute, insensible particles of matter (corpuscles), each of which is characterised exclusively by certain fundamental and irreducible properties—shape, size, and impenetrability.[12]

Simply put, all things reduce to *matter* and *motion*. There cannot be, in principle, anything else in the universe than these fundamentally inert and dead bodies that just lie there until something bumps them (the word "corpuscles" comes from the same root word as *corpse*). Aristotelian science, by contrast, had a more organic feel to it, for it saw everything—even rocks—pursuing a purpose or goal (i.e., a "final cause"; see Chapter Seven). Early modern scientists saw such purpose-driven (or teleological) description of things as unmechanical because it relied on a belief in "occult forces" or secret powers within the world, much like superstition or magic. A purpose or tendency within natural things to seek out their own goals was too alive, too creepy, and too much like Halloween for a mechanical science. By default, therefore, the inert particles of the mechanistic framework had to rule out the possibility of ghostlike powers in the machine. Spiritual forces are simply not allowed in proper science, or so we moderns are supposed to believe.

However, ruling out the possibility of occult forces is not demanded by empirical observation. A *worldview*, after all, is not something observed so much as a way of observing. An interpretive framework is a model of how you think the world *has to be*, which is not an empirical experience itself. The new philosophers, therefore, "generally recognised [sic] the hypothetical nature of their proffered explanations,"[13] confident that the nature of the universe was mechanical even though there's no way they could observe that it was.

The mechanistic worldview, then, depended on a number of empirically unverifiable assumptions without which mechanical explanations could not work. The first, in the words of Robert Boyle (one of the founders of modern chemistry), was that "there is one catholic or universal matter common to all bodies."[14] That is, there can only be one fundamental substance in the universe: matter. There might be immaterial substances (Boyle believed in God), but they cannot be accepted as possible reasons for why material things behave as they do.[15] Otherwise, there could be immaterial ghosts in machines, and they wouldn't be mechanical anymore. Of course, it's impossible to experimentally verify a statement about what everything is made out of, because nobody can empirically observe everything possible; a universal negative (e.g., there are no non-mechanical causes) cannot be proven.

The second unverifiable assumption required by the mechanistic worldview was that nature must be "uniform in its operations."[16] Matter must behave identically under identical conditions, regardless of where it is in time or space. If the law of gravity, for example, only operated on Tuesdays on Jupiter's moon Io, refused to operate on Wednesdays on Earth, and only randomly operated in the Andromeda galaxy, then the machine of the universe would not work very well; its system of governance would be erratic. Therefore, everything everywhere has to follow the same laws of motion for mechanistic explanations to work, even though—as with matter—this universal requirement cannot be evidentially proven to be true. If, however, we require (a priori) the universal homogeneity of matter and the universal uniformity of motion to be true, those assumptions allow modern technoscientists to explain everything in the same way: as clockwork.

Mechanical Problems

Of course, the mechanistic explanation of everything was (and still is) a work in progress. The early modern technoscientists faced two major difficulties, both of which sprung from the same fact: *power is not inert*. The mechanists had so thoroughly modeled the world in terms of inert simple particles that they were at a loss to explain what governed the machine, or how motion could be transferred between two fundamentally dead corpuscles. Mechanical particles were supposed to be just shapes and sizes; nothing about them said they could be able to *influence* each other. In their eagerness to get rid of occult forces, the mechanists were at a loss to explain impulse or change in momentum; their mechanical bodies were so dead that they had *no abilities whatsoever*, not even the power to affect[17] each other! They couldn't even explain how bumping worked. The mechanistic worldview suddenly didn't look very mighty at all.

The theoretical powerlessness of material particles produced two more specific problems: the first is known today in contemporary psychology as 'the hard problem of consciousness.' Even if we take it for granted that inert material particles can have the power to influence each other on contact, there appears to be a categorical difference between impacts on insensitive particles (like rocks) and impacts on sensitive beings (like ourselves, or cats). In his theory of perception, John Locke (1632–1704) noted that the experience of pain "hath no resemblance … to the motion of a piece of steel dividing our flesh.…"[18] If sensation was truly mechanical, and pain was imparted to our minds by the impulse of minuscule corpuscles in motion, one would expect the information of matter in motion to convey to our minds the same impression. But it doesn't. Raw pain caused by a knife cutting your finger feels *nothing like* matter in motion, even though it is obviously caused by certain forms of matter in motion. The problem is the same with color: the perception of blue or red has no resemblance at all to the wavelengths of light that cause it. The problem, then, is that there appears to be an irreducible gap between mental or emotional facts, and the mechanistic qualities of the mindless and unfeeling particles which supposedly cause those facts—even if those emotion-causing particles are cells in your nervous system.

In fact, the entire notion of consciousness appears irreducible to mechanical particles in motion:

> We are moreover obliged to confess that *perception* and that which depends on it *cannot be explained mechanically*, that is to say by figures and motions. Suppose that there were a machine so constructed as to produce thought, feeling, and perception, we could imagine it increased in size while retaining the same proportions, so that one could enter as one might a mill. On going inside we should only see the parts impinging upon one another; we should not see anything which would explain a perception.[19]

Whatever consciousness is, it appears to be *in principle* non-mechanical, by definition *not* inert, and therefore an exception to the supposed homogeneity of a single material substance across the otherwise mechanical universe. Even if your brain is a three-pound computer made out of meat, your mind appears not to be the same thing as your brain.

Even weirder than mechanical impacts causing unmechanical feelings is the fact of motion between (supposedly) inert bodies that is *not* caused by impacts (assuming that there is empty space between bodies, and not filled with some insensible "aetherial" matter maintaining contact between everything). This is the second problem that arises from the general powerlessness of inert material particles. For fun and effect, I call this 'the problem of seriousness.' If two inert corpuscles are at rest, at a distance from each other, there is no mechanical reason—in terms of mere matter and motion—why either of them should move at all. If there was motion at all, it would be in equilibrium (at least if aetherial matter collided with them "in every which direction ... on every side").[20] However, the two corpuscles will attract each other even though no other material particles push them towards each other.

Mechanical particles should not be able to do this to each other: "There is no action at a distance, no forces or powers which can exert their influence over empty space,"[21] any more than you should be able to bend a spoon with only your mind. And yet corpuscles do this all the time, just as Newton's famous apple fell from a tree. There appears to a be a system of governance in the universe pushing inert material particles around like a magician, as if there were ghosts or feelings in the machine. We have all kinds of unusual words for this phenomenon: weightiness and heaviness, for example, which have a familiar meaning in physics, but also psychological connotations of profundity, depth, or even sadness. That is why we call this phenomenon *gravity*; gravity literally means "weight, dignity, solemnity of deportment or character, importance ... heaviness, pressure."[22] Physical attraction between bodies is serious business! It's weird how dead corpse particles have feelings, isn't it?

Newton's Synthesis: The Force

A mechanistic worldview simply cannot accept the possibility of action at a distance or emotional motion because that would entail a return to occult qualities like those it found in Aristotle. This is why so much of modern psychology tries to explain the mind in terms of computers. If modern technoscience is going to be able to continue treating everything like a power tool, then the problems of consciousness and seriousness (let alone impulse) will have to be solved. The Newtonian synthesis offered a solution to the problem of seriousness (not consciousness), but in a sneaky way that annoyed the new philosophers because it allowed for the presence of feelings in the machine universe that we'd all rather forget about. He performed this sleight-of-hand magic trick by using *math*.

Newton was no slouch. In addition to his physics, he was also famous for inventing calculus by himself, before he turned thirty.[23] Newton was as aware as you are (now, having read Chapters Seven and Eight) that the premoderns had separated *theoria* and *technē*. For this reason, geometry and physics (as purely theoretical sciences) were considered to be completely independent of mechanics and astronomy (as practical arts involving moving objects). But if mathematical principles could be shown to be the fundamental basis of geometry *and* physics *and* mechanics *and* astronomy, then math really would be the tool for unifying theory and technique the way Francis Bacon always wanted. It might seem obvious to us now, but Newton revolutionized natural philosophy by showing that "the quantity of matter is the measure of the same" (i.e., mass = volume x density) and that "the quantity of motion is the measure of the same" (i.e., momentum = velocity x mass).[24] Newton's laws of motion used mathematics to synthesize *theoria* and *technē* "in a single science of matter in motion."[25] Mathematics became the philosopher's stone: the language of all things.

When it came to the mechanically insoluble problem of seriousness, Newton mathematized gravity. He simply *described* mathematically the creepy feelings of attraction between two supposedly inert material particles rather than *explaining* how their attraction worked mechanically. The ghost was still in the machine having feelings, only now (thanks to Newton) technoscientists could predict and control these forces using math. Kepler's laws of planetary motion were derivable from Newton's laws of motion, and so astronomy and physics were unified by mathematically "flinging gravity across the void" of space.[26] Francis Bacon had said that to command nature, she must be obeyed (Chapter Nine), and by unlocking the math of gravity, Newton allowed humans to eventually escape the gravitational field of Earth and enter space itself.

The fact that Newton's gravitational force was no less occult than the Force in *Star Wars* was not lost on his critics. By mathematizing the "most subtle spirit which pervades and lies hid in all gross bodies,"[27] Newton had by no means shown that gravitation was caused by mechanical contact between two inert particles of matter. To the contrary, *spirit*, *force*, *attraction*, or *love* comprised the evidently non-mechanical system of governance for the machine universe, and the new mechanistic philosophers found this to be extremely unscientific: "taken as an explanation of the universe, [Newton's] system failed—or rather it was no explanation at all, since no cause could be assigned and no mechanism imagined for its central principle," namely gravity as the force of attraction.[28] Mechanical functions aren't supposed to be based on how attractive bodies are to each other! However, the math was good and technoscientists eventually stopped complaining, imagining that they had stripped all the "mysticism" away from Newton's theories because the forces could be measured.[29] Today, we're all used to the serious attraction between particles being an unremarkable part of physics. What's remarkable is that we've forgotten that "force" is not a physical concept at all, but a metaphysical one, and that "gravity" (never mind "attraction"!) is an emotional one.

Newton, however, was fully aware that he was not using these concepts "physically, but mathematically."[30] As a good empiricist, he knew that technoscience is surface knowledge:

> In bodies, we see only their figures and colours, we hear only the sounds, we touch only their outward surfaces, we smell only the smells, and taste only the

> savours; but their inward substances are not to be known either by our senses, or by any reflex act of our minds....[31]

An empirical scientist really cannot know what's going on behind the scenes. All there are, are appearances of things; the reasons for things are forever hidden from us. Newton saved the mechanistic worldview, then, not by showing what force, energy, or power *are*, but by allowing technoscience to predict, quantify, and control force, energy, and power so that we can get more of them—even though they are, at root, *mythos* (Chapter Four) rather than *logos* (Chapter Six).

Welcome to the Machine[32]

Mechanistic philosophy spread widely across Europe, but weirdly only the English used it to make a lot of machines. The French and other European nations largely followed the rationalist science of Descartes (see Chapter Nine), excelling in mathematics and chemistry but believing in aetherial matter (because "nature abhors a vacuum"). The English were so proud of Newton, though, that they taught his mechanical physics in schools and churches. This meant that not only natural philosophers but also English craftsmen and inventors knew Newtonian mathematical science: the "science of scholars became the science of the educated layperson" in England.[33] This led to an "engine culture" there, where instruments of precise measurement and experimental mechanical apparatuses were the tools used by "enthusiastic ... craftsmen, artisans, and entrepreneurs" to know the world *as* a mechanical thing—and to transform it into a more useful, working thing.[34]

From this we get the modern definition of engineers: "craftsmen who specialized in turning the principles of natural philosophy into mechanisms useful to entrepreneurs for production, relying on precise measurements with scopes, graphs, and instruments...."[35] Moreover, Newtonians believed in empty space (even though Newton mentioned aetherial matter in his book, *Opticks*), which meant that creating a vacuum—and thus a piston-driven atmospheric engine powered by steam—was a conceptual possibility open to them.[36] Thus, while it was by no means natural or inevitable that this should happen, by "1850, England's 18 million inhabitants were using nearly as much fuel energy ... as the 300 million inhabitants in all of Qing China."[37] By 1900, all the British Isles (not just England) had less than 3 percent of the world's population but produced 25 percent of the world's fuel energy. You might have heard of this: some historians call it the Industrial Revolution.

This massive power differential would not only have sent Bacon into ecstasy, but came in very handy for continuing the process of colonialism and empire (see Chapter Ten). That's why the mechanistic framework of modern technoscience didn't only leave an indelible mark on Western culture, but continues to do so across the globe. It is also the reason why the usual story generally presents a "picture of a soulless, deterministic world machine,"[38] even though that wasn't necessarily Newton's point. As we have seen several times in this book, modern science and technology generally see the world as a complicated system of gears and levers (even the machine's feelings are mathematical), and now we've seen how that came about. But what about the impact of this worldview on human societies? What happens when we become "so used to the atomistic machine view of the world ... that we [forget] that it is a metaphor"?[39]

The 'social sciences' that were to spring forth from the mechanistic (and surreptitiously ghostly) philosophy each, in their own way, treated individual human beings or human social institutions as aggregates of inert material particles controlled by external forces. If the social sciences weren't mechanistic, then they wouldn't have been considered sciences at all. For example, Thomas Hobbes (1588–1679) saw the head of state (the king) as the literal system of governance controlling the inert material particles (individual citizens, incapable of making free choices) out of which the body politic (a mechanical monstrosity, the Leviathan) is composed. Auguste Comte (1798–1857) tried to discover the mechanisms by which society worked, so that they could be more efficiently manipulated to achieve the goals of Progress. Modern economics and psychology also view human individuals as isolated objects engaged in scientifically predictable behavior because they are simply determined by forces such as self-interest (Adam Smith) or desire (Sigmund Freud). Political *science*, soci*ology*, economics, and psych*ology* were all conceived of as sciences (aka, legitimate forms of knowledge) because they insisted on explaining their subject matter (humans and society) only in terms of matter (corpuscles) and force (techniques of governance). Doing so (alongside the rise of statistical mathematics) all feeds very nicely into the bureaucratic state that we know and love today.

So how has Newton and the mechanistic framework formed the picture of the world in which you and I were brought up? In Lewontin's words, "We no longer think, as Descartes did, that the world is like a clock. We think it *is* a clock."[40] Things bump into things, and you're one of those things. So am I. We're just bodies interacting with other bodies in ways predetermined by forces external to us. The world we're in is also just a bunch of things bumping around, into each other and into us. If we can figure out what makes those things tick, we can figure out how to control or redirect what they do, and that will make things better for us. That's pretty much it. Welcome to the dead machine. That's life.

What remains to be seen, now, is how we might respond to this usual modern, Western, Baconian, progressive, and mechanistic story that technoscience tells about us and the world. There are, after all, other ways to think about the world, and other ways to live. In the next section, we will look at several alternative conceptions of knowledge or power that pose problems for the supposed universality and obviousness of the usual story.

Notes

1 R.C. Lewontin, *Biology as Ideology: The Doctrine of DNA* (Anansi, 1991), 12.

2 Charles Coulston Gillispie, *The Edge of Objectivity: An Essay in the History of Scientific Ideas* (Princeton University Press, 1960), 143.

3 So-called simple machines have few if any moving parts (like levers or pulleys). In this respect, they're more readily classified alongside generic tools, and not as "mechanisms" with complex organization of parts.

4 Lewis Mumford, "The First Megamachine," *Diogenes* 14, no. 55 (1966): 3.

5 Mark Little, "Ferd Fteenthousand," YouTube, accessed 15 March 2023, http://youtu.be/F8P5vGcf-NU.

6 René Descartes, *Discourse on Method* (1637), in *Discourse on Method and the Meditations*, trans. F.E. Sutcliffe (Penguin, 1968), 56–57.

7 Mumford, "First Megamachine," 7.

8 René Descartes, *Meditations on First Philosophy* (1641), 104.

9 The "ghost in the machine" is Gilbert Ryle's phrase, from his book *The Concept of Mind* (Hutchinson, 1949), starting on p. 18.

10 Marvin Minsky, quoted in James Wright, "A Note from IESBS [International Encyclopedia of the Social and Behavioral Sciences] Author James Wright," *SciTech Connect*, last modified 4 May 2015, http://scitechconnect.elsevier.com/a-note-from-iesbs-author/.

11 Steven Nadler, "Doctrines of Explanation in Late Scholasticism and in the Mechanical Philosophy," in *The Cambridge History of Seventeenth-Century Philosophy*, ed. Daniel Garber and Michael Ayers (Cambridge University Press, 1998), 518.

12 Nadler, "Doctrines," 520.

13 Nadler, "Doctrines," 521.

14 Quoted in Nadler, "Doctrines," 521.

15 Nadler, "Doctrines," 527–28.

16 Nadler, "Doctrines," 521.

17 "Affect," by the way, is another word for *emotion*, and everybody knows that inert material corpuscles aren't allowed to have feelings. Even the word "impulse" has an emotional register, as in "impulsive" behavior.

18 John Locke, *An Essay concerning Human Understanding* (1693), Book II, Chapter VIII, §13.

19 Gottfried Wilhelm Leibniz, *The Monadology*, in *Philosophical Writings*, trans. Mary Morris (J.M. Dent, 1934), §17; original emphasis.

20 Nadler, "Doctrines," 534.

21 Nadler, "Doctrines," 528.

22 *Online Etymology Dictionary*, accessed 25 July 2024, https://www.etymonline.com/word/gravity.

23 So did G.W. Leibniz, and there's a whole STS case study to be had examining the nasty political and scientific bickering about who invented calculus first!

24 Sir Isaac Newton, *Philosophia Naturalis Principia Mathematica* [i.e., "the Mathematical Principles of Natural Philosophy"], quoted in Gillispie, *Edge of Objectivity*, 141.

25 Gillispie, *Edge of Objectivity*, 144.

26 Gillispie, *Edge of Objectivity*, 144.

27 Newton, *Principia*, quoted in Gillispie, *Edge of Objectivity*, 148.

28 Gillispie, *Edge of Objectivity*, 146.

29 Cliff Bekar and Richard G. Lipsey, "Science, Institutions, and the Industrial Revolution," *Journal of European Economic History* 33, no. 3 (2004): 719.

30 Newton, *Principia*, quoted in Gillispie, *Edge of Objectivity*, 141.

31 Newton, *Principia*, quoted in Gillispie, *Edge of Objectivity*, 149.

32 This is the title of a classic Pink Floyd song, in case you want to look it up.

33 Bekar and Lipsey, "Science," 728.

34 Jack A. Goldstone, "Efflorescences and Economic Growth in World History: Rethinking the 'Rise of the West' and the Industrial Revolution," *Journal of World History* 13, no. 2 (2002): 371.

35 Goldstone, "Economic Growth," 371.

36 Goldstone, "Economic Growth," 367.

37 Goldstone, "Economic Growth," 364.

38 Gillispie, *Edge of Objectivity*, 146.

39 Lewontin, *Biology as Ideology*, 14.

40 Lewontin, *Biology as Ideology*, 14.

V

Classical Modernity

Chapter Twelve

Rage against the Machine

OKAY FINE, THIS CHAPTER ISN'T REALLY ABOUT RAGE, ALTHOUGH THE opportunity to use the name of the best rap-metal band of the 1990s was too good to pass up. This chapter is, however, about a metaphorical (and sometimes literal) 'machine': the social system of modern Western technoscience. Our brief intellectual history in this book has finally resulted in a picture of science and technology that is familiar to the usual story: science as uncovering the truth of the material universe, always getting better at it, and always providing new technological powers. This vision of science is supposed to structure our society, and everybody is supposed to be okay with that. That's why our examination of Western technoscience in this book is becoming less historical and more thematic: now that we know where this usual story came from, we can focus more closely on problems that arise from it.

As you may recall from Chapter One, there are some pretty big issues with the usual story, and STS is the critical study of those issues. In the remaining sections of this book, therefore, we will survey how STS engages in that study. Some of these issues arose in the midst of modernity's heyday, what I'll call 'classical modernity.' Very roughly, this is that period of time from the end of the eighteenth century up to the two world wars of the twentieth century, when the Western world was explicitly and proudly modern. Even then, up to centuries before STS arose as a discipline, there was critical perspective on, and pushback against, the radical optimism about modern Western technoscience. The technoscientific system of governance wasn't something everyone believed in or agreed with. This chapter is about those perspectives where the usual story was resisted or rejected (or rather, *some* of those perspectives). We will look at some

Imperial Chinese approaches to technology and conquest, the Romantic movement in eighteenth- and nineteenth-century Europe, and the violent rebellions of the Luddites later in nineteenth-century Britain. In the latter case, we will certainly encounter some raging against machines, but in each case we will find alternatives to the hegemony of the usual story. Gaining critical perspective on science and technology is what STS is all about, and that happened even before STS did.

Imperial China

China is one of the world's oldest civilizations, but its efforts to modernize started little over a century ago. Its last emperor was deposed in 1912, after the dynasty proved itself incapable of fighting off numerous foreign incursions. Five centuries earlier, however, when modernity wasn't even a twinkle in European eyes, China was a technological superpower. You will remember from Chapter Nine that Francis Lord Bacon thought the purpose of modern science was to endow societies (especially his!) with new technological powers. With the right technology, he thought, you could be a like god in comparison to other cultures. He went on to say the following:

> Again, it is well to observe the force and virtue and consequence of discoveries, and these are to be seen nowhere more conspicuously than in those three which were unknown to the ancients, and of which the origin, though recent, is obscure and inglorious; namely, printing, gunpowder, and the magnet. For these three have changed the whole face and state of things throughout the world; the first in literature, the second in warfare, the third in navigation; whence have followed innumerable changes, insomuch that no empire, no sect, no star seems to have exerted greater power and influence in human affairs than these mechanical discoveries.[1]

With technologies like the printing press, firearms, and the compass, Bacon thought you could conquer the world (if not the entire universe). What didn't make much sense to him was that the Chinese invented all three of these technologies centuries before the Europeans did (that's why for him, their origin was "obscure and inglorious"), and yet China didn't proceed to spread out over the globe like Europe eventually did. From the perspective of the usual story, it's a mystery why the Chinese didn't conquer the world. Like, what else are you supposed to do with awesome technology, other than occupy other people's countries? The Chinese were oddly non-Baconian, at least from a Baconian perspective.

Bacon wasn't around to inform the greatest maritime officer of China, Zheng He (c. 1371–1435), what he should do with his technological superiority, because Bacon hadn't been born yet (plus they lived on roughly opposite sides of the globe). Zheng He's voyages occurred around the same time that king Henry V of England's longbows beat an army of French knights at Agincourt, about 150 years before Bacon's birth. Zheng He had a fleet of over 250 ships and 27,000 crew members, which is more than twice what the Spanish Armada had almost two centuries later. Sixty-two of his ships were the famous 'treasure ships,' huge vessels over 400 feet (122m) long. By contrast, Christopher Columbus—who sailed the ocean blue in 1492, 60 years after Zheng He—had only *three* ships in total, the largest of which was about 75 feet (23 m) long. Chinese naval technology eclipsed the Europeans in more ways than one! This is one reason why the empire for which Zheng He sailed was called the Ming dynasty; *Ming* means "enlightened."

Scale models showing size comparison between Zheng He's and Columbus' ships.

Zheng He's voyages were well recorded. He sailed all the way from China to Somalia, and did so numerous times. (Claims that he sailed to North and South America have been decisively debunked by both Chinese and Western historians.) But his purposes in these voyages was not conquest or colonization. Contrary to what a modern European would have wanted, Zheng He's voyages were intended "to gather treasures and unusual tributes for the Imperial court."[2] They weren't even intended to "develop trade [or] political influence with the countries visited,"[3] although if you want somebody to give you giraffes, ostriches, and zebras, then using a compass to arrive in East Africa with thousands of soldiers on hundreds ships (possibly armed with cannon) is an effective way to do that.

If not for conquest, colonization, trade, or political influence, what then was the point of these treasure ship voyages? Zheng He's displays of Chinese technology and military power were meant to acclaim the Ming emperor as the "Son of Heaven" and the rightful ruler of China.[4] Zheng He's emperor (Yongle) was a usurper, and by collecting costly treasures from across the seas, he confirmed his superpower status and removed all doubt of his legitimacy. As his soldiers reminded the inhabitants of the island of Pulau Sembilan, "We are the soldiers of the Heavenly Court, and our awe-inspiring power is like that of the gods."[5] After delivering this information, they cut a few incense logs and left. Thus the voyages weren't supposed to make China bigger or stronger: *it already was*. China didn't have to create an empire or an enlightened civilization. The tributes were simply proof that China was an empire and an enlightened civilization, and that Yongle was the right emperor for it. Later Ming emperors didn't have the same need for legitimation, and so canceled the voyages as an extravagant waste of money.[6] The massive treasure ships eventually rotted away in the harbor.

This is hard for the usual story to digest. After all, everybody is supposed to know that science and technology are for conquering more of the world (and eventually the universe), and that everybody everywhere has wanted to do this since the dawn of time. Evidently not. Zheng He's voyages are just one example among many that the modern framework of technoscience is an historically, geographically, and culturally unique way of interpreting the world. Zheng He and the Ming emperors were not trying to be modern or Western and failing at it. In fact, this decidedly non-modern, non-Western approach to technological power also continued into the Qing Dynasty, which maintained Chinese traditions and resisted European colonization up until its end in the early twentieth century.

The Qing emperor, however, had to deal with pesky modern European nations by the late eighteenth century. In a pair of letters to the king of Great Britain, Emperor Qianlong (1711–99) reiterated much the same policies that had influenced Zheng He's fleet 360 years earlier. The British envoy to Qianlong had been requesting new trade access to Chinese markets, including getting British merchants the right to move freely within Chinese borders. This request, noted Emperor Qianlong, was "contrary to all usage of my dynasty."[7] His respect of tradition was so strong that it was unthinkable for his "dynasty [to] alter its whole procedure and regulations, established for more than a century, in order to meet [the British crown's] individual views...."[8] Unlike the early modern Europeans, the repudiation of tradition was not a legitimate option for the Chinese.

Moreover, the Qing emperor was aware that he, like the Ming emperors before him, was the Son of Heaven and the rightful ruler of (one of) the world's greatest civilization(s): "Our Celestial dynasty possesses vast territories, and tribute missions from the dependencies are provided for by the Department of Tributary States...."[9] From Qianlong's perspective, China was the center of the world: "My capital is the hub and centre about which all quarters of the globe revolve."[10] As it had with Zheng He's fleet centuries earlier, China's "majestic virtue has penetrated unto every country under Heaven, and Kings of nations have offered their costly tribute by land and sea."[11] The "Celestial Empire" of Qing China "possess[ed] all things in prolific abundance and lack[ed] no product within its own borders."[12]

Therefore, the Chinese emperor saw no need for any technology or trade modern Europe could offer him: "Swaying the wide world, I have but one aim in view, namely, to maintain a perfect governance and to fulfil the duties of the State; strange and costly objects do not interest me.... I set no value on objects strange or ingenious, and have no use for your country's manufacturers."[13] Britain's industrial revolution was in full swing by this point, but China had no interest in the technological products of Progress or mechanism. Inversely, while there "was therefore no need to import the manufactures of outside barbarians [i.e., the British] in exchange for our own produce," Emperor Qianlong knew that "the tea, silk, and porcelain which the Celestial Empire produces are absolute necessities to European nations and to yourselves [the British]."[14] This is why the empire allowed limited foreign trade at Canton (modern Guangzhou): "so that your wants might be supplied and your country thus participate in our beneficence."[15] Like most other agrarian civilizations (see Chapter Five), in the Qing empire "a firm barrier is raised between my subjects and those of other nations";[16] the "distinction between Chinese and barbarian is most strict...."[17] Colonization of the barbarians, however, was not an option, for "it has never been our dynasty's wish to force people to do things unseemly and inconvenient" like "adopt[ing] the dress and customs of China...."[18] Rather, by generously providing access to Chinese technological wealth (like tea, silk, and porcelain), the empire could "exercise a pacifying control over barbarian tribes, the world over."[19]

"From every point of view," Emperor Qianlong concluded, "it is best that the regulations now in force should continue unchanged."[20] The perpetual change of the Doctrine of Progress was not desirable. Modernizing, mechanizing, and colonizing were uniquely Western and recent ideas that did not make sense within the framework of two millennia of Chinese civilization. As STS scholars ourselves, we can conclude that approaching science and technology in the context of repudiating tradition, unifying theory and craftsmanship, unending progressive change, and mechanistic atomism is not an inevitable result of the human spirit, any more than being British is. The examples of Zheng He and Emperor Qianlong show that the social dimensions of human variation extend to the role of technology in different cultures, even very powerful ones. Just because you have technological superiority doesn't mean you have to conquer the world like Francis Bacon thought you should. Being a modern European is *obviously* not the only way to be an intelligent human being, even though the usual story about technoscience implies that it is.

Romanticism

While Imperial China provides a powerful example against the universality and inevitability of the usual story, there was also resistance to modernity within Western culture itself. The Romantic movement in eighteenth- and nineteenth-century Europe stood in contrast to the power-seeking, ever-changing, and machine-industrializing early modern approach to science and technology. Romanticism is so broad and diffuse, however, that it is difficult (some say impossible) to generalize about it, but I will try anyway. It was a movement in art, literature, music, politics, and philosophy. One of its early heralds was Jean-Jacques Rousseau (1712–78), who forcefully disagreed with the progressive story about the State of Nature (see Chapter Ten). The State of Nature (according to

Rousseau) was a positive situation where humans enjoyed spontaneity, freedom, and the bounty of nature, rather than a negative situation of nasty, brutish, and short human life (according to Hobbes) out of which human ingenuity creates civilization to free us from nature (according to Locke). The Social Contract, according to Rousseau, created inequality, social oppression, and a decline in human access to the good life, instead of being the source of civilization, progress, and the other values that the usual story holds dear.

Jean-Jacques Rousseau (1712–78), portrait by Maurice Quentin de La Tour (1753).

Romanticism was characterized, therefore, by a general "disillusionment with some of the fundamental tendencies of modernity," namely "the growth of science and technology, the division of labour, and a competitive market economy."[21] The way in which mechanistic science had understood everything in terms of dead matter, and the way in which industrial factories had ravaged communities and the countryside, indicated a deep alienation or sickness at the heart of modern life. Romantic thinkers specifically rejected calculative rationality, the idea that everything could and should be reduced to a mathematical understanding. It may be more efficient to manage large populations of human beings (like students and employees at a university) by assigning each of them a number that can easily be crunched by a computing system, but we—you and I—are neither numbers nor computers. We are individuals with names and faces, unique histories and stories, and we have soul, baby! You can't reduce that to math. According to Romanticism, human beings—and the world we inhabited before industrial technoscience cut it into pieces—are alive with meaning, and there is always *something more* to life than just what can be plugged into an engineering problem—just like there is always something more to music or dance than what can be captured by an algorithm. Modernity saw everything through the lens of "stiff rationality,"[22] creating a world of straight lines, big boxes, and concrete, but Romanticism affirmed the incalculable, irreducible, and subjective dimensions of life: emotion, imagination, inspiration, and heroism.

These subjective dimensions of life were what the Romantics thought had been eliminated by the calculative rationality of modern technoscience. In seeing everything as dead mechanical particles to be counted, science and technology had killed the world and made humans misplaced strangers in what remained of it. Everything surrounding us was now dull, inert, gray, and tasteless. In the words of the English poet William Wordsworth, "We murder to dissect. Enough of Science and Art...."[23] Technoscience destroyed everything that made life worth living! This didn't mean that the Romanticists were hopelessly backwards looking, however. They highly valued modernity's emphasis on the freedom of the individual, but conceived of individuality in terms of love and community connection rather than atomistic isolation in a mechanistic universe. Some Romanticists even proposed an alternative scientific worldview, where matter consisted fundamentally of "living force or power" rather than inert corpuscles. They hoped technology could be transformed into a fully human endeavor that exalted "intuition, spirit, sensibility, imagination, faith, the unmeasurable, [and] the infinite...."[24]

Romanticism failed to win the battle against industrial rationality in Europe, however, which is why the name of the movement is now a bad word. For example, to "romanticize" history is to (supposedly) falsely portray the past by making it sound better than it actually was, the way that nostalgia conveniently forgets how bad things used to be, back in the day. But notice the callous nature of this anti-romantic advice: according to the usual story, we're not supposed to 'romanticize' reality because the reality which modern technoscience has made is *actually pretty horrible* (which is not what the usual story wants us to realize). To be "realistic" is to be honest about how nasty everything is. That's the "truth" we're supposed to accept by rejecting Romanticism, and yet it proves Romanticism's point: "science, technology, and industry" really "were building ... difficult and unlovely works" from which we need "a solace and an escape."[25] The best modernism was able to offer in contrast to Romanticism was a so-called realistic worldview where we have no choice but to accept that the world we live in is lifeless and dead.

The critical perspective for STS, then, is whether we actually have to accept "the reality" of the way things currently are thanks to modern Western technoscience, or whether there might be something more than that, something more important than what is currently "real." The point of the Romantic reading of history, after all, was not to misrepresent history, but to find something inspiring in it, "to see the infinite in the finite, the extraordinary in the commonplace, the wonderful in the banal."[26] This is what it really means to "romanticize the world."[27] To think critically about science and technology will involve asking ourselves if Romanticism tells us something important about "the spirit" of modern technoscience. Could we—*should* we—engineer more beautiful things than we currently do? Romanticism would say yes.

Luddism

Imperial China and Romanticism both offer important perspectives from which to reconsider the value of modern Western technoscience, but this can feel somewhat theoretical, as if we're simply contemplating science and technology at a relatively safe distance where everything is historical, abstract, or mental. The perspective of Luddism, however, is very hands on and physical, even dangerously violent. This is probably why you've heard of the Luddites before, and you probably already know that 'Luddite' is—like Romanticism—a bad word. A Luddite is generally defined as a technophobic regressive, someone who is afraid of and hates new technologies, even "one who stands in the way of progress, an ignoramus, an untutored primitive who resorts to instant violence."[28] This is not a particularly flattering description.

This negative connotation exists because the usual story doesn't like Luddism, and modern Western technoscience prevailed against the Luddites, just as it did against Romanticism and (eventually) Imperial China. At this point you might be wondering if there might be something inevitable or unstoppable about the usual story after all, because it seems to prevail against any resistance it encounters. It is important to remember, however, that there is nothing inevitable about the outcome of a battle or a war unless you really believe in an irresistible *force* working behind the scenes to manipulate world events to its liking. Should we really accept such a mythology about technoscience? Moreover, as we saw in Chapter Eight, just because something is powerful doesn't make it true. Even if Imperial China couldn't fight off the colonial powers forever, and even if Romanticism couldn't undo industrialization, that doesn't make colonization or industrialization right. The point is that the modern Western world didn't *have to* be colonialist or industrialist, even though it was. The same goes for science and technology. It didn't have to end up as modern or Western, even though it did.

One of the things modern Western technoscience did, though, was *create* the Luddites by mechanizing the textile industry of nineteenth-century England. As you'll remember from Chapter Eleven, the mechanistic framework thought that all "useful" explanations reduced to *matter and force*. To make an industry like textile production "efficient," therefore, meant to mechanize it, treating everything in the textile production process as if it were a particle of matter subject to the force of a system of governance. This included the people working in the textile industry, not just the spindles and looms they worked with. They were inert corpuscles too.

The mechanization of the textile industry meant the replacement of cottage industry crafting with a factory system, and the replacement of human powered weaving looms with steam powered ones. This meant that the skilled adult labor of the weaver (usually the father in a household) became unnecessary, because unskilled labor (usually children or women) was sufficient for greasing the wheels of the machines that did the weaving. The weaver was no longer a human being; the weaver was literally a machine, and the humans became attendants to the machines, rather than being the active agents in the weaving process. So not only was the head of the household out of a job, but the remaining family members were more easily exploited by the factory owners. In the nineteenth-century context, skilled men were able to command higher wages both because of male chauvinism, and because their skilled labor was harder to replace. Unskilled laborers who just took care of machines, however, were easy to replace and commanded less high wages. This meant that the women and children who filled those jobs were far easier to manipulate and control than the weavers they replaced. For example, many factories were staffed by indentured, unpaid, and malnourished child orphans working up to sixteen hours per day. This did not make the spinning and weaving mills very popular with the neighbors, even though it was explicitly defended as consistent with Adam Smith's laissez-faire capitalist economics. The mechanization of the textile industry in England thus disrupted the social arrangements of the time in ways that were beneficial to factory owners who could pay less in wages while having more power over their workers. This was not beneficial, however, to working-class people.

A Luddite, therefore, was someone who didn't believe that the mechanization of the textile industry was the right thing to do. The issue for them was not the newness of the technologies per se, but the *values* that came along with mechanization: "deskilling, depersonalizing, demoralizing, [and] degrading...."[29] The first Luddite Manifesto (1811) indicted any factory owner who "gain[ed] riches by the misery of his Fellow Creatures."[30] This refers to a fundamentally moral issue rather than one of economic or technological efficiency. The Romantic movement even came to the defense of the Luddites when the poet Lord Byron gave a speech to the British Parliament's House of Lords, condemning the idea that human beings should be "sacrificed to improvements in mechanism."[31] As David Linton argues, "it was not machines *per se* that were viewed as the enemy. The enemy was wage-cutting, speed ups, excessive employment of apprentices, unemployment, and high prices."[32] The machine was a symbol of what was wrong with the mechanistic framework that workers were forced to conform with. Moreover, the machine was a crucially important piece of capital, one where factory owners were particularly vulnerable. Lacking any legal recourse through labor unions or Parliament, the Luddites broke their weaving machines as acts of "industrial sabotage" or "collective bargaining by riot."[33] Therefore, a Luddite came to mean "anyone who willfully smashed his machine ... for political ends,"[34] even though they weren't afraid of technology just because it was new or powerful. In Brian Merchant's eloquent words, "the Luddites understood technology all too well; they didn't hate it, but rather the way it was used against them."[35]

It's not hard to see that the usual story wouldn't like machine breaking; it's tough to be more against mechanistic power progress than breaking machines! The English factory owners didn't like machine breaking either, nor did their government, which was afraid of a revolution like the one that had just happened in France. So the government solved their Luddite problem in a pointedly mechanical way: the application of force to matter. By 1812, 14,400 British troops had been sent to squash the Luddite revolts, and the Luddites did not survive. What started as a question of social values ended up with a technological 'solution' called *the military*. The Luddites were viewed as inert particles of matter that needed to be reorganized into a more 'workable' arrangement by the application of firepower.

At the moment, 'solution-based approaches' to problems are preferred in politics and commerce, not least because they are seen as ideologically neutral or objective—scientific, you might say. The Luddites illustrate, however, that technological solutions are not neutral. Problems are, by definition, negative—at least for someone. The Luddites were not a problem for themselves, of course; they were a problem for the status quo: factory owners, the industrial economy, even the mechanistic

> **A Solution-Based Approach:** a practical approach to problem-solving where problems are taken for granted, and solutions are seen as ideologically neutral and objective.

framework, progressivism, and the drive for more power. Problems don't just exist; they are always relative to a social context and a trajectory that is trying to go somewhere. The Luddites were 'a problem' for the British establishment because the former didn't like the trajectory of the latter. A solution to any problem, therefore, is a solution *for* whatever social arrangement sees it *as* a problem. The application of military force to the Luddites was obviously not a solution for the Luddites, but a solution for the direction the British establishment was taking against the Luddites. The technological solution to Luddism was in no way neutral, therefore. It saved and reinforced a particular, unique, and even subjective social framework and trajectory. All technological solutions are value-laden, therefore, as much as the usual story wants them to be seen as neutral.

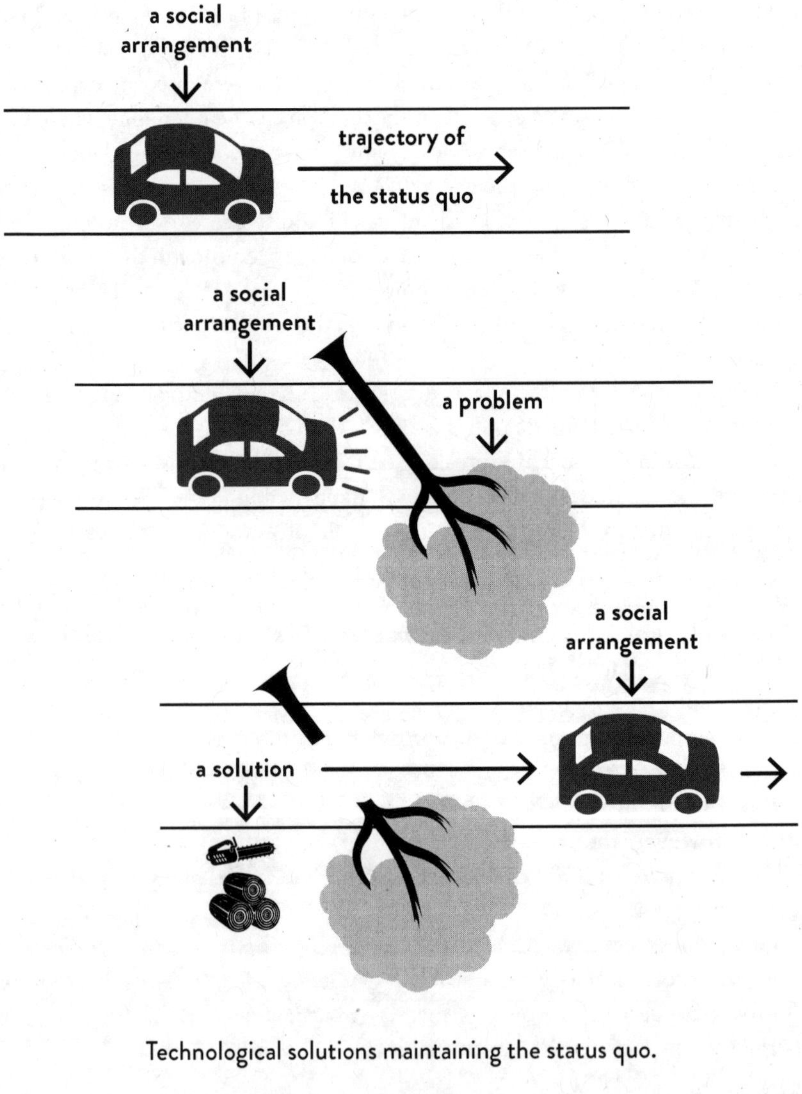

Technological solutions maintaining the status quo.

But what are we supposed to do with value-laden problems? How do we "solve" a "problem" of values? Whose interests should have been satisfied in 1812: the factory owners or the Luddites? This is not easy for technoscience to answer. As we saw in Chapter Ten, progress is a system of value that quite intentionally avoids talking about values, focusing on facts instead (facts are supposed to be neutral, remember). And yet moral questions are 'problems' for modern Western technoscience, specifically because modern Western technoscience doesn't know what to do about moral questions except treat them like questions of fact; they are arrangements of particles and force that can be 'fixed' by shooting bullets at them. The question for us as nascent STS scholars is whether modern Western technoscience is an historically contingent system of values (even if the usual story says it's objective and neutral) and, if it is, if it's a system of values worth having. Maybe shooting at everything isn't the best way to deal with our problems. Maybe we should figure out why we have these problems in the first place.

This is the critical perspective Luddism encourages us to take with respect to the usual story: we should be concerned about the value impact of technologies on society. We don't have to break steam looms to recognize that technoscience is a question of values, but such recognition makes us Luddites, possibly Romantics (although not Chinese emperors). Just because modern Western science and technology are the most powerful things on the planet doesn't necessarily make them good: might does not make right. So while it might be correct to say that mechanization or industrialization increase(d) technological or economic efficiency, that does not tell us if mechanization and industrialization are good things. Moreover, technoscience can't tell us what the good things are; it got out of the fuzzy business of answering value questions when it abandoned Aristotelian final causality at the start of modernity. It's tough not knowing how to technoscientifically answer questions about what things are good, but sometimes progressive optimism can be so strong that these questions are temporarily forgotten. In the next chapter, we will look at the most optimistic formulation of technoscientific absolutism in the modern period, and see what problems might arise from that.

Notes

1 Bacon, *Novum Organun*, §129.

2 *Oxford World Encyclopedia* (2008), s.v. "Zheng Ho."

3 *A Dictionary of World History* (Oxford University Press, 2000), s.v. "Zheng He."

4 Edward L. Dreyer, *Zheng He: China and the Oceans in the Early Ming Dynasty, 1405–1433* (Pearson, 2007), 3.

5 Dreyer, *Zheng He*, 65.

6 Dreyer, *Zheng He*, 4.

7 "Qianlong's Rejection of Macartney's Demands: Two Edicts," in *The Search for Modern China: A Documentary Collection*, ed. Pei-kai Cheng and Michael Lestz with Jonathan D. Spence (W.W. Norton, 1999), 104.

8 "Qianlong's Rejection," 105.

9 "Qianlong's Rejection," 105.

10 "Qianlong's Rejection," 107.
11 "Qianlong's Rejection," 105.
12 "Qianlong's Rejection," 106.
13 "Qianlong's Rejection," 105.
14 "Qianlong's Rejection," 106.
15 "Qianlong's Rejection," 106. In fact, at the time many Europeans "consider[ed] India and China more technically 'advanced' than Europe," because the industrial revolution had yet to produce the kind of economic growth in Europe that historians call "modern" (Jack A. Goldstone, "Efflorescences and Economic Growth in World History: Rethinking the 'Rise of the West' and the Industrial Revolution," *Journal of World History* 13, no. 2 [2002]: 365–66).
16 "Qianlong's Rejection," 108.
17 "Qianlong's Rejection," 109.
18 "Qianlong's Rejection," 105.
19 "Qianlong's Rejection," 106.
20 "Qianlong's Rejection," 108.
21 Frederick Beiser, "Romanticism, German," in *Routledge Encyclopedia of Philosophy*, ed. E. Graig (Routledge, 1998), doi:10.4324/9780415249126-DC094-1.
22 Simon Blackburn, "Romanticism," *The Oxford Dictionary of Philosophy*, 3rd ed. (Oxford University Press, 2016).
23 William Wordsworth, "The Tables Turned," in *Lyrical Ballads* (1798), lines 28–29. Don't forget that "art" means *technology* here (see Chapter Three). It is worth noting that this poem was published only five years after Qianlong's letters to the British king.
24 Crane Brinton, "Romanticism," *The Encyclopedia of Philosophy*, vol. 7 (Macmillan, 1967), 209.
25 Brinton, "Romanticism," 209.
26 Beiser, "Romanticism, German."
27 Beiser, "Romanticism, German."
28 Ian Reinecke, *Electronic Illusions* (Penguin, 1984), 12, quoted in David Linton, "Luddism Reconsidered," *Et Cetera* 42, no. 1 (1985): 32.
29 Kirkpatrick Sale, *Rebels against the Future: The Luddites and Their War on the Industrial Revolution: Lessons for the Computer Age* (Addison-Wesley, 1995), 200, quoted in Paul Lindholdt, "Luddism and Its Discontents," *American Quarterly* 49, no. 4 (1997): 867.
30 Kevin Binfield, ed., *Writings of the Luddites* (Johns Hopkins University Press, 2004), 72.
31 Speech of Lord Byron upon the "Frame Work" Bill, delivered in the House of Lords, 27 February 1812.
32 Linton, "Luddism Reconsidered," 34.
33 Linton, "Luddism Reconsidered," 34.
34 Linton, "Luddism Reconsidered," 33.
35 Brian Merchant, *Blood in the Machine: The Origins of the Rebellion against Big Tech* (Little, Brown, and Company, 2023), 4–5.

Chapter Thirteen

Positivism

IN THE PRECEDING CHAPTER, WE NOTED THREE DIFFERENT PERSPECTIVES that challenged the hegemony and legitimacy of the usual story. The Imperial Chinese were a technologically advanced civilization with many values that differed significantly from those the usual story claimed everyone was supposed to share. The Romanticists advocated for a different approach to the value and meaning of life than what they saw embedded in early modern, industrial Europe. The Luddites were driven to violently resist the values embedded in England's first Industrial Revolution, but succumbed to the technoscientific power wielded by the wealthy classes who benefited from the economic status quo.

But these perspectives weren't the only ones around, and they were certainly not the ones that came to dominate the modern world. The dominant one is the usual story! As we saw in the Chapter One, boosterism is a position you can take in STS, and doubling down on the usual story about science and technology is certainly one phenomenon that we see in contemporary societies. This is especially the case when large segments of the population deny the legitimacy and reliability of scientific research into things like climate change or vaccines. Quite often, popular and professional responses to denials of scientific authority like these are rooted in *positivism*, the most optimistic position ever taken towards science (technology usually gets left out of the discussion at this point).

Science as the Only Way to Think Properly

Positivism is a position that is so unabashedly and overwhelmingly positive about technoscience that critical pushback against it is commonly labeled anti-science or conspiratorial pseudoscience. Oddly, however, this positivity towards science is not why it's called positivism! We're accustomed to hearing the word 'positive' in terms of evaluation: to be 'positive' usually means (nowadays) to be good or happy. Positivism, however, is not defined in terms of goodness or happiness. Positivism is rather about science being the best and only way for human beings to think properly. Positivism is a way of saying that science is absolute: it produces knowledge that is the most true (even if never absolutely true), and is the absolutely best method for discovering the truth. In this sense, it is fair to say that positivism is scientific absolutism. Some scholars even call it "scientism,"[1] although I won't call it that because it's a pejorative term and not particularly easy to pronounce.

Why on earth would radical optimism and absolutism about science deserve the title 'positivism'? Well, a 'positive fact' refers to anything that has a *position*, a location in space. In that sense, positive means 'physical' because only physical objects take up space (as did the material corpuscles in Chapter Eleven). Moreover, to 'posit' something means to *propose* an idea, to suggest that something should be accepted as true, or to make a *proposition*. Something posited, then, is something that is said to exist or to be correct. The 'positive' in this sense refers to what we typically call 'the real world,' at least when we aren't thinking too philosophically about what constitutes reality. A 'positive experience,' therefore, means a sensory or empirical experience, something that happens to us in the visible, sensible, physical, or material realm (see Chapter Eight).

Any rigorous intellectual endeavor concerned with empirical facts, things that take up space, experimental procedure, material particles, or physical processes used to be called a 'positive science,' or what we now call natural science. Non-positive sciences would have been those systematic theoretical inquiries into things that don't take up space, like the meaning of poetry or the existence of a deity. So even though theology could be a science of divinity (from the Greek *theos* + *logos*), it couldn't be a positive science because the gods don't (have to) take up space. Positive sciences, therefore, don't investigate gods or poems, because gods or poetic meaning aren't 'there' occupying a geometrical position in space.

A *positivist* is someone who views positive science so highly that they make an 'ism' out of it. That is, positivism treats natural science as the *only* science, or scientific method as the *only* method. A positivist sees modern science as the only legitimate way to know anything about anything. Poetic or theological sciences simply cannot exist, for positivism. That's why positivism can be considered scientific absolutism: it sees non-positive sciences as pseudo-sciences and therefore illegitimate. Herbert Feigl, who (along with A.E. Blumberg) coined the term 'logical positivism' in 1931, said that "Questions which are in principle incapable of being answered by scientific method turn out, on closer analysis, not to be questions of knowledge."[2] That is, if you want to know something that cannot be answered via the scientific method, then (according to positivism) the thing you want to know is not a thing that can be known. It's not a thing at all, because (evidently) the thing is a non-scientific thing: it doesn't occupy physical space. Only things that occupy physical space are real.

For the positivist, therefore, there is no method "but the method of science" to solve any question at all.[3] Science is *it*. It will explain everything. Nothing else can, and there's nothing else to be explained anyway. This is why positivism is the most optimistic perspective on science possible. Even if the positivist knows that scientific knowledge is always changing and can therefore never be absolutely certain (see Chapter Two),[4] the positivist is still an absolutist about science as a method and a body of knowledge. There is absolutely no better way of discovering anything than science, and the discoveries of science are absolutely better than any (so-called) discoveries in non-scientific disciplines. Ultimately, there cannot be any disciplines that are unscientific. To a positivist, psychology and physics have to be fundamentally about the same kind of things: empirical, physical, natural facts. If you want to talk about anything other than matter and force, then you aren't thinking properly.

Positivism: the belief that (natural) science is the best and only legitimate way to know anything about anything; scientific absolutism.

Comte's Positivism

Positivism is a deeply progressive position. In fact, it is the usual story running rampant. For positivism, science is the final stage of human history. After all, what could possibly come *after* science to replace it? The positivist's view of humanity has its roots in the nineteenth-century social theorist Auguste Comte (1798–1857), who we briefly mentioned back in Chapters Ten and Eleven. Comte invented the word 'sociology' to describe the method by which scientists were supposed to discover the mechanisms by which society worked. With this knowledge of mechanical social processes, positivists could more efficiently manipulate societies and achieve progressive goals more quickly. According to Comte's sociology, there are three stages to historical and scientific development. The first he called the *theological* stage, where people are religious and believe in myths. Their myths are mistakes, of course (see Chapter Four), because gods don't actually operate in the world or explain why things happen. But hey, at least people back then were *trying* to be scientific, right? Comte also claimed that theological societies are characteristically militaristic and violent, which wasn't exactly a compliment (or correct).

The second phase of human social development, according to Comte, is the *metaphysical* stage. This is (supposedly) a more philosophical period, where the investigation of the laws of Nature are separated from religious questions about gods. As we saw in Chapter Seven, Aristotle's metaphysics were concerned with the ultimate principles of explanation that came after or beyond physics. Aristotle's metaphysical principles were supposed to explain how physics was possible. At least metaphorically, this kind of explanation is divine, as it's superhumanly difficult to explain the conditions for the possibility of physics. Comte argued that in the metaphysical stage, 'forces' replaced gods as the explanations for material events, and 'Nature' (with an appropriately respectful capital N) replaced deity as the supreme reality. Rather than being militaristic societies dominated by warriors, Comte thought metaphysical societies were dominated by lawyers and judges.

The third and final stage of human progress—Progress (with a respectful capital P) is what these social dynamics amounted to for Comte—is the *positive* stage. In the positive stage of human development, science is king: "the determination of social policy and the formulation of social goals" is given over to science rather than philosophy or theology.[5] Rather than explaining physical events in terms of natural laws or causation, only mechanical processes and statistical accounts of correlation can remain. All decision making would be 'evidence-based' rather than shaped by (so-called) ideology. Indeed, "the highest or only form of knowledge," according to Comtean positivism, "is *the description of sensory phenomena*."[6] The only thing that could possibly be true is the objective, neutral, physical description of facts. Positivism is an extreme form of the empiricism we saw in Chapter Eight, where all knowledge is *a posteriori*. It is also reminiscent of what Newton said about science's inability to know what the gravitational force really is: "In science, there are no 'depths'; there is surface everywhere...."[7] Because we can only describe our empirical experiences, positivists think a "scientific description can contain only the *structure* (form of order) of objects, not their essence.'"[8] In the positivist stage, humans don't look for the meaning of things; they just try to discover the facts by describing what they look like.

Auguste Comte (1798–1857).

Unsurprisingly, Comte thought that the positive stage of human 'social physics' (his other term for sociology) was industrialist, rather than militaristic or legalistic. Much of his progressivism was therefore a defense of the status quo or "an elaborate rationalizing of the dominant economic conditions" of his time (compare with the solution-based approaches in Chapter Twelve).[9] Recall also what we saw back in Chapter Two: science, rather than being objective and politically neutral, can replace religion as the central ideological weapon in modern cultures. Comte literally took this literally. Science replaced divine authority so completely that Comtean positivism culminated in a "Religion of

Humanity [that] was to be the supreme technology" used to create the ideal reformed society.[10] In the positivistic religion, there is no god other than scientific reason, but there is a pope or high priest (Comte himself, conveniently) and a list of reason-saints (including Adam Smith, Frederick the Great, and Shakespeare—all posthumously). Comte's scientific religion was an explicit attempt to replace Christianity in his culture, all in the name of Progress.

There is more than a bit of irony, however, in saying that science *replaces* metaphysics and religion with a supposedly non-metaphysical, non-theological *religion*. Comte's positivism may be fatally incoherent in this respect. Even so, positivism remains one of the prominent positions taken towards the role of science in society today. A society guided by science, evidence, and facts is the dream of liberal, progressive, and technocratic politicians the world over. You don't have to look far to find intelligent people who think that

> [hu]mankind achieves intellectual adulthood only with the scientific way of thinking. Our age [however] is still replete with remnants of and regressions to such prescientific thought patterns as magic, animism, mythology, theology, and metaphysics. The outstanding characteristics of modern scientific method are mostly absent or at best only adumbrated in those less mature phases of intellectual growth.[11]

The difficulty for positivism, however, is to avoid becoming those very things which it hopes to replace: mythology and metaphysics.

Scientific Mythology and Metaphysics

Positivism has a complicated relationship with mythology and metaphysics. Myth, according to the usual story, is supposed to be a set of mistaken beliefs about how things actually work, but STS suggests that positivism itself is mistaken about what "the history of thought from magic to science would reveal."[12] The archaeological, anthropological, and historical record reveals that believing that the human spirit has progressed throughout time to become modern, Western, and technoscientific is itself a "myth" (in the usual story sense). The unusual story of STS shows that the progressive usual story about the development of science and technology is a mistaken 'mythological' narrative. This wouldn't be a bad thing if myths were a legitimate form of knowledge or meaning making rather than mistakes (see Chapter Four), but the usual story is *opposed* to myth because it sees myths as mistakes, and so it comes to find itself opposed to *itself* (in addition to being factually mistaken, of course). No wonder Comte's anti-religious positivism culminated in a positivistic religion: it's tough for positivism to avoid being weirdly religious when the usual story is itself weirdly mythological. If the usual story is an anti-mythological myth and positivism is an anti-religious religion, then neither of them make sense.

The relationship between positivism and metaphysics is even more problematic. Positivists define metaphysics in the way that Newton's critics defined an occult force: pseudoscientific claims about some vaguely mysterious, mystical, creepy, substance (like a ghostly goo?) that "lies hid in all gross bodies."[13] Metaphysical forces or principles of causation might, in religion, be equated with the supreme being or divinity (as with Aristotle), or "subtle spirit[s] which pervade [everything]" (in Newton's own words).[14] On the usual story's understanding, metaphysics is otherworldly and invokes the uncanny, a lot like Romanticism did (see Chapter Twelve). That's why 'metaphysician' was listed as one of the "seven best jobs for paranormal enthusiasts" in the Halloween edition of one of my local newspapers![15]

Along with metaphysics, the usual story sees itself as opposed to other fuzzy things like Romanticism. The usual story of science is that it is hard-nosed rather than soft-hearted, discovering cold, solid facts and avoiding any question of feelings or the supernatural. This is what positivists call the "scientific outlook" on the world, sometimes equated with "naturalism" or "materialism."[16] The Romantic desire for 'something more' than just material facts is simply unscientific: "Only an approach that is resolutely guided by the question 'What is what?' will avoid reading mysteries into the facts...."[17] For a positivist, the "scientific world-conception knows *no unsolvable riddle*."[18] It will explain everything in terms of material particles and (non-occult) forces. Positivism is utterly "disengaged from the theological and metaphysical ideologies,"[19] because science makes no claims about weird creepy powers or explanations that go beyond our ability to simply describe empirical phenomena. Right?

In the twentieth century, 'logical positivists' distinguished themselves from what they called Comte's 'biological-psychological' positivism by focusing on the *meaning* (i.e., logic) of metaphysical and scientific statements. On their analysis, metaphysical statements contained a "content of feeling" that "express[es] a certain mood and spirit."[20] Scientific statements, to the contrary, contained no mood, feeling, or spirit, but merely described the empirical facts as they appear to us, like Comte's descriptions of sensory phenomena. This is why scientists were (supposedly) so much more objective than the touchy-feely metaphysicians who were only talking about romantic emotions. The logical positivists went on to describe their scientific outlook thusly:

> The scientific world conception is characterized not so much by theses of its own, but rather by its basic attitude, its points of view and direction of research. The goal ahead is unified science.... [F]rom this springs the search for a neutral system of formulae, for a symbolism freed from the slag of historical languages; and also the search for a total system of concepts. Neatness and clarity are striven for, and dark distances and unfathomable depths rejected.... Everything is accessible to man; and man is the measure of all things. Here is an affinity with ... all who stand for earthly being and the here and now.[21]

What's weird about this (so-called) scientific outlook is that the positivists didn't appear to see it as the expression of a mood or spirit, even though that's exactly what it was.

First, the positivists say that science isn't about the knowledge it possesses, but its scientific *attitude*. But an attitude is a feeling. So is an *affinity* with thisworldly concerns. Affinities are feelings of connection or closeness, like sympathy or kinship. Secondly, you'll recall from Chapter Seven that the highest level of Aristotle's metaphysics was "final causality," or the *purpose* and *goal* of everything. Here, the positivists lay out a lot of goals for their science: unity, neutrality, ahistorical symbolism, system totality, neatness and clarity. None of these goals are descriptions of sensory phenomena, however; by definition, goals never are. They're all statements about what scientists (supposedly) *want*, not what's already there. Desires or purposes are not objective facts, but rather teleological statements ("the goal ahead"), value statements ("man is the measure of all things"—see the discussion of anthropocentrism in Chapter Eight), and even subjective statements ("points of view"). For a hard-nosed "scientific outlook" that rejects metaphysics as too touchy feely, positivists sure do talk a lot about their feelings.

It turns out, therefore, that scientific absolutism is both mythological and metaphysical according to its own definition of those two (supposedly bad) things. We've already seen that the usual story's definition of myth isn't very good (Chapter Four), and now it's time to propose a better definition of metaphysics as well. If metaphysics literally means "that which comes after (or lies beyond) physics," then a metaphysical claim will simply be *any set of beliefs about what is real*. Is there a supernatural world that lies beyond the physical world? If you think so, that's your metaphysic. (Perhaps we could call it "supernaturalism.") Is there nothing that lies beyond the physical world, such that the physical world is the only thing that is real? If you think so, then that's your metaphysic. (Perhaps we could call it "naturalism.") Either way, you've got a metaphysic, or a picture in your mind of what you think constitutes reality.

Your metaphysic is your *conception of the world*. Your metaphysic is your *outlook on everything*, or what scholars sometimes call a "worldview." A metaphysic is a *model of existence itself*, an interpretive schema for all possible experience you might have. A metaphysic tells you what to expect reality to be like. It can tell you, for example, that unicorns won't walk into your room (because unicorns don't exist), or that "there is one catholic or universal matter common to all bodies" (as in Chapter Eleven).[22] The mechanistic framework is a metaphysic, just as much as Aristotelian final causality is. In postmodern terms, your metaphysic is a *metanarrative*, the story you tell yourself about everything that could possibly exist.

Expanded Definition of Metaphysics: any set of beliefs about what is real; a conception of the world; an outlook on everything; a worldview; a model of existence itself; an interpretive schema for all possible experiences you may have.

Once we understand metaphysics as world-modeling, it's clear why science cannot avoid it. If there is a "scientific world conception,"[23] then it's a metaphysic, plain and simple. Metaphysics are the assumptions that undergird the operation of physical science. They're the framework of unobservable requirements that make (for example) a mechanistic explanation of everything possible. There isn't any way for science to proceed if it doesn't operate on certain assumptions about the way the world has to be. Importantly, however, this doesn't mean that metaphysics and science are doing the same thing. Scientists don't have to *do* metaphysics; they just have to *have* one operating in the background so they can do their science. Physics and metaphysics aren't necessarily in competition with each other, although physics will always have a metaphysics that accounts for why physics is possible. STS is in the business of pointing out what science's metaphysics are (e.g., the usual story's metanarrative), and sometimes asking whether those metaphysics are good enough or if they need to change.

Unfortunately, the usual story wants to see science in terms of a progressing continuity that replaces mythology and metaphysics, but replacing them both leads to no end of trouble. If science is continuous with mythology and metaphysics, then they're all essentially trying to do the same thing. If they're all essentially trying to do the same thing, but mythology and metaphysics are bad, then it's not going to be easy to see why science isn't just as bad as they are, especially when the usual story ends up being mythological (bad) and metaphysical (bad). But that's what we get when (according to the usual story) "science is like magic, but *real*" (see Chapter Four). This brings us to the heart of the controversy about positivism.

The Verification Criterion of Meaning

Positivism sees religious absolutism as bad, but scientific absolutism as good, because science (unlike religion) knows what is really *real*. Positivism sees metaphysical modeling of reality as bad, but scientific modeling as not really modeling at all, because scientific models (unlike metaphysical ones) are supposed to actually correspond to what is *real*. Things are just like the models say they are, apparently. Science is the only way to think properly, according to positivism, because it's the only way to get in touch with reality.[24] The logical positivists express this radical optimism about science in a very clear, succinct, and famous way: *the verification criterion of meaning*.

As already mentioned, the logical positivists focused on the meaning of metaphysical and scientific statements. Because they thought science was the only way to think properly, they thought that only scientific statements had any meaning. For them, meaning wasn't about your sense of significance or belonging in the world, but rather about whether a claim could be verified, tested, or experimentally confirmed. If a statement was not in principle verifiable—like a mythological or metaphysical statement—then it was meaningless, which is even worse than being false:

> If someone asserts "there is a God," "the primary basis of the world is the unconscious," "there is an entelechy [final cause] which is the leading principle in the living organism," we do not say to him: "what you say is false"; but we ask him: "what do you mean by these statements?" Then it appears that there is

> a sharp boundary between two kinds of statements. To one belong statements as they are made by empirical science; their meaning can be determined by logical analysis or, more precisely, through reduction to the simplest statements about the empirically given. The other statements, to which belong those cited above, reveal themselves as empty of meaning....[25]

So either you make scientifically verifiable statements—claims that can be experimentally shown to either truly or falsely correspond with reality—or what you say is pure gibberish. Science isn't only the way to think properly; it's the only way to *make sense.*

Metaphysics cannot make sense, therefore, because metaphysics does not make empirically verifiable claims. Neither does mythology, theology, or the study of poetry. In one sense, then, these forms of expression really are doing something completely different than science: they aren't making empirically testable claims. In another sense, however, they *should* make empirically testable claims, because if they don't, they are meaningless forms of expression that don't talk about what's real. For a positivist, therefore, the only meaningful thing to do, when you're speaking, is to *try* to be scientific. This is exactly why the usual story thinks human beings have been trying to be scientific since the dawn of time.

Unfortunately, positivism's so-called scientific outlook is not itself an empirically verifiable set of claims. It is also not an obviously true set of mathematical statements, or a logical law. (The verification criterion of meaning allows for statements that are true by definition—like math or logic—to be meaningful too.) Therefore, if the "scientific world-conception knows only empirical statements about things of all kinds, and analytic statements of logic and mathematics,"[26] then the scientific world-conception is itself *not* "known." The verification criterion of meaning is not a scientifically verifiable statement ultimately reducible to the simplest statements about the empirically given. It's not a statement about "reality" that can be measurably true or false; it's a statement about what constitutes meaning. So on its own terms—by its own definition of meaning—it's *meaningless* because it cannot be scientifically verified to be either true or false. This is just the most extreme and obvious example of the problem we saw earlier: positivism is a metaphysic. It amounts to a self-referentially absurd paradox, because positivism thinks metaphysics are meaningless. Scientific absolutism cannot be absolute, therefore, because scientific absolutism is not itself scientific.

The logical positivists were aware of "the fact that the verifiability principle threatened to destroy not only metaphysics but also science."[27] Because they could not reduce their verifiability principle to a set of scientific descriptions of sensory phenomena—and could not show it to be mathematically true or logically necessary—their only real option was to construct a language wherein the verification principle was true by definition. Had they been able to do this, however, it would have been completely arbitrary, as nobody would have been obligated to operate within the stipulations of that new language. We could just use our regular languages instead. Logical positivism thus failed, and became "as dead as a philosophical movement ever becomes."[28]

What Now?

The failure of positivism at the start of our late modern era led to a crisis of faith in the sciences. If scientific absolutism cannot be logically sustained, then what happens to scientific authority or legitimacy? Indeed, it was in the context of such soul searching that STS arose as a discipline. We will look at this deepening epistemic crisis in our next chapter. Before we do, however, we should recognize that a "philosophical movement" doesn't really die. Positivism retains much cultural currency in the twenty-first century, in spite of the crippling problems it creates for itself. Many well-meaning defenders of science, be they popularizers or professional scientists, default to a kind of subconscious positivism in their battles against anti-scientific ignorance and prejudice. It's certainly the job of an informed citizenry to overcome ignorance and prejudice. However, it's also the job of an informed citizenry to spot positivism in our time and to be able to think critically about it, rather than forgetting either that it exists or that it is deeply problematic.

As an exercise, therefore, take a look at these quotations from some popular scientists and ask yourself if you see positivism, scientific absolutism, or even just the usual story implied in them. If you do, ask yourself if the quotation would succumb to any of the problems discussed in this chapter. Then ask yourself how people might do a better job of understanding the role or value of science as a social enterprise.

> For me, it is far better to grasp the Universe as it really is than to persist in delusion, however satisfying and reassuring.[29]
>
> –Carl Sagan, astrophysics

> The question of whether there exists a supernatural creator, a God, is one of the most important that we have to answer. I think that it is a scientific question.[30]
>
> –Richard Dawkins, zoology

> [Project Reason is a] nonprofit foundation devoted to spreading scientific knowledge and secular values in society ... with the purpose of eroding the influence of dogmatism, superstition and bigotry in the world.[31]
>
> –founded by Sam Harris, neuroscience

> Five hundred years of science have liberated humanity from the shackles of enforced ignorance.[32]
>
> –Lawrence Krauss, physics

> The good thing about Science is that it's true, whether or not you believe in it.[33]
>
> –Neil deGrasse Tyson, astrophysics

Notes

1 For example, Mikael Stenmark, *Scientism: Science, Ethics and Religion* (Ashgate, 2001).

2 Herbert Feigl, "The Scientific Outlook: Naturalism and Humanism," *American Quarterly* 1, no. 2 (1949): 145.

3 Feigl, "Scientific Outlook," 146.

4 As Feigl says, "The quest for certainty is an immature, if not infantile, trait of thinking. The best knowledge we can have can be established only by the method of trial and error. It is of the essence of science to make such knowledge as reliable as humanly and technically possible" ("Scientific Outlook," 143).

5 L.L. Bernard, "The Significance of Comte," *Social Forces* 21, no. 1 (October 1942–May 1943): 9.

6 Simon Blackburn, *The Oxford Dictionary of Philosophy* (Oxford University Press, 2008), s.v. "positivism"; emphasis mine.

7 Rudolf Carnap et al., "The Scientific Conception of the World: The Vienna Circle," in *Philosophy of Technology: The Technological Condition—An Anthology*, ed. Robert C. Scharff and Val Dusek (Blackwell, 2003), 89.

8 Carnap et al., "Scientific Conception," 91.

9 Lewis Mumford, *Technics and Civilization* (Harcourt Brace, 1934), 185.

10 Bernard, "Significance of Comte," 8.

11 Feigl, "Scientific Outlook," 137–38.

12 Feigl, "Scientific Outlook," 142.

13 Sir Isaac Newton, *Philosophia Naturalis Principia Mathematica* (1687/1726), quoted in Charles Coulston Gillispie, *The Edge of Objectivity: An Essay in the History of Scientific Ideas* (Princeton University Press, 1960), 148.

14 Newton, *Principia*, 148.

15 "Seven Best Jobs for Paranormal Enthusiasts," *24 Hours*, 29 October 2012, 15.

16 Feigl, "Scientific Outlook," 143, 148.

17 Feigl, "Scientific Outlook," 136.

18 Carnap et al., "Scientific Conception," 89.

19 Feigl, "Scientific Outlook," 136.

20 Carnap et al., "Scientific Conception," 90.

21 Carnap et al., "Scientific Conception," 89; emphasis removed.

22 Robert Boyle, quoted in Steven Nadler, "Doctrines of Explanation in Late Scholasticism and in the Mechanical Philosophy," in *The Cambridge History of Seventeenth-Century Philosophy*, ed. Daniel Garber and Michael Ayers (Cambridge University Press, 1998), 521.

23 Carnap et al., "Scientific Conception," 89.

24 Some positivists deny that they make any claims about reality, but that's unconvincing. If what's empirically given isn't what's exclusively real, then there's no point in restricting knowledge or meaning to statements about the empirically given.

25 Carnap et al., "Scientific Conception," 89.

26 Carnap et al., "Scientific Conception," 90.

27 John Passmore, "Logical Positivism," *The Encyclopedia of Philosophy*, vol. 5 (Macmillan, 1967), 55.

28 Passmore, "Logical Positivism," 56.

29 Carl Sagan, *Demon-Haunted World: Science as a Candle in the Dark* (Random House, 1995), 12.

30 Richard Dawkins, interview by David Van Biema, "God vs. Science," *TIME* (Canadian ed.), 13 November 2006, 35.

31 "Project Reason," https://web.archive.org/web/20100306140031/http://www.project-reason.org/. Archived from the original (http://www.project-reason.org/) on 6 March 2010.

32 Lawrence Krauss, *The Greatest Story Ever Told … So Far: Why Are We Here?* (Atria, 2017), 2, 15.

33 Neil deGrasse Tyson, Twitter, 11 April 2021, https://twitter.com/neiltyson/status/1381197292728942595.

Chapter Fourteen

A Crisis of Epistemic Proportions

THE FAILURE OF POSITIVISM CREATES A PRETTY BIG PROBLEM FOR THE usual story. The usual story wants to say that science—and technology, if we care to mention it—are the biggest and best things ever. Positivism tries to say this in the most absolutely confident way possible: only scientific statements have any meaning. As we saw in Chapter Thirteen, however, the statement that "only scientific statements have any meaning" is not a scientific statement itself, and so if it's true, it's also meaningless on its own terms. Scientific absolutism is therefore an inherently contradictory position and cannot be rationally maintained.

Now what? What is modern culture supposed to think about Western technoscience if the highest aspirations of the usual story literally cannot make sense? This situation of deep questioning about the status of scientific knowledge is an *epistemological crisis*. We don't know what we're supposed to think about science anymore. Indeed, epistemological crisis is characteristic of our late modern era (we are increasingly skeptical of grand narratives), and in that crucible of uncertainty STS is formed. In this chapter, we will examine three foundational proto-STS thinkers who both intensified and responded to the epistemological crisis of modern Western technoscience: David Hume, Karl Popper, and Thomas Kuhn. Each of them will help us better understand contemporary controversies about science in modern culture.

David Hume (1711–76)

David Hume is not a recent thinker. He died over 200 years ago, although he was kind of an intellectual rock star for his time. As a member of the Scottish Enlightenment, he was friends with Adam Smith, and he entertained Jean-Jacques Rousseau (see Chapter Twelve) as a surly houseguest for a while. Hume's writing overlapped with Newton at one end of his life, and Romanticism at the other. During his lifetime, he was most famous for his work as an historian, but his philosophical insistence on *rigorous empiricism* is the most relevant to our study here.

An empiricist (you will recall from Chapter Eight) holds that all possible knowledge comes to us via our five senses *a posteriori*. In Hume's own words,

> [i]t seems a proposition, which will not admit of much dispute, that all our ideas are nothing but copies of our impressions, or, in other words, that it is impossible for us to *think* of anything which we have not antecedently *felt* [by our] senses.[1]

This isn't too far off from the positivist definition of knowledge as the description of sensory phenomena, and it sounds eminently scientific, doesn't it? The only things we can know are the empirical things that science studies. Hume points out, however, that science *can't* be empirical because neither causation (the power which Baconian explanations seek) nor induction (the supposed method of modern science) can be reduced to copies of our sense impressions. If so, science will have to reconsider its relationship to the natural world it seeks to explain.

David Hume (1711–76), portrait by Allan Ramsay (1766).

Causation, first of all, is an idea crucial to modern technoscience. If science cannot determine the likely cause of an event, it loses its predictive power and becomes "useless" in a Baconian sense: you can't use science to make more powerful technologies if science doesn't know how to cause things to happen. However, as Hume points out, there is no idea "more obscure and uncertain" than causation.[2] It is equivalent to the ideas of "power, force, energy, or necessary connexion,"[3] all of which indicate the metaphorical guarantee or glue between two causally related events. The cause of gravity—the force or hidden power that actuates the cosmic machine—was what Newton said he was incapable of discovering (see Chapter Eleven). Technoscience wants—indeed, needs—to know what causes things to occur. Hume set out to clarify the whole issue of causation, because if one of the central ideas of science is unclear, then modern technoscience doesn't know what it's doing.

As an empiricist, Hume thought that the best way to clarify an idea in our minds is to discover the sense impression that is the origin of the idea:

> To be fully acquainted, therefore, with the idea of power or necessary connexion, let us examine its impression.... by this means, we may, perhaps, attain a new microscope or species of optics, by which ... the most minute, and most simple ideas may be so enlarged as to fall readily under our apprehension....[4]

We need to identify the empirical sense experience that gives rise to the idea of causation itself. So what sense impression or empirical experience gives us the idea of cause: an apple falling from a tree, perhaps? Think of any experience you've had of a causal connection, like touching a hot pan on a stove, or dropping something onto the floor. (Hume's example is a billiard ball hitting another one.) The more you search for the cause's originating sense impressions, the less clear it is what they are. Let's use the example of a falling apple: what you see is the color, the shape, and the motion of the apple; you hear the thud as it hits the ground; maybe you feel the smoothness of the apple, or the pressure and slight pain of the apple hitting you, if you were snoozing under the tree. You could even smell and taste the apple.

But it turns out that you don't know what *causality* tastes like. The causal connection which guarantees that a detached apple will fall to the ground has no flavor or smell, texture, sound, or visible appearance. Likewise, nobody could possibly know what *necessity* feels like, what *force* looks like, or what *energy* sounds like:

> When we look about us towards external objects, and consider the operation of causes, we are never able, in a single instance, to discover any power or necessary connexion; any quality, which binds the effect to the cause, and renders the one an infallible consequence of the other. We only find that one does actually, in fact, follow the other.... Consequently, there is not, in any single, particular instance of cause and effect, any thing which can suggest the idea of power or necessary connexion.[5]

Causes are not sensed.

But hang on, you might ask, why can't we say, as Newton does, that there *must be* a cause behind events—or a force of gravitation behind gross bodies—even if we can't see it and can only mathematically describe its effects? Well, on the one hand, who are you to say what the universe *must* be like? Scientists are only supposed to describe what's there, not give prescriptions for what should be there. On the other hand, how could you possibly have *an idea* of this unseen causal power if it's not a copy of any of your sense impressions?

Where did you get this unique idea from? Did you just make it up in your mind? If you did, that's not very empirical of you; it rather sounds rationalist, maybe even medieval.

Hume's point is that you simply *don't* have an idea of causation or its equivalents:

> It is impossible, therefore, that the idea of power can be derived from the contemplation of bodies ... because no bodies ever [uncover] any power, which can be the [origin] of this idea.... [E]xternal objects as they appear to the senses, give us no idea of power or necessary connexion....[6]

If you are an empiricist, you cannot have any idea of cause, power, force, or energy. Science will either have to give up on empiricism, or give up on the Baconian will to power. Either way, things don't look good for the usual story.

Or maybe not. Even if we can't have an empirical notion of causation, scientists can still statistically measure the probability of correlations between events and thereby make relatively reliable predictions about what is likely going to happen. We have dropped enough things to be effectively certain that the next time we let go of the milk carton midair, it will make a mess on the floor. Using *inductive generalizations* (see Chapter Eight) with sufficiently large and representative samples, scientists appear to make predictions about causal relations that are adequate for both explanatory and technological power.

The problem with induction, according to Hume, is that it's no more empirical than causation is. The idea that we can derive probabilities from the repetition of events is not itself reducible to any set of sense impressions. Dropping the milk carton on the floor once will produce the same sense data as dropping it a thousand times. There is no empirical information that you receive from a repetition of events that you don't receive from a single event. Of course, humans actually do

> draw, from a thousand instances, an inference, which we are not able to draw from one instance, that is, in no respect, different from them. Reason is incapable of any such variation. The conclusions, which it draws from considering one circle, are the same which it would form upon surveying all the circles in the universe.[7]

But even though we draw these inferences, there is no empirical reason—no sense impression—that tells us that with the right kind of inductive sampling, we can form statistically likely generalizations about causes that allow us to make robust predictions about the future. We just stubbornly think so anyway.

A truly empirical science—one based entirely on descriptions of sensory phenomena, like the positivists want—cannot, therefore, know the causes of any event or even use inductive reasoning to generate statistically likely predictions. Inasmuch as modern technoscience does these things anyway, modern technoscience is *not* an empirical enterprise! In a chronologically backwards way, then, Hume's response to positivism is to point out that science never was and cannot ever be the kind of knowledge that the usual story wants it to be. If science works (and it does), it works for reasons other than it simply being a true description of facts. Science is not and will never be the result of the physical world telling scientists what is real. Science will always be at least partially a result of humans *adding something to the sense data*; we add to it either an idea of causation or an inductive expectation that the future will be like the past.

What is the nature of these nonempirical additions that apparently make science possible? Hume calls them *customs* or *habits*. Human beings—including scientists—simply cannot tolerate withholding their judgment about the cause of an event or the predictability of the future. It is simply paralyzing to believe that the floor might not be there the next time you get out of bed: "We should never know how to adjust means to ends, or to employ our natural powers in the production of any effect. There would be an end at once of all action, as well as the chief part of speculation."[8] Hume's rigorous empiricism requires such strict skepticism that apparently we can't handle it. We eventually throw our hands up in frustration and say "of course every event has a cause!" and "of course the Sun will rise tomorrow!" We don't care if there's no empirical experience of causation; we'll just believe that causes exist anyway, even if that's just as mystical as believing in a god. We don't care that we have no empirical basis to expect the milk carton to fall on the floor the next time we drop it; it's just too boring to pretend that we don't know if it will, and then just wait to see what happens each time. We're in the habit of expecting the future to be like the past, and (we think) the future is also in the habit of being like the past. That's why we think we know what will happen in the future even though we haven't experienced the future (yet).

Moreover, Hume thinks we habitually respond to this lack of empirical data with credulity *together as a human community*. That's why the mystical beliefs which make science possible constitute a *customary* practice. Custom is, after all, the way a human community has (supposedly) "always done things," like the customs of the handshake or eating with chopsticks. Similarly, both the scientific community and the world science describes seems to follow customary practices:

> All belief of matter of fact or real existence is derived merely from some object, present to the memory or senses, and a customary conjunction between that and some other object. Or in other words; having found in many instances, that any two objects, flame and heat, snow and cold, have always been conjoined together; if flame or snow be presented anew to the senses, the mind is carried by custom to expect heat or cold, and to *believe*, that such a quality does exist....[9]

Modern technoscience's foundations are *traditionalist*, therefore, even if modernity defines itself by the rejection of tradition (see Chapter Nine). Tradition is, roughly, doing something because that's how our community has done it for a long time. But this means that technoscience can only work on the basis of mystical beliefs and customary expectations for which there is literally *no empirical evidence*, only human habit and patterns of previous behavior. Of course we can continue doing science and inventing new technologies, but if Hume's right, we just can't say that by doing so, we know what we're doing any more than the medieval scholastics, ancient Greeks, or prehistoric flint knappers and shamans did. Hume's rigorous empiricism chastens the aspirations of modern technoscience to the point where we wonder if it really has made any progress in the way the usual story has been telling us is inevitable.

Karl Popper (1902–94)

Just so you know, almost nobody paid attention at the time to the first book where Hume made his points about empiricism, causation, and induction, so we'll fast forward a few hundred years to the twentieth century when philosophers and historians were finally grappling with the inadequacy of positivism that Hume had foreseen. Foremost among these early STS scholars was Karl Popper, who was later misidentified as a positivist by none other than Stephen Hawking.[10] Popper's main project was to identify precisely what constituted the scientific method, so that science could be clearly distinguished from nonscientific pursuits. Going further than Hume (who only argued that scientists lack an empirical basis for the causal connections and inductive generalizations they make), Popper argued that scientific method is not—and cannot be—inductive at all. It's just not possible to abstract a theoretical explanation from a set of mere facts. Facts don't tell you what theories to have, and there are no such things as mere facts anyway (but we won't get into that right now).

Karl Popper (1902–94).

Rather, what characterizes a scientific method, according to Popper, is *falsifiability*. A science, in Popper's modern sense, is a set of theories which could be inconsistent with empirical observations. For example, "the moon is made out of green cheese" is a scientific statement because it's possible to show that the moon isn't made out of green cheese (say, by landing on the moon and eating a lot of the dirt). By contrast, "the moon is an aspect of the goddess." is a nonscientific statement, because it is not possible to show, through empirical observation, that the moon isn't an aspect of the goddess. The moon will continue to look (and taste) the same no matter how divine it is or isn't. Because Popper isn't a positivist, however, he doesn't claim that it's meaningless or worthless to assert the divine aspect of the moon; it's just not science if you do, and that's fine.

Notice, however, that Popper does not claim that scientific method is comprised of verifiability; rather, he claims that it is comprised of falsifiability. Verifiability and falsifiability are opposites: the former shows that something is true, the latter shows that something is false. According to Popper, scientific theories cannot be verified or proven; they can only be disproven (i.e., falsified). Scientific statements are universal and lawlike (see Chapter Two), and only an infinite amount of empirical confirmation could ever verify or prove a universal statement. However, all it takes is one contrary empirical observation to show that something is wrong with a universal statement. Exceptions don't prove a rule: they disprove them. This is a truth of deductive logic: trying to confirm an hypothesis as true commits the fallacy of 'affirming the consequent.' This is because there will always be an infinite array of logically possible explanations for why an event occurs. Just because you've never gotten the flu since you've been taking aspirin daily doesn't mean that it's true that daily aspirin doses prevent the flu. I could just as well prove my goblin theory of gravity (i.e., objects fall to the ground because invisible goblins are always trying to pull objects towards the center of the Earth) by dropping a cereal box on the floor. As soon as the box hits the ground, I could say, "See? I've proven my theory; the evidence shows that invisible goblins have pulled the cereal box as far as they could towards the center of the Earth." I could successfully repeat this experiment over and over again, verifying it each time. Isn't that convincing?

If, however, I dropped the cereal box and it *didn't* fall, but rather hovered in mid-air, then something would be wrong with my theory (perhaps my goblins were taking a nap). This is hypothesis disconfirmation or falsification, and it is a deductively valid logical form known as 'denying the consequent', or *modus tollens* (it's so respectable that it has its own Latin name). This is how scientific method works, according to Popper: first, scientists propose theories or hypotheses (they don't induce them from particulars); second, they test those theories or hypotheses by trying to falsify or disprove them empirically; third, the longer those theories or hypotheses resist falsification, the better they are; but fourth, no amount of resistance to falsification could ever show a theory or hypothesis to be correct or true. Science simply cannot be proven true.

If falsification is the best that scientific method can hope for, then science obviously cannot attain absolute truth. Scientific knowledge is fundamentally and unavoidably "provisional, conjectural, hypothetical—the universal theories of science can never be conclusively established."[11] But having a body of scientific theory highly resistant to falsification still sounds pretty good, doesn't it? It does, but a Popperian cannot be triumphal about this. Even falsifying a theory involves the scientist making judgment calls. Say you've made an empirical observation that contradicts a scientific theory. Which part of the theory is incorrect? Is the whole theory falsified (e.g., there are no invisible goblins) or just part of it (e.g., the goblins weren't awake)? Empirical observation can only provide factual statements inconsistent with a pluralistic theory-set, rather than giving an unambiguous "no" to a single theory. Disconfirming observations can just as easily result from observational bias, measurement error, or equipment malfunction as from a faulty theory, and so what is actually being falsified remains to be seen.[12]

Finally, there can be no ultimate end to falsification attempts. There is no point where a theory has been tested enough, because it could always be falsified by the next test. You just never know. Therefore, all hypotheses and theories must be subject to scientific testing for infinity. This, of course, is not humanly possible. At some point, scientists must decide when to stop testing a theory. There is, however, no empirical indicator that tells scientists when to stop; they have to make that call using "convention and intersubjective human agreement...."[13] That's about as subjective as custom and habit! Modern science, therefore, even when understood as a process of empirically testing theories for falsifiability, "does not ... rest on any foundational bedrock."[14] If Popper is right about scientific method (and many if not most scientists think he is), then modern technoscience cannot supply either truth, proof, or objective progress. This isn't how the usual story is supposed to go either.

Thomas Kuhn (1922–96)

With Hume, we saw that modern Western technoscience is not an empirical affair, at least with respect the two central aspects of causation and induction, and with Popper, we saw that science can never be known to be true. However, something of the usual story might be retained, perhaps, if we imagine the progress of scientific knowledge to involve a step-by-step increase in theories ever more resistant to falsification. The epistemological crisis of modern science is further deepened, however, by Thomas Kuhn's historical refutation of this modified version of incremental scientific progress. Kuhn is probably the most important early STS scholar, and he sets the groundwork for many contemporary and ongoing debates about science (see Chapter Fifteen).

Thomas Kuhn (1922–96).

What I have been calling the usual story, Kuhn calls "the textbook tradition" of science.[15] This is the received picture of scientific progress that most of us—scientists included—have been trained to accept by our science textbooks: "[s]cientific development becomes the piecemeal process by which [facts, theories, and methods] have been added, singly and in combination, to the ever growing stockpile that constitutes scientific technique and knowledge."[16] We're generally supposed to think that science (and technology) has been composed of a long history of *improvements*, regardless of whether we see those as verified truths or theories resistant to disproof. Once upon a time, we suppose, nobody knew that there were X-rays; then one day, Wilhelm Röntgen discovered X-rays and—voilà!—science advanced.

The trouble with this textbook model, Kuhn points out, is that scientific *revolutions* are not particularly incremental. They involve massive leaps or shifts in perspective that do not map onto the usual story of gradually progressing continuity over time. In fact, the more that historians look at the history of science, they find

> it more and more difficult to fulfill the functions that the concept of development-by-accumulation assigns to them. As chroniclers of an incremental process, they discover that additional research makes it harder, not easier, to answer questions like: When was oxygen discovered? Who first conceived of energy conservation? Increasingly, a few of them suspect that these are simply the wrong sorts of questions to ask. Perhaps science does not develop by the accumulation of individual discoveries and inventions.[17]

To really do a history of science, therefore, Kuhn thinks we have to distinguish between two kinds of science: what he calls *normal* science and *revolutionary* science. Let's look at normal science first.

Normal science is tradition-bound, Kuhn says, because most of the time, scientists are busy *not* having a scientific revolution. Rather, they are happily operating within the limits of their assumed worldview (i.e., their 'scientific conception of the world'; see Chapter Thirteen). Non-revolutionary normal science 'already knows' how the world must be (e.g., it's a big machine), and normal science generally assumes that these "ways of seeing the world and practicing science in it"[18] will *not* be disconfirmed:

> Effective research scarcely begins before a scientific community thinks it has acquired firm answers to questions like the following: What are the fundamental entities of which the universe is composed? How do these interact with each other and the senses? What questions may legitimately be asked about such entities and what techniques employed in seeking solutions?[19]

These are sets of methodological and metaphysical presuppositions that comprise what Kuhn calls a *paradigm*. They are "firmly embedded in the educational initiation that prepares and licenses the [science] student for professional practice,"[20] and constitute the tradition within which scientists recognize facts. A fact is a fact because it fits within the paradigm. Fitting into the paradigm is what makes the fact acceptable. This is why unicorns (probably) can't be facts: they (probably) don't fit into your paradigm.

So what does normal science normally do, then? *Normal science makes facts fit acceptably into paradigms.* Most of the time, science is problem-solving. When a piece of data or observation looks like it might *not* fit into the paradigm, this is a problem for the paradigm. Scientists then figure out how to "force nature into the conceptual boxes supplied by" the paradigm,[21] thus making the data or observation fit the paradigm after all. This solves the problem and generates a fact. Tradition-bound normal science thus maintains its own status quo, in the same way that the application of force to matter 'mechanically solved' the problem industrialization had with the Luddites. As we saw in Chapter Twelve, solution-based approaches are not neutral because solutions always solve problems as determined by non-neutral background assumptions. The same could be said for Newton's theory of gravity solving the problem of seriousness (see Chapter Eleven). Both of them propped up the status quo.

No 'fact' is neutral either. A fact is always *theory-laden*: loaded down with the background assumptions of the paradigm it has been fitted into. A classic example of this is geocentric astronomy. Astronomers had long thought that the (spherical) Earth was the center of the universe, at least as far back as Ptolemy of Alexandria, a scientist and mathematician of the second century CE. If the Earth is the center of the universe, the stars should appear to revolve around the Earth, which they do (thus far, we have hypothesis confirmation or verification). However, some planetary bodies seem to wander in their orbits (this is called 'retrograde motion')—and we call these bodies 'planets,' from the Greek *planētai*, meaning 'wanderers.' Ptolemaic astronomers could not explain the wandering planetary routes around the Earth using single circles, so they added *epicycles* to the theory: the orbit of a planet is a circle (or set of circles going around circles) going around another circle around the Earth. The introduction of epicycles allowed Ptolemaic astronomy to resist being falsified by the retrograde motion of the planets.

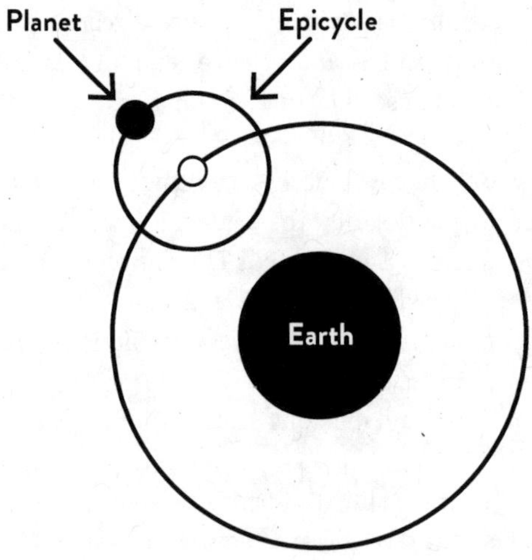

Epicycles as an explanation of planetary paths around the Earth.

Well, look at that! Ptolemaic astronomy is resistant to falsification. It turns out that planets do, in fact, rotate around the Earth, only that they do so in epicyclical rather than circular orbits. But wait: you might object by saying it's *not* a fact that planets revolve around the Earth in epicyclical orbits, because they actually revolve around the Sun in elliptical orbits. Okay, but why do you think that? Perhaps it's because you think that Copernican and Keplerian astronomy are true, and the facts fit those theories better. Maybe they do (see below for more on that), but even if they do, notice that what you think are facts are already embedded in a theoretical framework or paradigm, in this case heliocentric rather than geocentric astronomy. As such, "[s]cientific fact and theory are not categorically separable,"[22] which makes it hard to see how a *theory* could ever be falsified by contrary facts. Facts *assume* theories; they do not stand outside them.

Back to Ptolemaic astronomy, then. If epicycles allow geocentrism to resist falsification by the wobbly paths of planets around the Earth, how did the Copernican Revolution happen? If we follow a simple Popperian model, geocentrism should have been falsified and heliocentrism accepted because the latter was more resistant to falsification than the former. But this isn't what happened. As telescopes improved, astronomers were able to detect new wobbles in the planetary orbits, requiring more epicycles to be added to their paths, a process that can be continued forever without any loss of predictive power. In fact, Ptolemaic astronomy was able to make more accurate predictions of planetary positions than Copernican astronomy for several centuries thereafter. The shift away from geocentrism to heliocentrism thus involved a sacrifice of predictive power, a scientific revolution in favor of less accuracy!

Why? Because adding even more epicycles to save the Ptolemaic system looked *inelegant* in comparison to the relative simplicity of the Copernican system's heliocentric planetary orbits. The revolutionary switch from the Ptolemaic paradigm to the Copernican paradigm was made on the basis of a judgment call, not falsification, on what scientists thought was more beautiful, elegant, or parsimonious (see Ockham's Razor in Chapter Eight). It wasn't until 1838, nearly 300 years after Copernicus' death, that telescopes were sensitive enough to detect stellar parallax, producing an increase in the explanatory power of Copernican over Ptolemaic astronomy.

The Copernican Revolution was not 'normal science.' It was what Kuhn calls 'revolutionary science' because it entailed a *new paradigm*. Having the Earth and the Sun switch places in your model of the universe is not an incremental addition or accumulation of a new fact; it is the replacement of one picture of the world with a radically different picture of the world. This radical shifting of worldview is the general characteristic of revolutionary science:

> Sometimes a normal problem, one that ought to be solvable by known rules and procedures, resists the reiterated onslaught of the ablest members of the group within whose competence it falls. On other occasions a piece of equipment designed and constructed for the purpose of normal research fails to perform in the anticipated manner, revealing an anomaly that cannot, despite repeated effort, be aligned with professional expectation. In these and other ways besides, normal science repeatedly goes astray.... [These] extraordinary episodes in which that shift of professional commitment occurs are known as ... scientific revolutions. They are the tradition-shattering complements to the tradition-bound activity of normal science.[23]

Are the anomalies that cannot fit into old paradigms the objective, theory-neutral facts we couldn't find earlier? Are new paradigms, which can account for what was previously anomalous, more correct than the earlier ones, and so our paradigms can be said to progress towards the truth? Not really. To choose a paradigm because it solves problems better isn't objective or neutral, because problems are themselves theory-laden. Moreover, what constitutes a solution is equally theory-laden: just because something is simpler or more elegant (as Copernican astronomy purportedly was) doesn't mean simplicity or elegance are objective, neutral, or more likely to be true. Who are we to say that what is real cannot be complex or awkward?

A problem is not a sheer fact about the universe, nor is a solution to a problem. They are relative to the subjective frameworks of theory and paradigm. Similarly, preferences for simplicity over complexity, or elegance over awkwardness, are subjective rather than objective. The choice of one paradigm over another cannot be determined on the basis of any fact whatsoever, because facts are always *inside* paradigms, anomalies included. The choice of one paradigm over another is rather determined on the basis of *preferences* (for simplicity or elegance, for example), and as far as Kuhn can see, preferences are arbitrary:

Paradigm Shift: when a theoretical framework in science is replaced by another, incommensurate, framework; also known as a scientific revolution.

> [T]he particular conclusions [the scientist] arrives at are probably determined by his [or her] prior experience in other fields, by the accidents of his [or her] investigation, and by his [or her] own individual makeup.... For the individual, at least, and sometimes for the scientific community as well, answers to [subjective] questions like these are often essential determinants of scientific development.... An apparently arbitrary element, compounded of personal and historical accident, is always a formative ingredient of the beliefs espoused by a given scientific community at a given time.[24]

At this level, Kuhn would agree with Hume that science's fundamental basis is not empirical, factual, or objective. It is rather artistic and subjective, potentially as romantic as the simplicity of a poem or the elegance of a dancer. Science is an art, as aesthetic as getting a manicure or a haircut.

Scientific revolutions are why "the textbook tradition" of knowledge-by-accumulation does not actually fit the historical record and why the usual story of progressing continuity in science is false: either scientists engage in long periods of fitting facts into the received paradigm (aka normal science), or scientists engage in "the extraordinary investigations that lead the profession at last to a new set of commitments, a new basis for the practice of science" (aka revolutionary science).[25] In neither case can science tell us about what is really out there. All science can do is *make* what's out there *fit* the theories that *we prefer*. In this respect, knowing does turn out to be making after all, although not perhaps in the way Bacon thought (see Chapter Nine). Scientific revolutions cannot be mapped onto a model of progress towards truth, because modern technoscience consists of making, not knowing, what is 'out there.'

After the Epistemological Crisis of Science

It's unclear, then, how humans could ever break free from their theoretical frameworks and finally intellectually master the world the way the literate ancient Greeks had hoped to: seeing things as they really are (see Chapters Six and Seven). Scientific absolutism is a failure, and Hume, Popper, and Kuhn each (in their own ways) dig the hole deeper. Empirical facts are incapable of giving science an inductive method or knowledge of causes, and scientific facts are inextricably theory-laden and therefore incapable of determining which theories are better than others. The best science can hope for is a method of falsification that can never provide proof, but falsification is precisely what normal science works at preventing. Nor is it ever clear that falsification can decisively happen, given that it depends on arbitrary or aesthetic factors.

In this epistemological crisis, science loses its privileged knowledge status, and is no better (or worse) than investigations in the humanities or fine arts—which everybody knows is (supposedly) based on arbitrary or aesthetic factors. Science appears to be—at best—just one way of knowledge creation among many. Nevertheless, this appears to lead to a kind of skepticism about science that could justify all kinds of dangerous positions, such as creationism, climate change denial, and oppositions to vaccines in the middle of a global pandemic. Can some sense of scientific legitimacy or authority be salvaged from the epistemological crisis of science? The answer to that question is a central and ongoing issue in the discipline of STS, and I am not likely to be the one who answers it, least of all in this book. Rather, you and I both are caught up in the midst of trying to find a way forward through the controversies of our present time. All I can offer you is a broad overview of the ways in which STS has been grappling with these problems, which is the topic of the next section of this book. There we will examine some of the central controversies that the usual story faces in the unusual field of STS.

Notes

1 David Hume, *An Enquiry concerning Human Understanding*, ed. Peter Millican (Oxford University Press, 2007), 45; original emphasis.
2 Hume, *Enquiry*, 45.
3 Hume, *Enquiry*, 45; emphasis removed.
4 Hume, *Enquiry*, 46.
5 Hume, *Enquiry*, 46.
6 Hume, *Enquiry*, 47.
7 Hume, *Enquiry*, 32.
8 Hume, *Enquiry*, 33.
9 Hume, *Enquiry*, 33; original emphasis.
10 Stephen Hawking, *The Universe in a Nutshell* (Bantam, 2001), 31.
11 Stephen Thornton, *Stanford Encyclopedia of Philosophy* (2021), s.v. "Karl Popper."

12 Imre Lakatos, "Falsification and the Methodology of Scientific Research Programmes," in *Criticism and the Growth of Knowledge*, ed. Imre Lakatos and Alan Musgrave (Cambridge University Press, 1970), 129–31.

13 Thornton, "Karl Popper."

14 Thornton, "Karl Popper."

15 Thomas S. Kuhn, *The Structure of Scientific Revolutions*, 2nd ed. (University of Chicago Press, 1970), 8. In a twist of irony, Kuhn's book was published in a series edited by logical positivists, including Neurath, Carnap, and Feigl from Chapter Thirteen. The atomic physicist Niels Bohr was also on the advisory committee, as was John Dewey, Bertrand Russell, Ernest Nagel, and Arne Naess.

16 Kuhn, *Scientific Revolutions*, 1–2.

17 Kuhn, *Scientific Revolutions*, 2.

18 Kuhn, *Scientific Revolutions*, 4.

19 Kuhn, *Scientific Revolutions*, 4–5.

20 Kuhn, *Scientific Revolutions*, 5.

21 Kuhn, *Scientific Revolutions*, 5.

22 Kuhn, *Scientific Revolutions*, 7.

23 Kuhn, *Scientific Revolutions*, 6.

24 Kuhn, *Scientific Revolutions*, 4.

25 Kuhn, *Scientific Revolutions*, 6.

VI

Late Modernity

Chapter Fifteen

Social Construction and the Science Wars

THIS IS THE BEGINNING OF THE END OF THIS BOOK. WE HAVE COME TO see why STS views modern Western science and technology as a recent and unique social phenomenon worthy of critical examination. Technoscience is not a universal yearning of the human species, and was not combined into its currently powerful form until well after the Middle Ages. It is a product of early modern European culture, but was actively resisted as problematic in various times and social contexts. But what about right now, our time and contemporary social context? Are there similar or related issues with modern Western technoscience right now that STS studies and responds to? Yes there are, and the last two parts of the book will introduce us to some of these.

An annoying thing about such an introduction, though, is that it will be inconclusive. As noted in Chapter One, the history of our current moment is still being written. That is why the open-endedness of current STS research is itself characteristic of our time. As you can see from this section's heading, I have called our era "late modernity," but others might call it "postmodernity." Postmodernism is a controversial term, but it refers to how modern people no longer feel sure about modernity itself. In this late modern moment, we encounter *skepticism* towards the grand stories which have guided our cultures up to this point[1]—including the usual story about science and technology. The optimism about science and technology which characterized early modernity is far more muted now, if not entirely gone, in late modernity.

Therefore, as STS looks at contemporary technoscience, we find unresolved controversies rather than confident proclamations of truth. Late modernity is a difficult and uncertain time, to say the least, but that doesn't mean STS can't help. STS offers important discoveries and insight into science and technology, and these can guide us as we actively engage and respond to the scientific and technological issues we currently face, and will continue to face into the future. My hope is that you will be an active participant in the ongoing unusual story of STS.

Historically, the epistemological crisis of science in the 1960s was the furnace in which STS was formed. The resulting perspectives on science as a social phenomenon generated such controversy that by the 1990s, STS debate spilled out into contemporary culture as 'the science wars' between the (so-called) two cultures: the natural sciences, on the one hand, and the liberal arts, on the other.[2] In recent years, the conflict has faded somewhat, not because the STS issues have been resolved, but because the parties involved seem to have lost the taste for open battle. Nonetheless, the core issues of the science wars are still live, as seen in controversies about anthropogenic climate change, the use of falsehoods as 'alternative facts' by the Trump administration,[3] and hesitancy surrounding mRNA vaccines for COVID-19. In this chapter, I will introduce you to what STS scholars and their critics are (still) fighting about. In the process, hopefully we will find insights that offer a way forward for science in our time that is informed, critical, and reasonable. First, we need to understand the key flashpoint in the debate: social construction.

The Social Construction of Reality

Much STS theorizing about science can be categorized as social constructivist. Even before we're sure of what it means, social construction triggers alarm bells—because it sounds like science will be something people can make up out of thin air. If science is a social construction, it sounds like a human invention rather than a discovery of facts. But how can science tell us "what's really real in a world increasingly full of fake"[4] if it's just an invention of human beings? The seminal book on this topic is *The Social Construction of Reality* by Peter L. Berger and Thomas Luckman,[5] and a title like that seems to make the prospects of distinguishing the 'really real' from the 'fake' pretty dismal. If anything sounds like the laughable philosophy of denying the reality of what's right in front of our faces, the social construction of reality does.

However, social construction isn't a product of naughty philosophy: it's sociology. "Social construction," according to Wenda K. Bauchspies, Jennifer Croissant, and Sal Restivo, "is little more than basic sociological thinking."[6] This "fundamental theorem of sociology" states that "We humans have no other way to become human, to be in the world, and to find our way through truths and falsities than as social beings and in social interactions."[7] It's a sociological fact that human beings are social animals and understand the world socially, but what does this imply about the status of knowledge or the nature of reality? Why might *reality*—and eventually science—be 'socially constructed'?

Social construction means that humans—even human scientists—do not stand alone in front of objects in the world and passively receive knowledge from those objects through human powers of perception or intellect. Rather, we actively and communally build networks of meaning that constitute 'the world' we take for granted. The world that we all perceive as 'real' is at best a composite of what is objective and what is subjective. In a sense similar to the (usually banal) advice that 'life is what you make of it,' social construction means that *reality is at least partially a product of human action*. Francis Bacon said it first: *knowing is making*, but he probably didn't anticipate it turning into the human construction of reality (or did he?). Even if he didn't, that's what you get if truth and utility are the same thing (Chapter Nine). If knowledge is power and science is technology, then the human power to invent things and change the world is the same as knowing the truth about reality. What we know is real is something we invent too—including scientific knowledge itself.

Let's use an example to get a better sense of what this means. Consider, for a moment, the Great Canadian Ice Storm of 1998. Was it real, or am I making it up? You might hesitate to answer, because you might not have been alive in 1998, or you might not live anywhere near Canada, and therefore you might never have heard of 'the Great Canadian Ice Storm of 1998.' Maybe you want to look it up on the internet first? If you do, the first thing to note is that looking it up on the internet is a *social activity*. That is, you're relying on other people to tell you what may or may not have been real in 1998, and you're doing so within a network of trust, testimony, and authority. That's a social enterprise at the outset, and so it's already hard to say that the reality (or not) of the Great Canadian Ice Storm isn't at least partially a result of human interaction. The objective facts of the ice storm (if there are any at all) are *mediated* through human sociality.

But wait a minute! Doesn't that point fail to distinguish between the context of discovery and the context of justification? Sure, your *discovery* of the reality of the Great Canadian Ice Storm is socially mediated by 1) me asking you the question, 2) the internet's very existence, and 3) the expertise of the people who wrote what you read on the internet—but that doesn't mean that the *reasons* you have for believing it to be true (or not) are social constructions too, right? Surely the *reality* of the Great Canadian Ice Storm is independent of *how* you found out about it! A sociologist might be able to describe all the social factors that come into play as you investigate the ice storm, but the logical validity of your reasons for accepting the truth (or falsehood) of the ice storm is objective and universally true, right?

Not necessarily. What are the standards of logical validity, and who decides what they are? Sociologists? Philosophers of logic? The gods? The universe itself? Answering those questions is a social enterprise too, and if you think *that* answer can be split into the two contexts of (socially constructed) discovery and (ideally rational) justification, we can interrogate *that* distinction too, ad infinitum. None of the potential answers here rest on any self-evident bedrock. There's no end to the possibilities of justificatory standards being invented or contested by human beings, such that it's not clear that we ultimately can distinguish them from discovery contexts.

And we still haven't decided if the Great Canadian Ice Storm was real. Let's change tactics, then: what's an ice storm? Maybe it's when a lot of freezing rain coats everything nearby with ice. Okay, but so what? Good question. Ice storms are pretty unremarkable if that's all they are, objectively. The ground is slippery for a while and some tree branches break off. What makes this so Great and so Canadian? Well, we guess that it needs to happen in Canada and be really, really big. Okay, but so what? Does any of this even matter? Why are we even talking about this? (Why are you reading this book?) Do you even care if there was a great big ice storm in some part of some country called Canada sometime long ago? Probably not. You might not care at all whether it was real or not. For all you care, it might as well not exist. Its reality probably means nothing to you at all.

As far as you're concerned, the Great Canadian Ice Storm of 1998 doesn't even need to be real.

However, if sheets of ice knock out your electrical grid leading to 34 fatalities and the deployment of 16,000 military personnel to get everything back on line, then—to quote Martin Lawrence from 2003's *Bad Boys II*—"shit just got real." Reality, from a sociological perspective, is what we notice and care about: "A fact has to be named by the community, accepted by the community and practiced by the community."[8] The fairly narrow strip of east-central Canada that got hit by the Great Canadian Ice Storm in 1998 noticed it as such because *they got hit* by the ice. I wasn't there and you probably weren't either, which is why you and I have to rely on social networks of testimony to even begin taking notice of it.

But we also need to take note of *why* those Canadians noticed the ice storm and eventually we heard about it: it destroyed their power grid and killed people who depended on that electricity to stay warm enough to survive. Mainstream Canadian society is highly dependent on electricity, like many parts of the world. This dependence, however, is *literally* a human construction. We *build* electrical grids and the societies that depend on them. It is not necessary, however, that humans build electrical cultures: the Amish in the area, for example, did not use electricity. As a result, they did not notice the ice storm very much. According to Raymond Murphy, a sociologist of disasters,

> [w]hereas modern farms were devastated, Amish farms ... that refused to be connected to the grid were hardly affected at all. This apparently natural disaster, then, was in fact a technological calamity. There may have been intense freezing rain a century ago, but there was likely no disaster on account of it.[9]

> **Social Construction:** the theory that human experience or knowledge of reality cannot be separated from subjective and contingent social factors such as history, ideology, politics, language, religion, personal relationships, interests, systems, or artifacts.

Whether our societies are dependent on electricity is going to determine in large part whether an ice storm matters to us enough to notice it as a thing. Great Canadian Ice Storms don't really exist if you're Amish, even if you live in Canada and a lot of ice covers everything. Natural disasters only 'exist' in certain social contexts, therefore, and not in others. That's true for ice storms, as well as more recent disasters: hurricane Katrina (flooding New Orleans in 2005), the Japanese earthquake and tsunami (causing the Fukushima nuclear plant to melt down in 2011), or hurricane

Sandy (flooding the New York subway system in 2012). Each of these things existed as disasters because of the way those cities or systems were constructed.

To say that reality is socially constructed doesn't necessarily mean, therefore, that humans create the ice or the earthquakes (although with weather, there can be an anthropogenic component to its intensity or frequency), but it does mean that ice storms or tsunamis are only recognized as such when they interact significantly with contingent human social arrangements like electrical grids, subterranean railways, or nuclear power plants with backup generators on the ground floor. So-called objective reality goes unrecognized if it does not have a subjective or value-laden impact on people for whom it matters. In that respect, reality—as far as we know, inasmuch as we notice it at all—is a social construction.

The Social Construction of Science

So what are the implications of social construction for understanding science as a social phenomenon? Science can be interested in ice storms, of course, as a part of meteorology. But we've seen that the social construction conjecture isn't saying that the weather is a human invention, but rather that the importance of the weather as it impacts human interests is itself a function of human invention. Science is socially constructed because it's like meteorology: it's in the business of explaining to various groups of people why they should care about some set of physical phenomena. Science doesn't give us the cold, hard facts; it tells people what the importance of the facts are for them (which is, ultimately, *why* they're facts in the first place). Meteorology doesn't simply describe sensory phenomena, therefore; it describes the value-laden impacts those phenomena are likely going to have on our lives. Our lives, meanwhile, aren't just facts, but the complex results of history, culture, choices, and blind luck—all social factors. The weather is meaningful to us inasmuch as it impacts our complicated and non-neutral social reality. Meteorology is meaning, and that meaning is socially constructed and conditioned by the social context wherein that meaning is found.

Let's try another example: Pluto the ex-planet. Pluto used to be a planet, but in 2006 it got demoted to 'dwarf planet' status. So was Pluto ever *really* a planet? Did the facts about Pluto *change*? If reality is socially constructed, then maybe they did; maybe the astronomers constructed new facts about Pluto, because—as we saw above—"A fact is an idea or concept that everyone (or some subset of everyone such as a community or network) accepts as true."[10] What was accepted by astronomers as true before 2006 was not accepted by astronomers as true after 2006. But *Pluto itself* didn't change, did it? If meanings *but not objects* are constructed, then the orbit and rocky iciness of Pluto (etc.) remained the same; it's rather the *meaning* of that object which changed. Consider the definition of a planet: a planet is ... well, it is whatever astronomers say it is. Astronomers invent the definition of 'planet' and then see what objects will and won't fall under that classification. In 2006, the International Astronomical Union held several votes to define what a planet was, and after the votes were counted, Pluto no longer satisfied the definition of a planet (it had not gravitationally "cleared the neighbourhood around its orbit," which was the third part of the new definition).[11]

What this means is that a 'planet' is not a neutral or objective description of something; it is *not* a category that comes to us from the things themselves. Planets do not tell scientists what they are; *scientists tell planets what they are*. Facts, therefore, do not create the definitions that scientists use; scientists rather create the definitions that make facts facts. Whether Pluto is a planet is not a (so-called) factual question. It is a question of meaning. *Planets are meanings*, not mere objects. Meaning is applied to, not derived from, objects, and as such, facts are what *result* from meaning attribution. Scientific facts are what we get *after* the scientists have counted their votes: facts are whatever the relevant authorities agree on. Science is a *meaning system* comprised of socially mediated definitions and values, and as such does not—indeed cannot—give us objective knowledge of things as they really are. There are so far as human beings are concerned, no such things as 'mere objects.' 'Pluto'—whatever that means—is socially constructed, therefore (including the fact that Pluto is the name of the Greco-Roman god of the afterlife, and not just Mickey Mouse's dog).

So even though science, according to the usual story, is objective and doesn't talk about meaning or values, that's actually all it talks about. Unfortunately, several centuries of thinking that science doesn't talk about meaning or value has meant that scientists aren't very well trained in thinking or talking about meaning and value. This might be something STS would like to see improved in science education going forward. Similarly, science needs to do a better job in thinking and talking about reality. Rather than naively claiming that science tells us what's 'really real,' social construction wants us to recognize that naive realism about anything—science included—is false.

What is naive realism, though? It is what Greeks scientists (see Chapter Seven) thought they could get with *logos*: truth as an exact, non-symbolic, literal correspondence between an idea or concept and a real thing (see Chapter Six). Naive realism is why the positivists thought they had eliminated metaphysics; they thought their descriptions of sensory phenomena gave them a model of the universe that wasn't actually a model: it was a copy of the thing itself (see Chapter Thirteen). Naive realism is thinking that because you (a subject), perceive a tree, that tree (an object) is really there. A scientist's description of that tree is simply a mental picture of the tree, and the mental picture is true if it is an exact copy of the real, objective tree.

If social construction is true, however, then naive scientific realism is false. No picture of anything can ever be a perfect copy of the thing it's a picture of, even if that picture is a scientific description or model of your own perceptions. The Belgian surrealist René Magritte made this point with his famous painting "The Treachery of Images." The painting is a photorealistic representation of a tobacco pipe with a caption (in French) that reads "this is not a pipe." The philosopher Ludwig Wittgenstein (who we met in Chapter Four) made a similar point about the word 'cat.' You can write the word 'cat' on a chalkboard, but if you point to it, you have to admit that the word 'cat' is not a cat. Neither an image of a pipe (or a cat) nor the word 'cat' (or 'pipe') *is* a pipe or a cat. All representations of things are stand-ins for things, but are never the actual presence of those things. (That's why they're called *re*-presentations, not presentations.) Everything a human being could ever think, say, or model about something will always be a symbol for that thing, but no symbol is *literally* true. A symbol is always different from the thing it symbolizes. (Symbols don't even have to resemble the things they symbolize! The word 'cat' neither looks nor sounds like a cat.) But science is unavoidably a language of symbols about the world, and because of that, it can never be literally true. Only the actual things can literally be the things themselves; nothing else can be what they are, no matter how precisely we try to copy them with words or images or other products of the human mind. Science is a product of the human mind, and therefore the hope that goes as far back as ancient Greece—the hope that we can build "in the human understanding a true model of the world, such as it is in fact, not such as a man's own reason would have it to be"[12]—is a hope that can never be satisfied.

The fact that science is subject to the profound limitations of language and symbolism is what Bauchspies, Croissant, and Restivo call "rhetorical pathos," referring to linguistic or conceptual problems inherent in science itself.[13] Not only is naive realism in science inherently flawed, the status of otherwise solid and productive scientific models becomes uncertain. Scientific models, theories, or statements (like all languages) are social conventions; they are whatever the scientific community agrees upon. Those social conventions are loaded with all kinds of historical and philosophical baggage. Based on what we've already learned in this book, we could say that modern Western science is agrarian, literalistic, anthropocentric, sexist, colonialist, mechanistic—and that's just for starters.

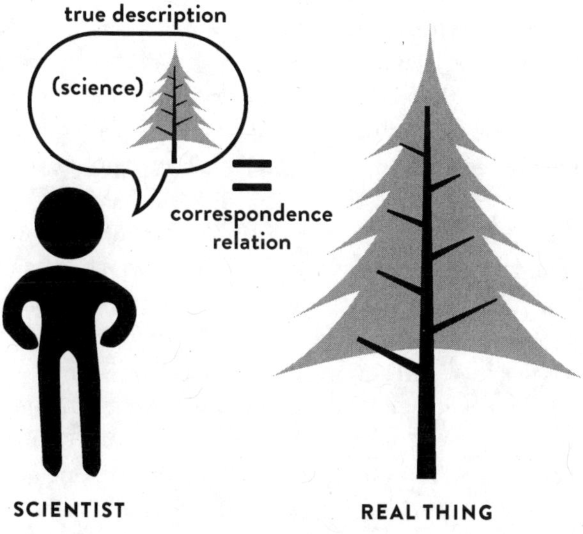

Naive scientific realism.

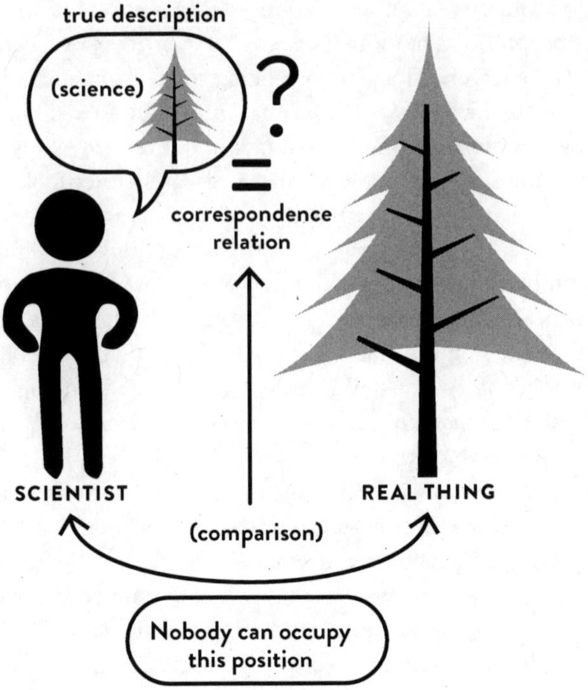

Nevertheless, why can't a symbolic system of meaning with all that baggage actually correspond with what is really out there? Even if it could resemble what's out there, epistemologically we could never know:

> A realist believes that there is an independent arbitrator called Nature that exists outside of humans, that facts are distinct from human thought and practice. A relativist believes that the situation or representations cannot be "sorted out" without an outside arbitrator, but there are no universal arbitrators not themselves grounded in a specific historical and intellectual position.[14]

To know whether our historically and linguistically conditioned scientific claims are true, someone would have to stand outside of all human symbolic frameworks, having no historical, cultural, or ideological perspectives whatsoever, and compare the limited constructions of scientists with the world as it actually is in itself—just to see if they line up. The problem is that there isn't anyone who could do that. That person would have to be more than superhuman; they'd have to be an all-knowing, omnipresent deity to do that. Disappointingly for the usual story, divine science is humanly impossible. Modern Western technoscience may have *wanted* to know the truth like God does,[15] but we have seen that kind of absolutism to be self-defeating (in Chapter Thirteen).

Science, therefore, is the socially constructed process of assigning meaning to an external reality we can never verify as actually being what we think it is, because to do

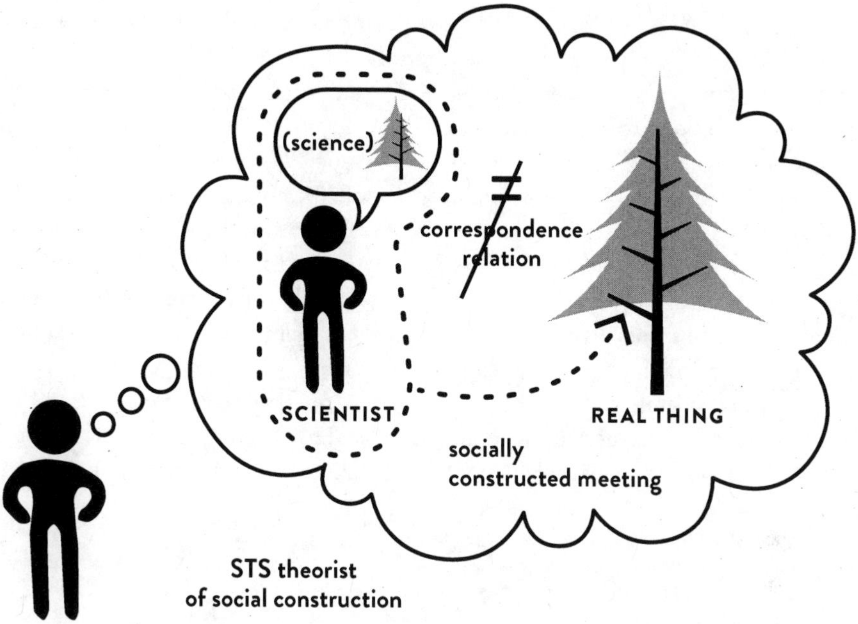

so we would have to somehow step outside our own systems of meaning, symbolism, and representation—our science—and see things as they actually are, neutrally, as if we were gods. The external world (if there is one, ha ha) cannot possibly tell us—or scientists—what it is in itself, because we will unavoidably conceptualize it in terms of socially constructed systems of meaning. We don't even have an 'external' world until *after* we've noticed things via our socially constructed symbolic categories: "'Reality' cannot be used to explain why a statement becomes a fact, since it is only after it has become a fact that the effect of reality is obtained."[16] The whole point of saying that science is socially constructed, then, is to say that science *cannot* tell us what is 'really real.' Reality doesn't tell scientists what it is; scientists tell reality what it is. Science constructs reality as we know it. This is *not* how the usual story goes. Unsurprisingly, a lot of scientists (and philosophers of science) found this suggestion offensive! That's where 'the science wars' came from.

The Strong Program and Scientific Anarchism

I'm not going to give you a play-by-play of the science wars, although that could be something fun to read about in another book. Rather, I'm going to raise the intensity of the "relativistic trends" in social constructivist theory to show you how heated things can get in STS.[17] In a weird way, some sociologists came to criticize other social constructivists by pointing out that social constructivism is itself socially constructed. This sociological movement was called the 'strong program'—because it was an extra strict form of constructivism—but it also became associated with the sociology of scientific knowledge (SSK) as a whole. Their basic point was that when social constructivists point out how science is a socially constructed system of meaning that could not correspond to reality, those same social constructivists thought that their own theories—their claims about how science is socially constructed—were in fact *not* socially constructed! To the strong program of SSK, this appeared to be a double standard. Why should social constructivists get to say that their theorizing about science corresponds to reality, whereas the theorizing of scientists about nature could not? Why aren't the social constructivists themselves socially constructed? Why are their factual claims about socially constructed science not themselves social constructions or any less problematic than naive realism?

The strong program called this double standard "ontological gerrymandering." The word 'ontological' means pertaining to the study (-logical) of reality or existence (ontos), while 'gerrymandering' is a word from politics. It means to re-draw electoral boundaries to

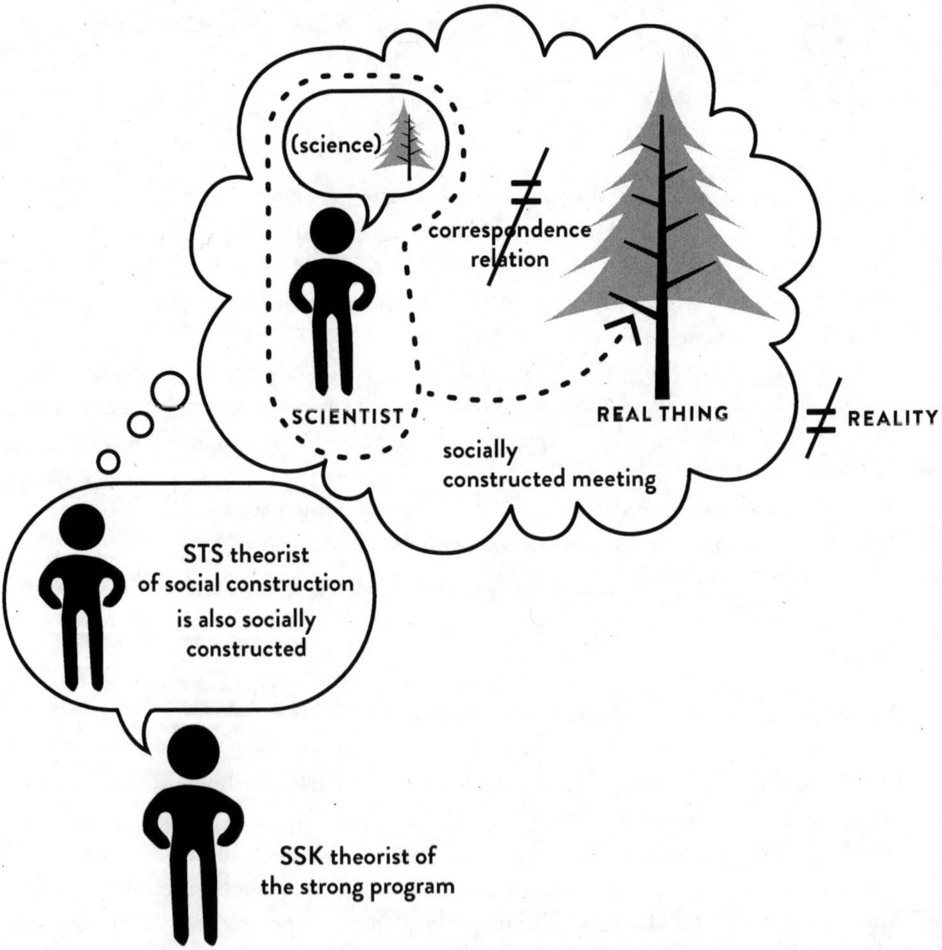

make it easier for a specific political party to win an election. Ontological gerrymandering, therefore, means redefining what's real so that it's easier for you to win an argument about what's real. If social constructivists think that they aren't socially constructed themselves while scientists are, then they're ontologically gerrymandering. They're being unfair to scientists by self-servingly redrawing the boundaries of what's real: social construction is really real while scientific descriptions are not. Pretty convenient, hey?

According to the strong program, social constructivism which ontologically gerrymanders is being unfair to science. To overcome this unfairness, social constructivists need to apply social constructivism to themselves. But if being socially constructed means that your theorizing cannot correspond to reality, then social construction theory cannot be said to correspond to reality either: "just as sociologists recognize that the language and concepts of their subjects are social constructions, so too are the language and concepts used by the sociologists themselves."[18] If sociologists recognize this, then they avoid ontological gerrymandering—and that's good, because avoiding ontological gerrymandering levels the playing field between the social constructivists and the scientists they study. That's being fair. However, avoiding ontological gerrymandering also means that social construction itself cannot be any more true than socially constructed science is. How could anyone step outside of social constructivism itself to see, by comparison, if *it* is a correct model of reality? We can't, and therefore everything's relative, including social construction. We can't know if science is true, and we can't know if social construction is true. Same with this book: we can't know if it's true either.

If that's the case, then maybe we do just make everything up. Science is an invention of the human mind, and so is STS (and this book). Nothing is true, nothing is real. This might lead to what some scholars call 'scientific anarchism,' because it appears that nothing has any authority, not even science. That might sound worrisome to you. The original scientific anarchist—Paul Feyerabend (1924–94)—didn't find this worrisome at all, however. Strictly speaking, he wasn't even a social constructivist; he thought that science was actually, not just possibly, false:

> We have now discovered that science has no solid results, that its theories as well as its factual statements are *hypotheses* which are often not just locally incorrect but entirely false, making assertions about things that never existed.[19]

Feyerabend knew that modern science was a goal-oriented enterprise of value and meaning attribution, but he refused to assume that the goals of modern science were better than other ones (say, for example, the goals of Aristotelian science), or that modern science has achieved its goals better than other meaning systems have. As we see from its history, science is "a fight between ... opposing schools [that] is a power struggle, pure and simple." Therefore, "the case of science vs. witchcraft (for example) is still entirely open."[20] There's no way to way to rationally resolve a power struggle any more than you can do diplomacy with monkeys throwing turds. In the end, science has equal authority—i.e., none at all—alongside poetry, baking cookies, naturopathy, exorcism, or drinking bleach (etc.).

Again, Feyerabend doesn't think this is a bad thing. Shouldn't we *want* to be able to make anything up that we want to? Isn't knowledge power and power knowledge, after all? If humanity is the measure of all things, then why should nature—or reality—be the measure of anything?

> Taste, not argument, guides our choice of science; taste, not argument, makes us carry out certain moves within science.... There is no reason to be depressed by this result. Science, after all, is our creature, not our sovereign; *ergo*, it should be the slave of our whims, and not the tyrant of our wishes.[21]

As far as Feyerabend is concerned, there's simply no way to determine which scientific theory or description is better than another, and what a relief that is! We no longer have to worry about scientists being right or wrong; at long last, Western technoscience has finally achieved the power Bacon always wanted: it gives us the power to do anything we please. In this respect, radically relativist constructivism fulfills the deepest yearnings of modern Western technoscience. Why then complain about relativism?

The Science Wars

If we do complain about the relativistic social construction of science, perhaps that's because we're not as modern as we think we are. Maybe we don't want knowledge as power after all; maybe we want the truth for its own sake, and not just as a tool for our own devices. Maybe there's a way for the world to tell us something true about itself through science, even though we're not yet sure how that's possible.[22] Maybe there's a lot more in common between us moderns and our medieval and classical forebears than the Enlightenment—or even social constructivism—would have us believe.[23] But this wasn't the way most scientists responded to relativistic social constructivism about science. They argued that science is a true description of reality *because it works*, which only made their problems worse. Richard Dawkins, for example, argued that

> Western science works.... When you take a 747 to an international convention of sociologists or literary critics, the reason you arrive in one piece is that a lot of Western-trained scientists and engineers got their sums right. If it gives you satisfaction to say that the theory of aerodynamics is a social construct that is your privilege, but why do you entrust your air-travel plans to a Boeing rather than a magic carpet? As I have put it before, show me a cultural relativist at 30,000 feet and I will show you a hypocrite.[24]

Dawkins doesn't understand that if science's truth and authority come from its practicality (like making reliable airplanes), then knowledge is power rather than truth. Technoscience is just our ability to force the world to do what we want it to do, which reveals nothing of what it might be like in itself, objectively. Dawkins doesn't realize that he's a nominalist, not a realist (we'll say more about this at the end of this chapter). Moreover, if science is true and authoritative because it gives us the power to fly (etc.), then the (supposed) authority and truth of science derives simply from its role in a power struggle between incompatible meaning frameworks and goals (as if flying is obviously a better goal than knowing the mind of God). This is *exactly* what Feyerabend says about modern science. Dawkins walks unwittingly into the arms of scientific anarchism.

The fact is, however, that social construction never denied that technoscience works. It simply denies that we are in a position to say science is true, even if it works. Social construction's point is like Hume's point about induction: inductively, we like to think that we 'know' the Sun will rise tomorrow, because we've seen it rise a million times before. But there is no 'objective' indication from anything in the world that the future *has* to be like the past. We have no *empirical* data of the Sun rising tomorrow until it does, actually, rise tomorrow (and that hasn't happened yet). So just because something (like induction, or aerodynamics) works doesn't mean we have an actual indication that it is true or tells us what is real (see Chapter Fourteen). Similarly, theorists of social construction do

> not believe that artifacts, such as airplanes, engineered in the light of scientific knowledge, usually fail to work. Constructionists are creatures of Humian habit. They expect airplanes to get you there, and know that science, technology, and enterprise are essential for air travel.[25]

What social construction denies is the belief that because something is powerful, that thing is therefore true and gives us access to reality as it is in itself. Power tells us more about ourselves, our values, and our culture than it does about the world. Recognizing the social construction of science refocuses our inquiry on these social realities, realities that science is otherwise inattentive towards because it's busy trying to discover the (supposedly) objective truth. While science may be preoccupied with *that*, social construction is at least an inducement to consider other dimensions of science—including *why* modernity fetishizes the achievement of absolute knowledge.

Moreover, social constructivists aren't necessarily scientific anarchists. Even the strong program of SSK only holds that social construction is just as socially constructed as science is. If meanings but not objects are socially constructed, then scientists like Dawkins don't have to worry that sociologists are denying the absolute objectivity of *the external world*; they're just denying the absolute objectivity of *science*: "relativism was not meant to oppose realism but rather to oppose absolutism. Somehow, critics and colleagues too easily forget this."[26] All they're saying is that science and the external world are not (and cannot be) the same thing. It's clear that there are "many science-haters and know-nothings" out there who would love to latch onto relativistic social construction to delegitimize science,[27] but it's not clear that relativistic social constructivists are science-haters and know-nothings:

> They [relativistic social constructivists] may query some self-serving images of science that are in circulation, and exalted pictures of what scientists do, why they do it, and how they do it. That is very different from doubting the truth or applicability of any propositions widely received in the natural sciences. If they are social constructionists, they are so at 30,000 feet.[28]

The science wars did, however, force STS scholars to defend social construction theory in order to distance it from science-haters and know-nothings. They tried a variety of ways to do this. Some scholars argued that constructivist relativism isn't radical skepticism, but rather methodological disinterestedness: "two of the founders of 'relativistic' sociology of science, Barry Barnes and David Bloor, defined relativism as 'disinterested inquiry,' a classical definition of science."[29] Just as natural scientists aren't supposed to assume that their theories or hypotheses are true before they experimentally test them, social scientists aren't supposed to assume that natural science is automatically true while they're studying natural scientists. Methodological disinterestedness is pretending that you don't care what the outcome of your inquiry is going to be, so that you can be more objective or neutral in the process of discovery. In this respect, STS (arguably) has "scientific traditionalism at [its] core," and its "core researchers" are explicitly committed "to the scientific worldview."[30] In the context of objectively examining natural science as an object of study, STS scholars should withhold judgment about the truth or falsehood of claims made in the natural sciences, right? Rather than being strict constructivists (like the strong program insists), STS could simply be *contextually* constructivist.

However, if all social constructivism amounts to is bracketing our beliefs about the truth of science, then most everything social constructivism has said about the social construction of meaning and the limitations of symbolic systems turns out to be irrelevant to social constructivism. If you're just talking about methodological disinterestedness, then you don't have to get into issues about meaning, symbols, realism, representation, or cats. Contextual constructivism misses the 'constructivism' part. Other STS scholars argued, therefore, that even though all knowledge claims are socially constructed, some social constructions are *better* than others. That way, modern technoscience can be (sort of) true and authoritative, even while it's socially constructed:

> [T]he evidence brought to bear to support the chemistry professor's claims takes very different forms than does the evidence presented by UFO advocates: Chemists can make and test all manner of hypotheses that might be disproven, we can see that their claims have predictive value, and we can therefore have considerable confidence that their description of the world is accurate. To be sure, chemistry has evolved through centuries of research; we can see this as a long process of social construction, guided by (socially constructed) principles for scientific thinking that allow chemists to develop knowledge in which we have considerable confidence. Modern chemistry

> is a social construction, as was Aristotle's model of a world composed of four elements, but they are not equally useful for making nylon, and Berger and Luckman would not have understood them as having equal standing as descriptions of empirical reality.[31]

Therefore, just because social construction applies equally to chemistry, UFO conspiracy theories, Aristotelianism, or witchcraft doesn't mean that they all have equal value. Some actually produce useful outcomes (like nylon or airplanes), some don't.

These better or useful outcomes, however, are not objective, neutral, or obviously the right ones. They're values masquerading as facts. Is the usefulness of nylon universally recognizable as an obviously uncontroversial value in all possible historical and cultural contexts? Or does nylon only 'work' in some particular contexts where some particular interests consider nylon to be a 'solution' to some 'problems' those interests face? For whom is nylon 'useful' (see Chapter Twelve)? This returns us to the problem of saying science is true because it works: if some social constructions are 'better' than others on Baconian grounds of 'usefulness,' that's like saying nylon is better than formal and final causation because nylon is more useful for making hot air balloons. But who's to say that hot air balloons are better than knowing the purpose and goal of all things? This brings us right back to the power struggles of scientific anarchism with no resolution in sight.

Finally, some sociologists simply resorted to arbitrary fiat and declared that they were going to ontologically gerrymander anyway because if they didn't, they couldn't do sociology anymore. If sociologists were to avoid ontological gerrymandering, they'd have to admit that their social constructivist analyses of science couldn't be true. Conceding this point would rob sociology "of its empirical grounding and significance. This reflects a widespread resistance to the idea of sociology as a discovery science."[32] If sociologists of science are held to a strict constructivist standard of relativism about their own claims, that would make STS incapable of "contributing to our understanding of the world as we have traditionally understood that pursuit."[33] Clearly, it would be unacceptable (for STS) if STS couldn't continue with its traditional ways of operating, and so

> analysts accepted the basic constructionist insight that people assigned categories and meanings to the empirical world, and they sought to observe and understand this process by locating claims in their larger context. They conceded that ontological gerrymandering occurred but saw it as a necessary evil, something to be aware of, to be done with caution.[34]

In the name of maintaining tradition, therefore, the (not strong) program of social constructivist STS must be allowed to point out that natural science does not contribute to our understanding of the world (as we have traditionally conceived that pursuit) while STS does. How convenient! And what a mess.

Contingency vs. Conditional Inevitability

As mentioned earlier, this book is not trying to win the science wars. I can only hope to explain what the wars were fighting about, and maybe—just maybe—prepare us a little bit for getting to a place where we might be better informed about science as a social activity, even if we don't have all the answers. This is what the philosopher Ian Hacking can help us do. He boiled down the core of the disagreement between social construction and the usual story into what he called basic "sticking points," potentially irresolvable bones of contention between scientific realists and relativists. His hope was that if we can be clear about what those sticking points are, at least we won't be confused about what we're fighting about when we're wrestling with social construction. At least we'll know what the disagreement is fundamentally about, even if we're not sure (yet) how to resolve those fundamental disagreements. So let's finish this chapter by looking at those sticking points.

The first sticking point is *contingency*. Socially constructivist STS sees scientific progress as contingent, where the usual story sees progress as conditionally inevitable. What does this mean? According to Hacking, social constructivists don't just say that scientific *ideas or concepts* are socially constructed. They also say that the *objects* which scientists discover are socially constructed, but not literally. Rather, the reasons why some objects are discovered by science, and not other objects, are *sociological reasons*. Unique and particular historical, cultural, and even philosophical factors lead to the discovery of some facts and not others. For example, not only is the notion of a dwarf planet (like Pluto) socially constructed, it's also an historical and social accident that Pluto was ever discovered at all. Of course scientists will agree that nobody *had* to discover Pluto. But *if* there were astronomers and *if* they had the right telescopes, *then* they'd discover Pluto at some point, right? This is what 'conditionally inevitable' scientific progress means: *if* you're going to do science, then *eventually* you'll discover all the same things that Western technoscience has discovered so far. If you replaced all of humanity in the past, present, and future of Earth with velociraptors who liked to do astronomy, they'd eventually discover Pluto too and eventually recognize it as a dwarf planet. Western technoscience is therefore an inevitability *if* anybody does any 'real' science at all.

Social constructivists are *not* conditional inevitablists about science. They do not think it's inevitable that all scientists will eventually turn out to be like modern or Western ones, or discover the things modern Western technoscience has discovered, like Pluto or quarks or whatever. The discovery of Pluto or quarks is dependent on things like social factors that didn't have to happen. That's what *contingency* means: something that didn't have to happen. It's historically, culturally, even philosophically *accidental* that we have modern scientists, or telescopes, or clear skies, or a thousand other things besides that make finding Pluto possible. Therefore, there's no reason why Pluto would *have to be* discovered when people (or dinosaurs) do science for long enough.

This means there are a thousand different ways to do real, legitimate, solid science *other than* the way Western modernity actually happened to do it. Even if your culture

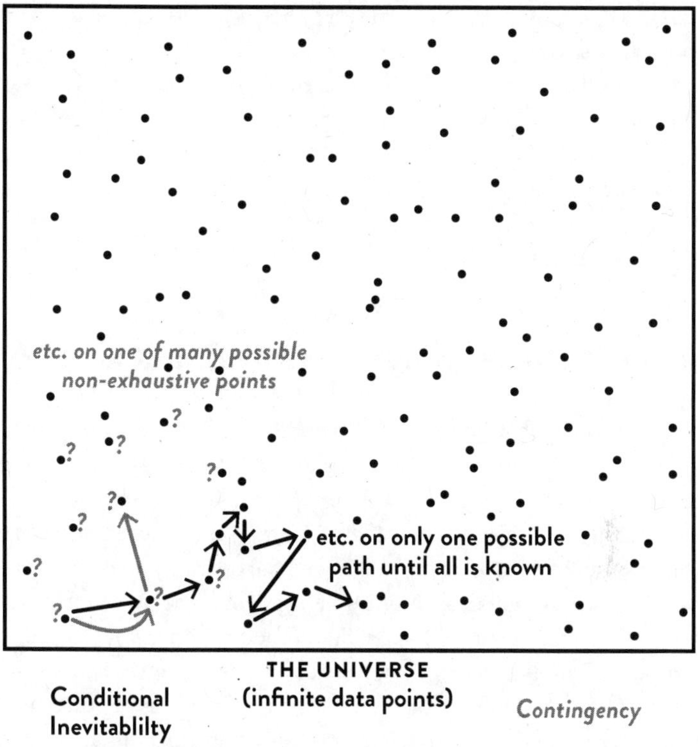

has astronomers, they don't have to do astronomy in a telescopic way. Telescopes might not get invented—and even if they do, astronomers might not want them—and even with them, telescopes might be pointed at other things. Science can discover all kinds of things, but different science under different conditions might discover different things. That's why social constructivists believe that it's contingent that modern Western techno-science exists and has turned out the way it has. And so quarks and dwarf planets may not exist to a particular culture even if that culture is highly scientific.

In 1984, Andrew Pickering wrote a book called *Constructing Quarks*.[35] He's both a physicist and an STS scholar, having doctorates in both disciplines. He argues that quarks, the basic building blocks of reality according to the standard model of subatomic physics, are socially constructed. The discovery of quarks happened because, in the 1970s, physicists stopped using bubble chambers to track particle decay and started using new detectors. The new detectors are what they discovered quarks with. But Pickering thinks the physicists *could have* kept on using the old bubble chambers instead, and physics *would have* kept on advancing and discovering new things without ever discovering quarks: "Physics did not have to develop in a quarky way."[36] This means that the *context* of discovery has an effect on the *content* of discovery. There are an infinite number of possible discoveries we can make about the universe, and finite human scientists cannot discover them all. The paths that science charts through these possibilities is contingent upon all kinds of social factors, and each path might reveal something different.

This does not mean that quarks aren't real or that scientists invent them out of nothing. It means that discovering quarks isn't inevitable: "alternative 'successful' science is in general always possible."[37] If that's what you think, then that's one reason why you're a social constructivist. If you think all other possible ways of doing subatomic physics (including the original bubble chamber way) will eventually hit dead-ends until someone discovers quarks, then that's one reason why you're not a social constructivist. Hacking doesn't attempt to resolve this difference of opinion. He just points out that, hey, at least we now know what these people are arguing about.

The End of the Medieval Synthesis All Over Again

I'd like to add, however, that I don't understand how anyone could possibly *know* that there's only one possible path for physics to take. You'd have to have some serious faith in a particular metaphysical picture of the universe, and some serious faith in a particular epistemology, to be able to know that modern Western technoscience is the only way the universe can be truly known. This is why the other set of major sticking points takes us back (again) to the European Middle Ages and early Western modernity. As you may or may not recall from Chapter Eight, medieval natural philosophers—aka *scientists*—disagreed about whether there were any universal truths in the abstract realm of the absolute (the mind of God), or whether the truth was just a powerful tool to get the results that we wanted, and not a reflection of what was ultimately real. We used the terms *rationalism*, *empiricism*, and *nominalism* to map this dispute, which ranged from Franciscan monks to Francis Bacon (Chapter Nine) and David Hume (Chapter Fourteen). In this chapter, we've added the term (naive or scientific) *realism* to the mix as well. My apologies if you were hoping to be finished with all that history, but there's a reason why we learned it: Hacking's point is that "the present science wars, especially as they hook up with social construction, have strong resonances with traditional philosophical issues."[38] That is to say: the problem of social construction in science is older than Western technoscience itself.

Maybe a distinction will help: the issue here that divides the social constructivists from the usual story can be understood both metaphysically (pertaining to the nature of reality) and epistemologically (pertaining to the nature of human knowledge). That means there are four possible combinations in this fundamental disagreement between constructivists and anti-constructivists: 1) metaphysical anti-constructivism, 2) metaphysical constructivism, 3) epistemological anti-constructivism, and 4) epistemological constructivism. It would be nice to give each of these a simpler name, so let's use terms we've already encountered: 1) realism, 2) nominalism, 3) rationalism, and 4) empiricism, respectively. The problem with 'realism,' though, is that nobody can agree on what it is supposed to mean (unless it's qualified in a lot of different and cumbersome ways, like adding 'naive' or 'scientific' to it, as we've already done). Instead of 'realism,' therefore, Hacking suggests calling metaphysical anti-constructivism "inherent-structurism," because that word is so ugly that no one else will want to use it in another way.[39] It is more common in STS, however, to call it *essentialism*, for the idea that science gives us access to the *essence* of reality.

With these four terms in play, we can make a nice chart of these sticking points:

	Anti-Constructivism	Constructivism
Metaphysics	essentialism	nominalism
Epistemology	rationalism	empiricism

Another problem, however, is that the connection between rationalism, empiricism, and social construction isn't yet clear. What I will do, therefore, is explain *essentialist rationalism* as one side of the disagreement (the first column), and then *nominalist empiricism* on the other (the second column). First, essentialist rationalism.

An essentialist believes that there is an objective, external reality outside of our mental representations (the *logos*), and that external reality is solid, stable, and discrete: these ultimate realities are the essences of things. (This is a metaphysical position.) If you want to know what a zebra *is*, you find out what makes a zebra objectively a zebra and not a horse or a quagga. If it starts to look like there might not be a clear distinction between a zebra and a horse, or a zebra and a quagga, then you might have to conclude that the two share the same essence—an abstract universal identity—in spite of any initial appearances to the contrary. The essence of a thing, then, will be its ideal and absolute definition in terms of necessary and sufficient conditions, and when science discovers what these are, science is true. (This is an epistemological position.) Scientific truth is obtained when a scientific statement correctly describes the essence of something objectively real. (This is a combination of correlative metaphysical and epistemological positions.) This is *not* what social constructivists think, however; this is what the usual story thinks.

Why is metaphysical essentialism epistemologically 'rationalist,' though? Well, in the first place, essentialism believes in an abstract order of truth that is beyond what we can see with our senses. For example, nobody can sense the genome of a zebra with their eyes or ears, but we can—through reason, presumably—figure out what it is anyway using a variety of intellectual methods. Secondly, scientists of this usual sort think that the universe *has to be* this way. If the universe *wasn't* full of objective facts ready to be discovered, then how could science even be possible? The scientific essentialist, therefore, has to believe (or think they know) that the universe is organized in a particular way *in advance* of their experience of it, if science is going to be possible. That is, they believe 'the truth is out there.' The truth *has to be* out there, somewhere, even if they haven't found it yet: "The whole point of [scientific] inquiry is to find out about the world. The facts are there, arranged as they are, no matter how we describe them."[40] We may discover what those particular facts are only after we experience the world *a posteriori*, but we know *a priori* that there are facts there for us to discover. The truth has to be out there, every event has to have a causal explanation, and properly disciplined rationality has to be able to reveal these truths to us (otherwise, what's the point of science?). This is why the usual story is, in Hacking's view, epistemologically rationalist even though science's experimental method is inductive and its knowledge of individual things is empirical. A rationalist, remember, knows *most* things empirically, but also believes they can know *some* things through reasoning alone.

This essentialist-rationalist way about thinking about science is so deeply embedded into modern culture that it probably seems like common sense to you. Science is true (epistemologically) because science correctly describes the way things really are (metaphysically). And how do we know science correctly describes the way things are? Because science works—not only in Dawkins' sense of making airplanes that reliably fly, but also by producing reasons and evidence that conclusively prove scientific statements:

> Rationalists think that most science proceeds as it does in the light of the good reasons produced by research. Some bodies of knowledge [like science] become stable because of the wealth of good theoretical and experimental reasons that can be adduced for them.[41]

The enduring *reliability and stability* of modern scientific knowledge is supposed to be a testament to its truth because that stability presumably comes from *inside* the things themselves, where the essences are: from inside the technoscientific enterprise and the objects it manipulates. Reality is *telling* us, in a sense, that science is true. This is what Hacking means when he says essentialist rationalists give *internal explanations for scientific stability*. The reason why essentialist rationalists think science (as a body of knowledge and a method for discovery) is so authoritative and effective—i.e., why it is solid instead of being a random and unreliable source of obvious bullshit—is because the universe actually *is* the way scientific reason says it *has to be* if science is going to work (and it does work). Scientific rationalists believe, therefore, that the reasons for believing science is true come from *within* science and objective reality itself.

Social constructivists do not believe this, even though they believe that science is a stable, reliable, and authoritative body of knowledge and method of discovery. Metaphysically, Hacking thinks, social construction holds that what is real cannot be known *a priori* to have an objective structure which finite human language and modeling can accurately mirror:

> The world is so autonomous, so much to itself, that it does not even have what we call structure in itself. We make our puny representations of this world, but all the structure of which we can conceive lies within our representations.[42]

Of course we might want to believe that our scientific statements are true because they accurately reflect what is objectively real, but social constructivists argue that nobody is in a position to know what is objectively real, or to compare that objective reality with what people think it is and then see if the two correspond. It would be extremely unlikely if reality actually turned out to be exactly or even mostly the way some intellectual primates on a randomly hospitable rock in the corner of a galaxy in some small part of the universe think it is (although we can never be entirely sure that it isn't).

It's far more likely, the social constructivists say, that our technoscientific theories are themselves just helpful tools that make navigating our lives on the aforementioned space rock a little easier. But survival value is not truth value, and so the social constructivists are best understood as *nominalists* about what is real: there are no 'absolute forms' out there, no essences, and science (like everything else) is the *projection* of useful but human categories of order onto an unknowable chaos. (This is a metaphysical position.) Science is in the business of projecting *names* onto things because that helps *us* make sense of *our* world. But just because science is successful in this endeavor does not mean *it* tells us the truth about any *objective* reality in *the world itself*. We are a part of the world we are making manageable for ourselves, and we can never get outside of our world—and ourselves—to see if our success at managing our world means that our management models are correct.

Therefore, social constructivists can and do believe that science is a powerfully stable and authoritative way of describing things, but this authority and power does not come from within the described things themselves; it rather comes from outside those things and is imposed onto them by powerful and authoritative external forces. This is what Hacking means by an external source of scientific stability. Social constructivists believe that science is a stable body of knowledge and an authoritative method of discovery, not because of the way objective reality actually is, but because of the way social human beings are. With some difficulty, Hacking calls this an "empiricist" orientation. An empiricist, remember, doesn't believe that a human being can know anything through reason alone; everything that can be known must come to us through our five senses (this is an epistemological position). David Hume, the arch-empiricist, pointed out that nobody can possibly perceive the cause of an event empirically (see Chapter Fourteen). Your five senses never show you what the cause of anything is. A rationalist believes that there has to be a cause anyway (they think they know this a priori) and reason will show them what this is even if their senses don't. Empirically, however, we can't ever know what the cause of anything is, even if it's 'human nature' to want to believe there's one anyway. We just habitually assume that all events will have causes. This customary and traditional expectation, however, does not originate from within the events we observe; it originates in us. An honest empiricist, therefore, won't claim to know what has to happen because the universe told them so. An honest empiricist will just wait and see what happens, all the while noting that our powerful confidence that things will happen the way we predicted comes from the human power to construct explanations that work for us. *We* are the powerful external force that creates scientific stability. Our traditions, customs, and habits (etc.) are what supply effective causal explanations. Social constructivists are the actual empiricists, not the scientific 'realists' of the usual story.

Arguably, actual scientific practice more closely resembles the nominalist-empiricist account than the essentialist-rationalist account. Take those zebras again. If we want to know what a zebra is in general, we are usually referring to the *species* which makes any particular animal a zebra and not a walrus or a cabbage. Nobody has ever seen (or heard, tasted, smelled, or touched) a species before, but a rationalist thinks that the abstract reality of the species can still be known, and an essentialist thinks that the species is an abstract essence that is actually 'out there' in objective reality. Therefore, the notion of species—and particular species, like quagga or plains zebra (never mind the fancy Latin names)—are stable categories because they reflect objective essences that really exist.

Unfortunately, biological essentialism is bunk, and not just because it turns out that (whoops) quagga are an extinct variety of plains zebra. No species have "fixed essences," because they are all "thoroughly heterogeneous collections of individuals whose phenotypic properties [change] over time, and [vary] across the population at any given time."[43] Indeed, that variation within species is the reason why they can evolve. If species were precise essential definitions, evolution would be impossible, because essences do not change. Therefore, nothing in nature—not even genetics—can tell you exactly, essentially, or absolutely what a zebra is. Only a scientist can. As one STS scholar once said (I'm paraphrasing), a species is whatever scientists say it is.[44] That's nominalism. But it doesn't mean that species are just random labels that biologists invent, as if they could just make quaggas and plains zebras be separate species. Nominalists have to make do with the empirical world, just like everyone else: "The scientific nominalist [has] to accommodate, constantly, to the resistance of the material world."[45] The point is rather that the strong and rigorous reasons for why quaggas and plains zebras are the same species ultimately come from biologists, not from quaggas or plains zebras.

This leads into a particular school of thought within STS, and perhaps the most influential contribution of STS to scholarship in other disciplines: *actor-network theory* (also known by its initials ANT). Actor-network theory was most famously proposed by the theorists Bruno Latour and Michael Callon as a kind of rebuke to social constructivism. Social constructivists assume that the human or social realm is distinct from the objective or physical realm, and that we are ultimately cut off from knowing what the objective realm is in itself. Actor-network theory denies this distinction; after all, we're just as much a part of the world as the other objects in it. But this means that the distinctions between society and nature, subject and object, active and passive, and even modern and medieval break down. However, ANT shares with social constructivism an emphasis on anti-essentialist explanations for scientific stability: in general, it sees successful technoscience[46] as the result of creating extensive and effective networks of power between a wide variety of actors.

> **Actor-Network Theory (ANT):** Technoscience is the arrangement and association of human actors and nonhuman actants into robust networks that achieve a goal for a set of interests.

Some of these actors, of course, are scientists, but many are not. In fact, many of the actors in technoscientific networks aren't even human: they're nonhumans, like lab animals or chemical compounds. Traditionally, Western technoscience has thought of these nonhuman actors as 'objects,' inert particles that are subject to the force and manipulation of people (aka 'subjects'), as we saw with the mechanistic paradigm in Chapter Eleven. But animals and chemicals (etc.) are no more inert than people are; all of them have capacities and abilities, and in that sense, they even express kinds of interests. (Nitroglycerine, for example, has an 'interest' in exploding when related in a precise way to heat, shock, or flame.) If you really want to insist on a distinction between humans and nonhumans, you can call humans 'actors' and nonhumans 'actants' (as ANT does), but either way, the point is that everything in the universe has potentialities and powers, and technoscience is just one of the ways of arranging these various items into useful networks.

A scientist will interact with other scientists, graduate students, lab technicians, university administrators, governments and corporations, and even their own historical and ideological commitments. Through these various networks of meaning, a (techno) scientist or group of (techno)scientists will seek to establish a new configuration of material actants that will be seen as valuable by a certain group of people. Technoscience may seek to create a self-driving car, for example, and to do this it will create a new network out of a great many different things, including the tiniest electronic bitty-bits all the way up to the biggest frameworks of transportation management and oversight. If all these actors and actants can be made to do the thing as intended, then the technoscience succeeds: "a stable network ... is the result of managing all of these actors and their associations so that they contribute towards a goal."[47] This goal-oriented success does not mean, however, that technoscience is true. It's just useful (for someone), which is presumably nice.

Nevertheless, once the stable technoscientific network is in place, it becomes incredibly hard to dispute or dismantle it. Any particular technoscientific configuration of the world—from quarks to quantum computing—is held together by materials, measurements, modeling, manipulations, translations, laboratories, conferences, publications in peer-reviewed journals, royalty agreements, funding structures, advertising, market share, and everything in between. If you want to deny that there are any quarks or quantum computers (for example), you will have to pull apart all those actors and all those actants and undo all those relationships, which is incredibly difficult. That's why technoscience is stable: not because these "new alignments of forces" are inherently true reflections of the ultimately real,[48] but because very few people are in a position to say (or act) otherwise. Science is stable because 'external' factors (factors other than the 'essence' of things themselves) make it stable.

So there, that's (arguably) what the fuss is all about. Either there's only one way to do science, or there's many. Either the world is of an essence that allows us to know what it really is, or it isn't. Either the world tells us what is true, or we do. It's not my job—indeed, it can't be—to tell you which of these options is the right one. That's far beyond the scope of a book like this. What's within its scope is getting *you* to wrestle with these questions rather than ignoring them. There are plenty of very public disputes about the role of science and technology in our society, but we aren't going to resolve these if we ignore the reasons why those disputes exist in the first place. STS at least brings those reasons into the light. Whether catastrophic climate change is anthropogenic, or hydroxychloroquine is a good therapy for COVID-19, is going to depend at least partially on some metaphysical and epistemological position you take regarding the status of science. The science wars of the twentieth century already are and always were part of the culture wars of the twenty-first century. It's up to you (singular and plural) to figure out how we're all going to get through it together. That's why reading this book doesn't conclude STS for you. It's here to *start* STS for you. You're smart too.

Notes

1 This is Jean-François Lyotard's famous definition of postmodernism: "Simplifying to the extreme, I define *postmodern* as incredulity toward metanarratives." From *The Postmodern Condition: A Report on Knowledge*, trans. Geoff Bennington and Brian Massumi (University of Minnesota Press, 1984), xxiv.

2 The notion of the 'two cultures' is notoriously controversial, not least because in the science wars, many philosophers of science were opposed to social construction, just as much as the natural scientists were. You might even say that the science wars were between the philosophers and natural scientists against the social scientists, but at any rate, that's not how the public perceived the debate at the time.

3 Kellyanne Conway, *Meet the Press* interview, 22 January 2017.

4 This is the subtitle of Steve Novella's book *The Skeptics' Guide to the Universe: How to Know What's Really Real in a World Increasingly Full of False* (Grand Central, 2018).

5 Peter L. Berger and Thomas Luckman, *The Social Construction of Reality: A Treatise in the Sociology of Knowledge* (Doubleday, 1966).

6 Wenda K. Bauchspies et al., *Science, Technology, and Society: A Sociological Approach* (Blackwell, 2006), 24.

7 Sal Restivo and Jennifer Croissant, "Social Constructionism in Science and Technology Studies," in *Handbook of Constructionist Research*, ed. James A. Holstein and Jaber F. Gubrium (Guilford, 2008), 221.

8 Bauchspies et al., *Science, Technology, and Society*, 20.

9 Raymond Murphy, "Opinion: Lessons of the 1998 Ice Storm," *Montreal Gazette*, 14–15 January 2013, https://montrealgazette.com/opinion/opinion-the-lessons-of-the-1998-ice-storm.

10 Bauchspies et al., *Science, Technology, and Society*, 19–20.

11 One concern during the voting process was that demoting Pluto from planetary status would "weaken the popularity of astronomy, and a decline of popular support could translate into a decline in research funding" (Bauchspies et al., *Science, Technology, and Society*, 21). This is an example of why STS theorists who argue that science is socially constructed conclude that science is *not* "independent of social, economical, political, and subjective influences" (Bauchspies et al., *Science, Technology, and Society*, 23–24).

12 Bacon, *Novum Organum* §124.

13 Bauchspies et al., *Science, Technology, and Society*, 24.

14 Bauchspies et al., *Science, Technology, and Society*, 25.

15 "Thinking God's thoughts after him" is a phrase attributed to Johannes Kepler in reference to his role as an astronomer, but historians of science have not been able to locate an original source where he says this in exactly those (translated) words. The closest candidate is "Those laws [of nature] are within the grasp of the human mind; God wanted us to recognize them by creating us after his own image so that we could share in his own thoughts." Johannes Kepler to Bavarian chancelor Herwart von Hohenburg, 9 and 10 April 1599, in *Johannes Kepler: Life and Letters*, ed. Carola Baumgardt (Victor Gollancz, 1952), 50.

16 Bruno Latour and Steve Woolgar, *Laboratory Life: The Social Construction of Scientific Facts* (Sage, 1979), 180, 182; quoted in Joel Best, "Historical Development and Defining Issues of Constructionist Inquiry," in *Handbook of Constructionist Research*, 46.

17 Bauchspies et al., *Science, Technology, and Society*, 24.

18 Best, "Constructionist Inquiry," 47.

19 Paul Feyerabend, "Theses on Anarchism," in *For and Against Method: Including Lakatos's Lectures on Scientific Method and the Lakatos-Feyerabend Correspondence*, ed. Matteo Motterlini (University of Chicago Press, 1999), 114.

20 Feyerabend, "Theses on Anarchism," 117.

21 Feyerabend, "Theses on Anarchism," 117–18.

22 Nathan Kowalsky, "Science and Transcendence: Westphal, Derrida, and Responsibility," *Zygon* 47, no. 1 (2012): 118–39.

23 This is, very generally, Bruno Latour's point in *We Have Never Been Modern*, trans. Catherine Porter (Harvard University Press, 1993).

24 Richard Dawkins, "The Moon Is *Not* a Calabash," *Times Higher Education Supplement* no. 1143 (1994): 17.

25 Ian Hacking, *The Social Construction of What?* (Harvard University Press, 1999), 67.

26 Bauchspies et al., *Science, Technology, and Society*, 25.

27 Hacking, *Social Construction of What?*, 67.

28 Hacking, *Social Construction of What?*, 68.

29 Bauchspies et al., *Science, Technology, and Society*, 25.

30 Restivo and Croissant, "Social Constructionism," 216, 226 note 7.

31 Best, "Constructionist Inquiry," 45.

32 Restivo and Croissant, "Social Constructionism," 226 note 4.

33 Steve Woolgar and Dorothy Pawluch, "How Shall We Move beyond Constructivism?" *Social Problems* 33, no. 2 (1985): 162; quoted in Best, "Constructionist Inquiry," 48.

34 Best, "Constructionist Inquiry," 49.

35 Andrew Pickering, *Constructing Quarks: A Sociological History of Particle Physics* (University of Chicago Press, 1984).

36 Hacking, *Social Construction of What?*, 73.

37 Hacking, *Social Construction of What?*, 69.

38 Hacking, *Social Construction of What?*, 62.

39 Hacking, *Social Construction of What?*, 83.

40 Hacking, *Social Construction of What?*, 83.

41 Hacking, *Social Construction of What?*, 91.

42 Hacking, *Social Construction of What?*, 83.

43 Rob A. Wilson et al., "When Traditional Essentialism Fails: Biological Natural Kinds," *Philosophical Topics* 35 (2007): 193.

44 Gordon McOuat, "The Political Origins of Natural Kinds," public lecture, University of Alberta, Edmonton, Canada, 28 April 2009.

45 Hacking, *Social Construction of What?*, 84.

46 ANT also collapses the distinction between science and technology, making the term 'technoscience' doubly appropriate here.

47 Sergio Sismondo, *An Introduction to Science and Technology Studies*, 2nd ed. (Wiley-Blackwell, 2010), 82.

48 Sismondo, *Science and Technology Studies*, 85.

Chapter Sixteen

Not Just a Neutral Tool

ENOUGH WITH THE SCIENCE WARS, YOU MIGHT SAY, ENOUGH WITH ALL those annoying metaphysical and epistemological questions about science. Wouldn't it be better to turn our attention towards technology instead? It's a concrete and practical subject that we all have immediate, hands-on contact with, and (as we saw in Chapter Three) the critical study of technology in STS is less well developed compared to the critical study of science. That means there's more to be said about technology, in addition to it being more immediately relevant to our lives, than the abstract study of science. Examining technology might have been what interested us about STS in the first place, so let's get on with it.

Well, if you're *not* a social constructivist about science, you probably *are* a social constructivist about technology. I regret to inform you, therefore, that we're not yet done with social construction. Moreover, unlike the social construction of science, the social construction of technology is *part* of the usual story that you've probably grown up with! How so? We've already seen that the usual story says that science is objectively true, but it also thinks that technology is just the practical application of objective science to solve problems. In that vein, Herbert Feigl (a positivist; see Chapter Thirteen) claimed that

> [i]t is the social-political-economic structure of a society that is responsible for [the] various evils [of] ever more powerful weapons of destruction [and] the misery, physical and mental, of the multitudes [accompanying the] employment of scientific techniques in the machine age.... Scientific knowledge is itself

> socially and morally neutral. But the manner in which it is applied [technologically], whether for the benefit or to the detriment of humanity, depends entirely on ourselves.[1]

If science is objective, then technology is value-neutral. It's *entirely up to us* whether technology has good or bad effects, just like relativistic social construction thinks it's entirely up to us whether science is true or not (see Chapter Fifteen). We construct the goodness or the badness of a technology. It's a neutral tool, but how we use it is socially relative, and therefore so is its value. The social construction of technology isn't just that societies construct tools; that's obvious. It's also that societies get to decide if technologies are good or bad. There's nothing objectively good or bad about them.

Look, nobody said the usual story had to be consistent: it's anti-constructivist about science but constructivist about technology. Even so, there's a consistency of intent: in either case the usual story doesn't want you to worry. The neutrality of technology and objectivity of science go hand-in-hand because neutrality and objectivity suggest that there's nothing to think critically about. If science is objective, then everything has to be fine—it's not like we can change 'the facts' if we think critically about them. And if technology is neutral, then the only thing we need to think critically about is how we use it. Technology *itself* cannot be an issue of concern, because it's 'just a tool.' We'll invent and use technologies for one socially constructed reason or another, but that process has everything to do with *us* and nothing to do with *it*. In itself, therefore, technology is not a meaningful object of study. At least, that's what the usual story says.

Technological Liberalism

I'm going to call the usual story's position on technology 'technological liberalism.' Not only is liberalism the go-to position in Western societies for democratic politics, it's also the go-to position in Western societies for the value of technology: technology is both *free* (liberated) from bias, and we humans are free (at liberty) to use technology in any way we want to. This (presumed) double freedom with respect to technology dovetails nicely with the democratic freedom extolled by classical political liberalism: you can do whatever you want in a liberal society, unless it harms another person.[2] Everyone gets to choose for themselves what they think is good, and that freedom of choice is itself supposed to be good.

Liberty, therefore, is paradoxically both neutral and good. It's neutral, because liberalism does not care which vision of the good life you choose for yourself; it's good, because such open-ended freedom of choice is intrinsically good for individuals and generally leads to good outcomes overall. Same with technology: technological liberalism holds that technology is both value-neutral and usually good. Value-neutrality in this context means that technology does not tell us what we should do with it (e.g., "The computer does not impose upon us the ways it should be used"),[3] and this freedom to do whatever we want with it is intrinsically good and generally leads to good outcomes (unless it's used to hurt other people). "To our accustomed way of thinking," says STS scholar and political scientist Langdon Winner, "technologies are seen as neutral tools that can be used well or poorly, for good, evil, or something in between."[4] Technologies are objects, but *as objects*—in themselves, intrinsically, alone, and objectively—they have no value at all: "technical *things* do not matter at all.... We all know that people have politics, not things."[5] The only way technology is going to have value is *if we want it to*, and so again, in that case there's no reason to think critically about it as a thing.

Yet liberalism also sees this objective neutrality as good, so much so that it forgets to be neutral about technology: "Scarcely a new invention comes along that someone does not proclaim it the salvation of a free society."[6] Technologies are (in addition to being neutral) *inherently biased towards goodness*, which is why it's good that we should always have more of them—and more powerful ones. Science and its (presumptive) application as technology and industry are "the best guarantees of democracy, freedom, and social justice ... [they are] democratizing, liberating forces."[7] This means that boosterism (see Chapter One) is just as much a part of the usual story as technological neutrality is. Technological progress is *automatically good* (see Chapter Ten) and there's really no point in thinking critically about technology here either: everything will be fine, even if there are a few kinks (or side-effects) we'll need to iron out first.

This form of optimism about technology—especially technologies on the cutting-edge of innovation—is evident across the spectrum of politics and economic class. For example, Mark Zuckerberg (inventor-ish of Facebook and CEO of the Meta technology conglomerate) opined that

> [w]hen people are connected, we can just do some great things. They have the opportunity to get access to jobs, education, health, communications. We have the opportunity to bring the people we care about closer to us.[8]

Here, it's not up to you or I whether the Metaverse will bring all these beneficial connections to us. The Metaverse will (supposedly) just do that itself, because it's the Metaverse, and that's what the Metaverse does: it connects people. Zuckerberg's technologies will just make good things happen, because connecting people is good and Instagram, WhatsApp, and Facebook (etc.) are good at connecting people.

Similar things are said about guns in America. Individuals having the right to possess firearms, and the actual possession of those same firearms is an automatic guarantee of political freedom and personal safety. The professor of neurosurgery, Miguel A. Faria, Jr., claims that

> [c]ivilian disarmament is ... harmful to one's freedom and potentially deadly to one's existence.... While the United States and Switzerland have more guns per capita than any of the other developed countries, they also have more freedom in general than countries with draconian gun laws.[9]

Dr. Faria sees a causal link between gun ownership and "governments that sustain and affirm individual freedom" instead of being "despotic and tyrannical."[10] The value of a gun then, is not entirely up to the person who uses it and how it is used; it has intrinsic, political, and valuable effects just by virtue of being there. Guns make America (and Switzerland, we should suppose) the most politically free country(ies) in the world.

But surely you've heard the National Rifle Association's famous slogan, right? "Guns don't kill people, people do." If *this* is true, then even if guns *do* have an intrinsic value of generating political liberty, they *don't* have an intrinsic role in killing people. The good things about technology are automatic and inherent in them, but the bad things are incidental or accidental to the technologies themselves (that's why we call them 'side-effects,' by the way). This is how technological liberalism has its cake and eats it too: it remains relativistic about the potential *negative* effects of technologies by emphasizing contingent social factors and neutrality, but emphasizes inevitable progress and intrinsic value when advocating a technology's *positive* values. It's as if technology is super great, but we're in charge of that greatness somehow. As long as we're all responsible, law-abiding citizens, technological advancement is just going to make life better for us.

Neutrality and Ladenness

Technological liberalism evidently has a lot of issues to work through. So does STS, however. The preponderance of constructivist accounts in STS makes it less of an alternative to the usual story about technology than it is with respect to science. If the value of a technology is socially constructed, then there is no value in the technology itself. 'Objectively,' it's neutral. Perhaps only a naive realist would think they could identify the value of a technology in itself, 'out there' in 'the world' where 'the truth' is. Therefore, a sophisticated STS approach to the value of any technology should probably say that "What matters is not technology itself, but the social or economic system in which it is embedded."[11] To understand technology in a social context, we would only have to understand its social context and not the technological item per se. In terms of the 'externalist' form of STS (see Chapter One), the arrow of influence would always go from society towards technology and science (ST←S), and not the other way around (ST→S). Technology is not active, we are. The meaning or value of a technology must be found in what we project onto it, not what impact or influence it has on us.

One trouble with this socially constructivist approach to technology, though, is that it means that "there is nothing distinctive about the study of technology in the first place."[12] It's as if by doing constructivist STS, we eliminate the T as an object of inquiry. And yet STS actually does study technology and not simply the social context it is embedded in. As we saw in Chapter Fifteen, actor-network theory treats *everything* as actors (actants), even technologies, and not just human beings. So while it's likely true that many different social contingencies factor into what technologies we develop or how we use those technologies, constructivist analysis can and should be supplemented by consideration of how technologies might be "political phenomena in their own right."[13] So what happens when STS looks at technologies themselves?

First of all, it is clear that technologies have to be *good for something*. If they aren't good for anything, then nobody will want to use them. To call a useless thing a 'technology' or a 'tool' would be meaningless. At the level of mere use, though, technologies don't even have to be *designed*; we can just *find* objects that *are* good for something. This goodness for something is called an 'affordance' (see Chapter Three). An affordance is something valuable that we find embedded in an object, something that the object is good at doing—even if you don't want to do that thing. A fist-sized rock, for example, is good for throwing at glass windows (among other things), and it's good for that purpose even if you don't want to throw things at glass windows. It's also good for that purpose even though nobody designed or shaped the rock to be that way, and even though nobody designed windows to be rock targets. This affordance of this size of rock is what it is, therefore, regardless of what you want, project onto it, or socially construct it to be.

Of course, that affordance (throwability at windows) only exists in the rock because of its potential relationships (networks) with other things, like human hands and arms (for holding and throwing), panes of brittle glass (for breaking), and desires (wanting to throw rocks). This is a socially constricted context. But within that social context, the rock has that potential power inside itself, just waiting for someone to actualize it. So objects can embody a value even without anyone *making* them that way. Embodied value can be (in) a *found object*.

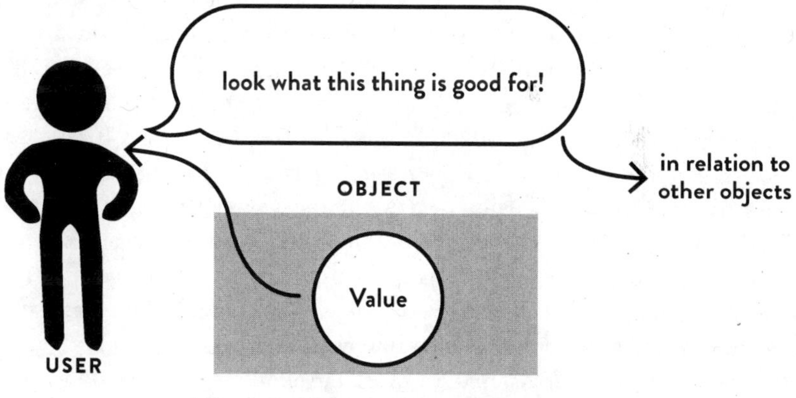

Affordance: found value embodiment.

Furthermore, this means that all technologies are *embodiments of power*. This is a specification of the 'good for something' aspect of affordances. If a tool doesn't have the *power* to do anything, it's useless, and as seen above, a useless tool doesn't make sense. If a tool is useful or good for something, the tool's ability to help obtain that good is its power, and that power is deemed a good by anyone who wants to

Reiterated Definition of Affordance: a power or potential use that is embodied or embedded in an object; see p. 47.

achieve what that tool helps them achieve. But what does 'embodiment' mean? It means to be in a body, a physical thing. An affordance, therefore, is a power or potential use that is embedded in a physical body or an object, like a rock. That powerful good is deemed to be *in* the tool, not in the user, otherwise the user wouldn't want or need to use the tool to help achieve their goals. In that case, they could just do the thing as easily by themselves, without using the tool at all. So the good—this value—is embedded in the object. It is 'objectively' valuable, at least from the perspective of the user. That's what value-embodiment means: a power that is present in an object, available to achieve a purpose or a goal.

Power embodiment is the opposite of value neutrality. Any object with an affordance is *value-laden*: it is *loaded* (laden) with values—or uses, goals, purposes, ends, or goods—whether or not anyone wants it to be. It embodies a set of powers willy-nilly. This is reminiscent of final causality in Aristotle (see Chapter Seven): all things, even inanimate, natural objects, have something good about them in relation to other things, and humans (among other beings, like otters and chimps) can often figure out what those goods are. Technologies are intrinsically value-laden, otherwise we wouldn't use them.

But now recall efficient causality in Aristotle: this is the agent cause, like an artisan. An artisan designs and builds an artifact. A *designer* can *make*—construct!—a tool to do something out of materials that wouldn't have been able to do that thing on their own. This is what an artifact is: an intentional arrangement of materials that contains the power to do something that the materials could not do in their original configurations. Let's say you want to make a gun, so that you can propel chunks of lead at supersonic speeds through panes of brittle glass (it's like throwing a rock, except 'better'). It's tough to make a gun out of rocks—and rocks won't do this for you, even if you could—but in some rocks, you'll find metallic ores like lead, iron, copper, and zinc. If you place these fancy rocks in a particular system of governance with other things and relations (like plastics, woods, explosives, smelting, and machining), you can get a gun that fires bullets. A designer—or rather, a whole community and history of designers and patrons—will produce an artifact (in this case, an object like a gun) in order to actualize a purpose or a goal. The resulting technic therefore embodies a value because it was *designed to*. The design process constructs a useful object by embedding values into—or bringing values forth from—a set of materials. In these cases, the affordances were put into the object *on purpose*, but taken from the materials.

These values, moreover, remain in the object even when the designers go away. Mikhail Kalashnikov may have designed (and given his name to) the AK-47, but you don't need him nearby for the AK-47 to work, or for you to know what it does. The AK-47 makes it fairly clear what its affordances are, even if it's just sitting there on a table. Mr. Kalashnikov designed it very well and you're smart; you'll be able to figure out the rifle eventually, even if you don't want to use it for anything. So the value-ladenness of a designed artifact can lay dormant in an object long after the designers are gone, and for as long as that object is functional, its design purpose will be 'out there' in the 'thing itself' even if nobody uses it, and regardless if it's used for good or evil.

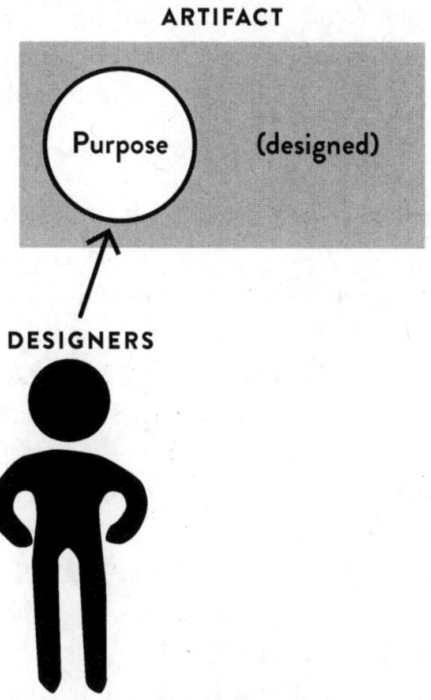

Design: intentional value embodiment.

All technologies are, therefore, value-laden embodiments of power, independent of the meaning or values we might want to use the thing for or project onto it (which we also do, of course). Those values are so fundamentally embedded in almost every aspect of our material world that we hardly even notice they're there. As Winner points out, "we see the details of form as innocuous, and seldom give them a second thought."[14] Paying attention to the values embedded in technical objects is not something we usually do. But paying attention to values is one of the things STS does, because "seemingly innocuous design features ... actually mask social choices of profound significance."[15] If designers can embody values into technological arrangements that exist even when the designers are absent, then things like artifacts, tools, or even techniques can be out there in the world achieving certain social effects even if we're generally ignorant of that fact. Blissfully believing that everything's neutral, good, and fine won't help us if there really are technological issues that demand our attention. STS aims to find out if there are any that do. Let us continue, then, to expand our unusual examination of the non-neutrality of technologies to see what issues might come to our attention.

However, you might have noticed already that it's not easy to talk about technologies unless we use examples of particular items. So far, we've used examples like Mark Zuckerberg's internet empire, rocks, and guns. STS scholars sometimes claim that the best way to study technology is to study individual technologies with "qualitative methods ... such as interviews, case studies, and ethnography."[16] This close attention to empirical detail can sometimes miss out on the larger theoretical questions about technology, but this doesn't have to happen. In the next two sections of the chapter, I will survey a number of classic examples of value-laden technologies in Winner's seminal though controversial paper on technological politics, both so that we have a better idea of what STS scholars are arguing about with respect to technology, and so that we can expand our understanding of the unusual possibility of non-neutral technology.

Political Artifacts, Prior to Use

Winner argues that technologies can sometimes embody political values even when most of us don't notice or want those values to be a part of our social context. He does not deny that we can and do socially construct the meaning of technologies: "Those who have not recognized the ways in which technologies are shaped by social and economic forces have not gotten very far," he says.[17] Nor does he deny that technologies embody the value-laden intentions of their designers. In addition to these dimensions, however, he advances the shocking idea that technologies can achieve political effects independent of—and logically prior to—them being used in one or another way. The social implications of a technology "precede the *use* of the things ...,"[18] meaning that in some cases it doesn't matter *how* you use a technology; it only matters *that you use it at all.* Winner's most famous examples of this are *racist overpasses* in Long Island, New York.

Apparently (so the story goes), there is a parkway that connects New York City to Jones Beach on Long Island, and the parkway is intersected by a number of overpasses. These overpasses are unusually low, in some cases leaving between 8 and 9 feet of clearance at the curb. If you want to use the overpasses, you can drive over them in your with car no problem. But using them to drive over the parkway was only part of the point. The low overpasses were also designed so that certain vehicles *couldn't go under* them, specifically public transit buses. It doesn't matter how you drive over a low overpass—you can drive over it nicely or rudely, for example—but underneath it, it will present an obstacle to tall vehicles like buses approximately 12 feet high, no matter what you do. Tall buses would smash into the low overpasses, if driven underneath. These are objective facts about certain material things, it seems easy to say.

It's a bit tougher to say what's political or value-laden about these material facts. According to Winner (who relied on journalist Robert A. Caro's biography of Robert Moses), "Poor people and blacks, who normally used public transit," were supposed to be discouraged from going to Jones Beach.[19] Moses, the designer of the parkway and the low overpasses, only wanted "whites of 'upper' and 'comfortable middle' classes" who owned cars to be able to get to the beach.[20] Moses also vetoed an extension of the Long Island Railway to ensure "limit[ed] access of racial minorities and low-income groups to Jones Beach...."[21] In Moses' time (at least), racialized communities were more likely to be economically marginalized, and their levels of income meant they depended more on public transit than private automobile ownership to get around. Preventing public transit from getting to the beach meant that certain social classes were indeed discouraged from getting to the beach, thanks to Moses' overpasses. Given a number of social and technological facts about race, economic class, public transit, bus height, overpass height, and the way that metal and glass react when smashing into concrete, Moses was able to embody racism in his overpasses such that they became "just another part of the landscape," regardless of how anybody used them.[22]

Winner's example of Moses' low overpasses has become something of a legend in STS. One journal devoted nearly 50 pages of discussion to what amounts to only three paragraphs in Winner's original.[23] There is dispute about whether Moses was an intentional racist or a casually systemic racist. There is dispute about whether *all* commercial vehicles were prohibited on parkways as a matter of course, making taller (and more expensive) overpasses superfluous. And there is dispute about whether the low overpasses prevented buses from getting to Jones Beach at all. Even though buses average between 118 and 144 inches high while Moses' overpasses average 107.6 inches of clearance,[24] you can take public transit buses to Jones Beach today (Google Maps says it takes at least an hour and a half from JFK airport, if you take the subway for the first part of the route). But all this misses Winner's point.

Let's consider this counterfactually: not about what does happen on Long Island, but what could happen. *If* Moses was an intentional racist, *if* economically marginalized and racialized people couldn't afford cars, and *if* public transit was prevented from accessing the beach by low overpasses, *then* Moses' racism was baked into the technological material of Long Island. Even if someone else who wasn't racist at all designed those low overpasses, the exclusionary effect would have still been achieved so long as those people rode buses of the usual size. All of this can change, of course, and that's the point of contingency in social construction: we didn't have to build the world in this way, and we can (and indeed should) build it in another way. But if all these actors and actants exist with all these potentialities, then if you put them together in the way Winner thought Caro thought Moses did, then you get discriminatory political effects embedded in a long-lasting technology that is altogether unaffected by how anyone uses it. The racism of Moses' overpasses would be embodied in the landscape of New York in addition to all the various ways in which society might use or project meaning onto those material things.

Another example of political artifacts are the broad boulevards of modern Paris. The old city of Paris followed a street plan rather like a rat warren. Like most medieval European cities, the streets were narrow, winding, and complexly intertwined. This made it comparatively easy to barricade a few streets with furniture tossed out of the windows, and thereby block off whole neighborhoods for revolutionary purposes. (If you've seen *Les Misérables*, then you've seen this represented on stage or screen.) However, Emperor Napoleon III of France wasn't too keen on revolutionaries barricading large sections of Paris during his reign. Therefore, he hired Baron Georges-Eugène Haussmann as a city planner, who bulldozed many old and poor neighborhoods to create wide, straight, and difficult-to-barricade boulevards like the famous Champs-Élysées. Winner claims that this was "to prevent any recurrence of street fighting [as in] 1848."[25]

As with Moses' overpasses, there is debate over whether the prevention of street fighting was the intent of Haussmann's boulevards or a convenient side effect. Again, however, that misses Winner's point. You don't have to *design* the wide boulevards in a particular way for them to have the strategic effect of making revolution more difficult, nor do you have to *use* the wide boulevards in a particular way for them to have that

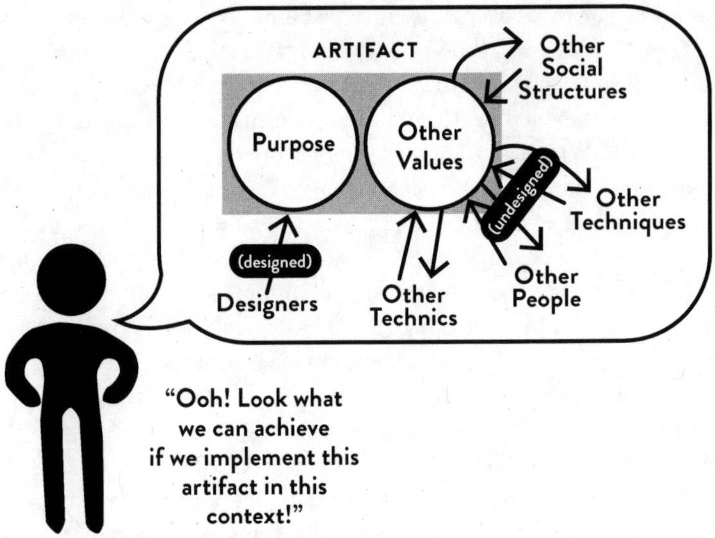

Intentionally political artifact.

effect. Rather, once those boulevards are in place, it's simply a fact that it's harder to fill them up with chairs and kitchen tables than it was previously; they simply are more efficient at moving troops around the city. These values are not in the technological arrangement because someone put them there; Haussmann simply actualized these values by bringing the boulevards into being. His intent, or anyone's use of them, doesn't change their political impact, although their political impact was very much aligned with the interests of the emperor.

A third example of political artifacts are the pneumatic molding machines used, for a time, to manufacture Cyrus McCormick's agricultural reaping machines. In the 1880s, pneumatic molding machines were introduced into his Chicago factory at a cost of half-a-million dollars (about 15 million in inflation-adjusted dollars at the time of my writing). These machines could be operated by unskilled laborers, which put the skilled laborers who previously manufactured the McCormick reapers out of work (compare with the Luddites in Chapter Twelve). McCormick said mechanization was a good way to "weed out the bad element among the men,"[26] even if it made lower quality products at a higher cost to his company, because that "bad element" was the skilled workers who were organizing a labor union. McCormick didn't want a unionized workforce, and his technological modernization helped him get rid of it.

However, whoever designed the pneumatic molding machine didn't design it to be a union-busting machine. That would be a very unlikely objective for any mechanical inventor working on pneumatics and moldings. Moreover, nobody implemented or used these molding machines in a uniquely union-busting way. If you implement such machines in a factory, you may or may not have anti-union sentiment. If you're an unskilled laborer, you may or may not have anti-union sentiment. Either way, however, the pneumatic molding machine will privilege unskilled over skilled labor, which makes it harder to organize or maintain a union shop. McCormick knew this, but he didn't construct the technology so that it did this. Nobody did. He just saw what it was capable of doing politically, and he arranged for that to happen.

Unintentionally Political Artifacts

But artifacts can be laden with value even if nobody notices or wants that to be the case. Moses, Haussmann, and McCormick may have been aware that technologies can be "thoroughly biased in a particular direction" and used that knowledge to their political benefit,[27] but the social effect of a technology can be *undesired* and still actualized by the very fact of its existence or use. One example Winner gives is the super-advanced technological marvel of *stairs*. Nobody designs, implements, or uses stairs to "systematically exclude [mobility impaired persons] from public life,"[28] but that's what happens anyway if there aren't any other options for getting into a building. Politically, the value implications of stairs vis-à-vis things like wheelchairs and the people who use them are objectively built into the structure of stairs without any relevant intention at all: "designs unsuited for [the mobility impaired] arose more from long-standing neglect than from anyone's active intention." Nevertheless, "justice requires a remedy."[29] If we leave our stairs—or our wheelchairs, or both—just the way they are, no amount of meaning attribution or socially constructed value-projection is going to change the value-laden fact that some buildings exclude some people for entirely unjust reasons. Something's got to change, and that something is objective, material, and technological.

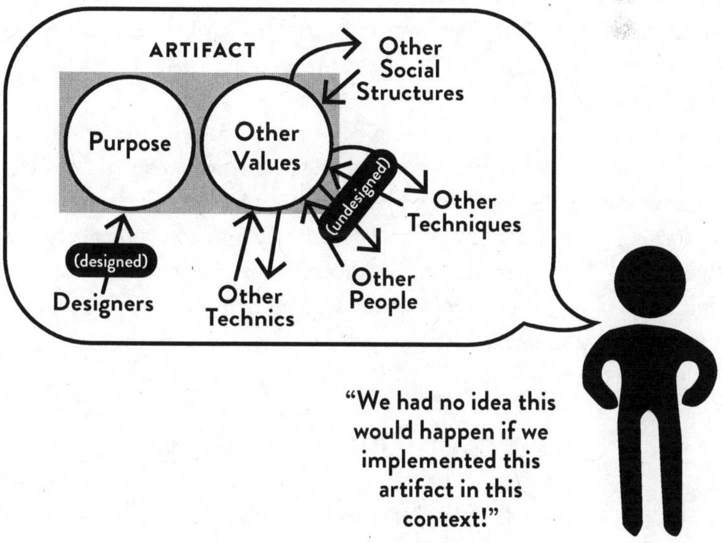

Unintentionally political artifact.

The last example from Winner I'll cover is the tomato-picking machine developed at the University of California Davis. According to the university's website, a plant breeder (Jack Hanna) and an engineer (Coby Lorenzen) worked together in the 1950s to create a mechanical tomato harvester.[30] The engineer needed to collaborate with the plant breeder because machines had real trouble picking the previously soft and juicy Rutgers tomato without squishing it. It's unlikely that Hanna and Lorenzen had a vendetta against Rutgers University and their tomato (at one point it was the most popular variety of tomato in the world), but their machine effectively did. Their machine needed a new kind of tomato. Their new tomato, called the 'square tomato' because it wouldn't roll off the machine, was "hardier, sturdier, and less tasty."[31] Today, the delicious Rutgers tomato is all but extinct. How so?

The tomato-picking machines substantially reduced the cost of picking tomatoes, which was otherwise done by hand. The machines were expensive, however, and so only larger farming operations could afford the initial costs up-front and then go on to benefit from the savings per ton of tomatoes. Therefore, those businesses that could mechanize their tomato operations outcompeted the smaller ones that couldn't. As a result, the ownership of tomato farms became concentrated in the hands of a few, tens of thousands of manual farm laborers lost their jobs, and most of us eat hard tomatoes that the machines can handle instead of the juicy ones common in the Campbell's soup and Heinz ketchup of a century ago.

None of this appears to be "the result of a plot," however.[32] Hanna and Lorenzen probably didn't have vendettas against small farms and migrant workers any more than they did against the Rutgers tomato. Nevertheless, it appears that their machine was value-laden in these ways. Winner concludes that

> [w]hat we see here instead is an ongoing social process in which scientific knowledge, technological invention, and corporate profit reinforce each other in deeply entrenched patterns that bear the unmistakable stamp of political and economic power.[33]

STS won't tell us if it's *better* to have more small farms, more manual farm laborers, or tastier tomatoes, but it will make us ask why those questions of value should be answered by two guys with a penchant for tinkering at a publicly funded university instead of by a deliberative process that involves all of us democratically. If technologies are unavoidably value-laden, then maybe our political process should apply to technological changes and not just electing politicians.

Democratizing Technology

The implications of technological value-ladenness are pretty significant for the usual story. If the usual story is right, then technologies are value-neutral and there's nothing particularly important for us to do about them. We can let the usual experts—scientists, engineers, inventors, investors, etc.—continue doing their usual thing, which is making life awesome, of course. But if technologies are value-laden—and in some or even many cases, loaded with values that we might not want to introduce into our societies—then technologies may need to be regulated in ways similar to how we regulate politicians: democratic accountability. Winner thinks "the same careful attention one would give to the rules, roles, and relationships of politics must also be given to such things as the building of highways, the creation of television networks, and the tailoring of seemingly insignificant features on new machines."[34] Just as much as we can ask ourselves if we want a particular politician to be in charge of our country, we can also ask ourselves if we want a particular technology "to join society as a piece of its operating equipment."[35]

We have some options with respect to our technologies, of course. We can utilize instrumental considerations of *how* to implement a particular technical arrangement (like an oil pipeline's route), but those choices become ossified once the artifact has been built: political "choices tend to become strongly fixed in material equipment, economic investment, and social habit...."[36] We have more freedom with respect to technologies *before* they've been implemented. In those cases, we can ask questions of intrinsic value: "Are we going to develop and adopt the thing or not?"[37] Some technologies (like oil pipelines per se) may have embodied values (like an infrastructural commitment to carbon emissions) that would be better to not actualize at all.

Either way, *technologies build values into the world*. Of course, much of this can and will be socially constructed, but there's more to technology than what we bring to it. Technology brings baggage to us too, especially once it's in use. Because of this,

> [t]he masters of technical systems, corporate and military leaders, physicians and engineers, have far more control over patterns of urban growth, the design of dwellings and transportation systems, the selection of innovations, our experience as employees, patients, and consumers, than all the electoral institutions of our society put together.[38]

"Consciously or not," Winner says, the structures of our technologies "influence how people are going to work, communicate, travel, consume, and so forth over a very long time."[39] STS makes political citizenship more difficult thereby, because we have to add technology to the list of things voters need to be informed about. It also makes engineering (etc.) more difficult, because we have to add politics, sociology, and moral philosophy to the list of things engineers (etc.) need to be educated about. And that, just like the previous chapter, is just the start of doing STS.

Notes

1 Herbert Feigl, "The Scientific Outlook: Naturalism and Humanism," *American Quarterly* 1, no. 2 (1949): 145.

2 The classic formulation of this principle of liberty is from John Stuart Mill, *On Liberty*, originally published in 1859.

3 Quoted in George Grant, *Technology and Justice* (University of Notre Dame Press, 1986), 19.

4 Langdon Winner, "Do Artifacts Have Politics?," *Daedalus* 109, no. 1 (1980): 125.

5 Winner, "Do Artifacts Have Politics?," 122; original emphasis.

6 Winner, "Do Artifacts Have Politics?," 122.

7 Winner, "Do Artifacts Have Politics?," 121.

8 Mark Zuckerberg, interview by Emily Chang and Sarah Frier, "Mark Zuckerberg Q&A: The Full Interview on Connecting the World," *Bloomberg*, 19 February 2015, https://www.bloomberg.com/news/articles/2015-02-19/mark-zuckerberg-q-a-the-full-interview-on-connecting-the-world.

9 Miguel A. Faria, Jr., "America, Guns, and Freedom: Part II—An International Perspective," *Surgical Neurology International* 3, no. 135 (2012): 6.

10 Faria, "America," 7.

11 Winner, "Do Artifacts Have Politics?," 122.

12 Winner, "Do Artifacts Have Politics?," 122.

13 Winner, "Do Artifacts Have Politics?," 123.

14 Winner, "Do Artifacts Have Politics?," 123.

15 Winner, "Do Artifacts Have Politics?," 127.

16 Anabel Quan-Haase, *Technology and Society: Social Networks, Power, and Inequality*, 2nd ed. (Oxford University Press, 2016), 51.

17 Winner, "Do Artifacts Have Politics?," 122.

18 Winner, "Do Artifacts Have Politics?," 125.

19 Winner, "Do Artifacts Have Politics?," 124.

20 Winner, "Do Artifacts Have Politics?," 124.

21 Winner, "Do Artifacts Have Politics?," 124.

22 Winner, "Do Artifacts Have Politics?," 125.

23 *Social Studies of Science* 29, no. 3 (1999): 411–57.

24 Thomas J. Campanella, "Robert Moses and His Racist Parkway, Explained," *Bloomberg CityLab*, 9 July 2017, https://www.bloomberg.com/news/articles/2017-07-09/robert-moses-and-his-racist-parkway-explained.

25 Winner, "Do Artifacts Have Politics?," 124.

26 Winner, "Do Artifacts Have Politics?," 124–25.

27 Winner, "Do Artifacts Have Politics?," 125.

28 Winner, "Do Artifacts Have Politics?," 125.

29 Winner, "Do Artifacts Have Politics?," 125.

30 "Remaking Tomatoes for Robotic Harvesters," University of California Davis, accessed 10 June 2022, http://ucdavis.edu/research/discoveries/agricultural-discoveries/tomato-harvester.

31 Winner, "Do Artifacts Have Politics?," 126.

32 Winner, "Do Artifacts Have Politics?," 126.
33 Winner, "Do Artifacts Have Politics?," 126.
34 Winner, "Do Artifacts Have Politics?," 128.
35 Winner, "Do Artifacts Have Politics?," 127.
36 Winner, "Do Artifacts Have Politics?," 127–28.
37 Winner, "Do Artifacts Have Politics?," 127.
38 Andrew Feenberg, *Questioning Technology* (Routledge, 1999), 131. The title of this section of my chapter, "Democratizing Technology," is also the title of the sixth chapter of his book.
39 Winner, "Do Artifacts Have Politics?," 127.

Chapter Seventeen

Technology with a Mind of Its Own?

IT MIGHT SEEM THAT STS HAS VERY LITTLE REMAINING TO DISCUSS regarding technology, at least in general. Once it becomes clear that technologies are unavoidably value-laden, scholars and students of STS can busy themselves with identifying the values embodied in particular technologies and figuring out (democratically) what to do about those embodied values. The thing about embodied values, though, is that they restrict our freedom. We aren't as free as we might like to be with respect to our use of value-laden technologies or their effects on us. Rather than the value or impact of a technology being simply *contingent* on any number of social factors (many of which are within our control), technology may be an *actor* in its own right—an active force with agendas of its own that could interfere with ours. That is, technology could limit our freedom to choose what to do with it or what impacts it will have on our society—as if technology were telling us what to do. Rather than contingency being the key to understanding it, technology might be a form of *necessity* that makes it difficult or impossible to deliberate, individually or collectively, about it. If technology has that powerful an influence on us, we'd have to say that it affects us *deterministically*.

Working Definition of Technological Determinism: the belief that technology is an external force that causes or determines social change, and is itself determined along a unidirectional trajectory by preordained social factors.

Technological determinism, therefore, is the idea that technology is an external force limiting or controlling the rest of what we call society and nature.

Moreover, technological determinism is a deep concern within STS and subject to considerable debate. Sergio Sismondo links it to the outdated notion of technology as applied science, but he doesn't say why.[1] Anabel Quan-Haase defines it as "view[ing] the effect of technology as unidirectional and ... not consider[ing] social factors" in technological development, which she further associates with "simple approaches" to technology like "Utopian and dystopian visions of technology...."[2] Sally Wyatt delineates four types of technological determinism: justificatory, descriptive, methodological, and normative,[3] only one of which holds "that technological developments take place outside of society, independently of social, economic, and political forces ..., following an internal, technical logic that has nothing to do with social relationships."[4] Wyatt notes that some STS scholars define technological determinism even more broadly: "that technological change causes or determines social change."[5] By this last definition, seeing technology as value-laden (see Chapter Sixteen) would be technological determinism—and STS tends to see technological determinism as bad. So maybe STS *shouldn't* view technology as value-laden after all?

One way to understand technological determinism is to consider petroleum-based automobility. Gasoline was originally discarded as a useless by-product of the distillation of kerosene. With the development of the liquid-fueled internal combustion engine and the automobile, however, the value of gasoline as a fuel became apparent. By the time Ford developed the Model T car, there were already nine million gas-powered automobiles on American roads.[6] However, to drive these vehicles, infrastructure had to be in place, including fueling stations, repair shops, refineries, transportation networks, even road surfaces (tarmac was invented to tolerate the higher speeds of automobiles in comparison to bicycles and carriages). The automobile also fueled the growth of suburbia, a form of development outside the urban core where the middle classes could feel like they were living in the country while commuting by automobile to their jobs in the city.[7] As of 2005, approximately 1 percent of the surface area of the United States was covered by a road, about the total size of South Carolina. Nowhere in the lower 48 states are you ever more than 12.4 miles (20 km) away from a road.[8]

It's unlikely that anyone's grand plan a century ago was to cover the continents in a network of asphalt and strip malls, but that's part of what internal combustion engines mounted on wheels unleashed. Looking back, however, it's not easy to imagine things going in a different direction. Looking forward, it's easy to imagine things going in the same direction, with more roads, more cars, and more suburbs. There's a trajectory and a momentum embodied in the petroleum systems of transportation, housing, and commerce that seem impossible to change or redirect. At the moment, the biggest alternatives to this status quo are just automobiles fueled by something else: electricity or hydrogen, and both of these have different infrastructure demands of their own. Looked at in this way, it can feel like because petro-mobility makes certain requirements of the societies it's in, those societies have no choice but to fulfill those requirements. We just have to build more roads and fueling stations. That can be understood as technological determinism: liquid fueled vehicles are an external force that causes social change, and liquid fueled vehicles were caused by earlier external forces that we neither have nor had control over.

STS scholars fear that technological determinism doesn't appreciate "how deeply social the processes of technological development are."[9] Technological boosterism can be a type of determinism that assumes that technological progress will inevitably make society better; boosters don't rely enough, therefore, on sophisticated sociological and historical analyses of technology and society. Technological determinists allegedly leave "no space for human choice and intervention" regarding technology, and as such it "absolves us from responsibility for the technologies we make and use."[10] Determinists conceptualize technology more as an abstract force than a collection of concrete objects, and as such it is mythologized as an autonomous entity with a mind of its own (which is, of course, what artificial intelligence is intended to be). STS—and especially the socially constructivist strands within it—tends to think of itself as having "abandoned this [deterministic] view" by demythologizing the idea of technology as a monolithic, threatening or liberating, and possibly self-aware being in its own right.[11]

There is, therefore, a tension within STS's investigation of technology. The technologically liberal usual story is inconsistent, being both constructivist and anti-constructivist, seeing technology as both neutral and positively value-laden. Critical responses to the usual story have to be accordingly complex. Against the progressivist boosters, STS sees technological development as contingent and socially conditioned; against neutrality, it sees technology as objectively value-laden. It's not yet clear that these two strands of analysis are compatible, however, especially if the first strand sees the second as pessimistically deterministic. If technology is objectively value-laden and limits our freedom to do whatever we want with it, how can it also be a contingent, socially conditioned reality that we are free to change however and whenever we want? Wouldn't we just end up with the same inevitabilism of the boosters, only without the optimism?

These are highly abstract issues, but that is usually how theory goes. Early scholars of STS had a tendency to talk about technology in a grand sense, as if it was "a universal force,"[12] whereas more recently scholars have focused more on individual, concrete instances of technology as a way to keep STS grounded in day-to-day practicalities.[13] And yet our day-to-day experience of hyper-modern technology does suggest that something is out of our control. We seem to be living through a transition into a potentially new epoch, facilitated by rapidly changing and radically powerful technologies that virtually none of us have any say over. From the ubiquity of social media platforms mediating and shaping human relationships, to digitally enabled gig economies with temporary and flexible contract work replacing permanent salaried careers, to weaponized drones making twenty-first-century warfare as static as the trench fighting of World War One, to large language models in generative artificial intelligence making writing seem obsolete, it can feel like forces are at work in technological change that follow their own direction and goals, and ignore any goals that we might have. In this context, perhaps STS should return to a larger-scale analysis of technological forces. One such STS theorist is the French sociologist of technology, Jacques Ellul (1912–94). In this chapter, we will consider his accounts of technological society and technological autonomy in the hopes of gaining some traction on our prospects in a technological future.

La Technique

We noted way back in Chapter Three that many European STS scholars find the English word 'technology' to be imprecise and problematic: does it refer to tools and techniques themselves or to the scientific study of tools and techniques? Ellul is one of these scholars. He carefully defines his object of study as *not* "machines, technology [in the English sense of manufactured objects], or this or that procedure for attaining an end [in the English sense of 'technique']."[14] What Ellul calls *la technique* (in French) is rather *the entire modern technoscientific system.* In this respect, Ellul's definition of technique has much in common with what people often mean when they complain about 'technology' because the wifi is slow. "Ugh, technology!" is an exclamation that refers to technology as a broad, interconnected abstraction encompassing a myriad of individual objects. For Ellul, *la technique* is a "high-altitude and 'transcendental'" perspective of technology as a society-spanning system.[15] Ellul thinks, moreover, that this system characterizes our society as a whole, so much so that we should call our modern societies 'technological' ones.

Therefore, *la technique* for Ellul is more of a way of life than it is any particular set of objects or trendy gadgets. *La technique* is rather the social system that produces those gadgets (etc.). As such, *la technique* is a sociological framework: a paradigm (see Chapter Fourteen), an interpretive schema, a mindset, a worldview, an ideology, even a form of consciousness. *La technique* is the way in which people think in order to produce things like artificial intelligence, self-driving autonomous vehicles, or neural links between brains and computers. As a result, *la technique* is not something external to our culture, having impacts on it. "It is incorrect to say," according to Ellul, "that economics, politics, and the sphere of the cultural are influenced or modified *by* Technique; they are rather situated *in* it, a novel situation modifying all traditional social concepts."[16] *Technique is our culture*, on Ellul's analysis, which is why the title of his magnum opus is *The Technological Society* (1954).

Jacques Ellul (1912–94).

Of course, you'll have noticed that his book title says 'technological' instead of 'technique,' which raises an issue of translation from French into English. It is fair, however, to speak of this technical form of social organization as 'technology' so long as we remember the oddity of the word pointed out in Chapter Three: techno*logy* can be a *logic*, a form of rationality or thinking about the world, and one with a long intellectual history stretching back to the literacy of ancient Greece and the agriculture of Near Eastern civilizations. Unifying that form of scientific rationality (*logos*) with the practical manipulation of objects to gain power (*technē*) is what this book argued was a characteristic of modern Western culture. Ellul calls this technoscientific equation of techniques with scientific logic 'the technical phenomenon,' the momentous unification of the theoretical sciences and the practical arts through mathematization starting in the sixteenth century.

This is why Ellul thinks that "the technique of the present has no common measure with that of the past."[17] *La technique* is *modern* technoscience, which means 'techno-logic' or 'techno-rationality,' rather than "an instrument" or simply "applied natural science."[18] *La technique* is modern technological thinking, and that means trying to put every particle of matter under our control the way Francis Bacon wanted us to (see Chapter Nine).[19] In our techno-logical society, therefore, to do science *is* to manipulate matter. Technology isn't 'applying' science at all. We're *doing* technoscience because science and technology are functionally the same thing. We can hardly even imagine a form of knowing that doesn't give us power over the physical world. If we try to imagine knowledge that doesn't embody technical power, all we can think of is something 'useless' and therefore not valuable. But this means that *technology as a way of thinking* is also *a way of valuing.*

Technological Values

As a sociological framework which produces a particular suite of unique artifacts, technology is a value system. It conveys a set of preferences and goals which we are supposed to actualize in our actions: designing, manufacturing, purchasing, using, etc. One such technological value is a preference for artificiality over naturalness. After all, artifacts *are* technologies; it would be odd if techno-rationality didn't prefer technological development over raw, natural material. This is exactly what the myth of the State of Nature has been telling us for several centuries: nature might be a nice place to visit, but it's a bad place to live (see Chapter Ten). Highly advanced technology is better than raw, savage nature. The whole point of the Doctrine of Progress is to use increasing technological sophistication to extract us from the brutal existence nature would otherwise force on us. Of course, many people enjoy nature and so-called natural products, even wilderness, but the usual story points out that nature lovers don't *really* want to live in wild nature without technology. They just want to visit nature, and they do so technologically:

> Much of the experiencing of nature ... that is hailed as a reaction to the so-called reductionist technological approach to reality paradoxically has arisen as a by-product of modern technological culture. Some recent studies, for instance, have discussed the experience of mountain climbing as a product of modernity, related to other experiences involving the search for and exceeding of limits—be they outer or inner.... Mountain climbing presupposes not only a modern

> technological infrastructure, but also a very technical manner of locomotion using many kinds of high-technological devices, and its professionalization can only be understood as an accessory of technological culture itself.[20]

And if you *still* want to escape technology by making recourse to nature, don't forget that 'nature' is a social construction: nobody can access what 'real' nature is without social construction, and constructions are *artifacts*. Therefore, 'nature' and 'the wilderness' are artifacts too, and not 'natural' in any non-artificial sense. There is no way to escape the technological mediation of nature. This is also a good reminder that social construction is a techno-logical theory!

A second technological value is calculation: artifacts are easily and preferably *measurable*. Calculative rationality seeks to break things down into their constituent parts so that they can be quantified and so that mathematical relations between the parts can explain how they work. This is the intellectual legacy of Newton's solution to the mechanical problem of 'seriousness' (see Chapter Eleven): technological value is so deeply embedded in our culture that we simply see it as 'scientific' to understand material particles with math. Calculative rationality is also where the lawlike universality of Mertonian science (see Chapter Two) is supposed to come from: "standardization of tools, metrics, units, and frameworks creates a kind of universality and eliminates subjectivity," we are led to believe.[21] If everything is made out of the same parts, then everything will follow the same mathematical laws and allow for uniform explanations and linear predictions that are independent of our subjective perspectives and personal preferences. Modern science works, we think, because it's a highly advanced information technology that gives us useful and objective truths.

Calculative Rationality: the belief that everything is best understood by breaking it down into quantifiable parts and explaining their relations mathematically, in the hopes of eliminating subjectivity and guaranteeing objectivity.

These are the values implicit within calculative rationality, then: atomization, quantification, universality, linearity, and predictability—and without them, we couldn't have technical sciences like economics or engineering. For example, modern economics uses cost-benefit analysis to establish the best outcomes of a decision by adding up or subtracting aggregates of preference units like money or "utils," which are basic units of happiness or pleasure.[22] Using techno-logical thinking like calculative rationality, we can optimize useful outcomes (which are better than 'natural' outcomes, of course) for as many people as possible.

Efficiency: maximizing outputs (of anything) with as few inputs as possible.

This optimization leads to the highest techno-logical value of all: *efficiency*. A good, optimally functioning technology should operate as smoothly as a well-oiled machine.

Machines are artificial arrangements of smaller parts of matter that follow the same uniform laws of motion, behave in linear and predictable ways, and ideally maximize outputs with minimal inputs (maximizing outputs with minimal inputs is the definition of efficiency in general). Technological rationality therefore sees everything in terms of machinelike efficiency: everything from politics to economics to energy to transportation should be streamlined to produce the most ______________ (fill in the blank however you like) with the least expenditure. Techno-logic seeks to optimize everything as *a means to an end*. That is, everything is a tool that should efficiently provide something else; everything is a machine geared towards efficiently producing the valuable goal of whatever the machine was designed to do. Surely this is obvious, for who would want an inefficient machine, or an inefficient society? You'd have to be crazy to want inefficiency for anything. That's how deeply technological thinking is embedded in our society: we think that thinking any other way than this is crazy. And *this* is why Ellul defines technology as "the totality of methods rationally arrived at and having absolute efficiency ... in every field of human activity."[23] Technology is the logic of our society that evaluates everything like it should be a perfectly functioning tool.

Importantly, however, techno-rationality does not specify what ends it is to be used for, or what goals or values it should efficiently pursue. Efficiency is the only named goal, and all other goals are blanks to be filled in. After all, there are a zillion different kinds of machines, and each kind has its own unique goal (like making ice cream, or driving fast). But in addition to all those unique goals, all machines have the *general* goal of achieving their specific goals efficiently. Technology is supposed to provide practical ease for *whatever* purposes you have, and nothing more. Indeed, there can be nothing more, for efficiency is the highest value of a technological culture. If I were to ask you the question "what is technology for, in general?" you might answer "I don't know, whatever." We generally have no idea what technology per se is supposed to be doing (other than being awesome), any more than we know where progress is supposed to be going (other than utopia).

Our society is therefore ordered "according to a process which is causal but not directed towards ends; It is formed by an accumulation of means which have established primacy over ends...."[24] Asking what other goals or values a technology should pursue or achieve is not a technical question; if our rational framework is predominantly technical, we will dismiss such value-questions as meaningless (see Chapter Thirteen, where only empirically verifiable statements had meaning). The purpose or goal of anything cannot be determined or decided by any mechanical process or analysis. We can quantify and measure the value of an artifact's efficiency, but no other value can be fit into techno-science's quantitative method because all other values are qualitative. Therefore, values cannot be the topic of 'rational' consideration because rationality itself is technoscientific and can refer only to the description and control of material particles. All ends other than efficiency itself are considered merely subjective or completely relative because they neither describe nor manipulate material particles. If a non-technoscientific end or goal (a qualitative good like justice or beauty) is ever subject to appraisal, therefore, our society generally assumes that the only way to evaluate it is by the democratic aggregation of individual preferences—i.e., taking a vote.

Technology *is* good for something other than just being good for something, however. Machinelike efficiency grants us *power*. We might think it grants us power for its own sake—especially if efficiency does not specify what it is to be used for—but if so, technological power would be as impractical and self-contained as Aristotelian divine science (see Chapter Seven). But we know that technology is good for more than itself; it is fundamentally geared towards the mastery of material particles! At first, Ellul says:

> Technical growth leads to a growth of power in the sense of technical means incomparably more effective than anything ever before invented, power which has as its object only power, in the widest sense of the word. The possibility of action becomes limitless and absolute.[25]

But think about this in terms of money. Most of us want more money, even if some of us have a lot of it. The usual answer to the question of "how much money is enough money?" is "more money."[26] And yet we don't want money simply because we want money. We want money because we can buy something other than money with it. We want the power that money gives us, and that is the power to purchase *other things* that we want. Money is a means to other ends, and we want as much of these means as possible, preferably without limit.

Similarly, techno-logic might *desire* power for its own sake, but that's just what it tells itself. Power *actually* only makes sense technologically if it is power over the physical world. The purpose of power for its own sake is actually power to control the world. Techno-logic is Baconian (see Chapter Nine). Technological rationality wants this power to be absolute, limited by no other value than power itself. Technology as a social system wants nothing more than the unfettered ability to manipulate matter in every possible way, whichever way we like. If we were to subordinate technical efficiency to any other values—like customs, traditions, or even ethics—that would make the system an inefficient machine and less powerful, which is technologically irrational.

Efficiency, then, is both the highest value that is rationally possible and an observable, objective fact that can be measured. Techno-logic appears to be an unassailable standard of value, one that is supposed to be so obvious that we don't even recognize it as a system of value which dominates our way of thinking and way of life. That is why Ellul calls it a "closed circle."[27] His point is that modern societies don't know how else to think about themselves. Since industrialization, everything should be like a factory; since Henry Ford, everything should be like an assembly line. From Amazon warehouses to cubicle farms to health care systems to construction projects to electrical power generators, we are all supposed to be parts of a larger system of governance that efficiently produces valuable outcomes from our labor.

This is what 'human resource management' is all about: "aim[ing] to increase worker efficiency through the application of scientific method to the understanding of labour" (technically known as 'Taylorism,' after the mechanical engineer Frederick Winslow Taylor who proposed the idea).[28] You and I are resources to be mined, in principle no differently than a coal seam underground is mined. We each have 'energy' to be harnessed, and the goal is to do so efficiently. Blue collar workers are a human resource that use their energies to sometimes literally mine coal, whereas white collar workers are a human resource that use their energies to maintain and improve bureaucracies. Bureaucracies are supposed to guarantee the efficient functioning of other workers, and the computer and the internet are indispensable tools for the smooth functioning of bureaucracies. In both cases, white and blue collar jobs generally involve the endless, monotonous, and invariant repetition of a single task—which, of course, is exactly what machines do.

Taylorism: the application of scientific method to human labor to increase worker efficiency.

Welcome to the rest of your life in the technological society.

Enframement, Determinism, and Autonomy

This is our culture, then: we think technologically, which is why we live in a technologically dominated world. Ellul says that technology is our *milieu*; it is our environment, that which surrounds us, rather than nature. It is the cultural context we exist within, the air that we breathe, the water we swim in. Techno-logic is our nature, our community, our society. We take it entirely for granted, as if its existence and presence were necessary or second nature, and because of this implicit acceptance, we don't even notice that it's there, functioning in this way. We assume that our modern societies are just necessarily efficient assemblages of technological methods (operating unobtrusively in the background) while we're busy pursuing our own subjective goals (like getting a job or going out drinking). Technoscientists, meanwhile, are busy providing us with 'objective facts' (like the needs of the market) that structure our lives and the frameworks for what we see as possible and important, while they constantly change our lives because progressive improvement is inevitable, unstoppable, and limitless.

We aren't supposed to notice that these 'facts' are socially constructed or value-laden. They aren't supposed to be harmful to us, reducing job security or purchasing power, having any effect on our ability to buy a house or purchase groceries. If we do notice that these artifacts (like the economy, or our transportation networks) are value-laden, we are supposed to think that those values are automatically good because technoscientific progress makes them so. If we do nevertheless think critically about those embodied technological values, our own critical thinking will be criticized for not being 'fact-based' but 'emotional' instead, which is (of course) assumed to be the opposite of 'reason' (because technological rationality cannot work with anything unmeasurable, like feelings). There aren't supposed to be any reasons why we should think twice about our technological society; there isn't supposed to be any way out of

Enframement: the condition of believing oneself to be incapable of thinking or living outside the framework of calculative rationality.

this conceptual trap. That is how technology as a system of value maintains a hegemony in modern culture.

Other STS scholars call this restrictive form of thinking and living 'Enframement,' the condition of being incapable of thinking or living outside the box provided by calculative rationality.[29] Enframing is not knowing how to think or imagine living outside the framework of modern technology. Rationality itself is seen as inapplicable to anything not reducible to strictly physical components, so critical thinking—which is not reducible to strictly physical components—is irrational. No matter what we or our culture might want to deliberate about, therefore, the only legitimate options will be the 'pragmatic' ones, those which make technological sense within the general trajectory of the status quo (see Chapter Twelve). We cannot possibly envision a world different from this one, because this is the only one that makes techno-logical sense.

Ellul isn't worried that this technological system will have 'negative impacts' or 'harmful effects' on people. He is confident that "we will doubtless succeed in averting certain technically induced crises, disorders, and serious social disequilibrations; but this will but confirm the fact that Technique constitutes a closed circle."[30] Techno-logic will take a solution-based approach to contemporary techno-social problems, and solve those as technicalities by mitigating potentially negative consequences with more technological solutions. Ray Kurzweil, a futurist, inventor, and Google engineer, claims that whenever a technology encounters an obstacle, a new technology will be invented to overcome that obstacle. This leads to what he thinks is the "Law of Accelerating Returns," the exponential growth of technology (eventually) leading to unimaginably fast economic progress, human cyber-evolution, and the Singularity, where biological life will merge with supercomputers and colonize the universe with "intelligence."[31] Ellul's concern with this pseudo-religious view of the future is that "these results will come about through the adaptation of human beings to the technical milieu." Technological solutions to social problems "result in the *modification* of [people] in order to make them happily subordinate to their new environment, and by no means imply any kind of human domination over Technique."[32] The problem with Enframement, therefore, is that we do not have control over technology. It means that technology has control over us, thus changing what it means for us to be human.

Ray Kurzweil (born 1948).

Many STS scholars find this type of analysis problematic. If techno-logic is "a continual move towards rationalizing all aspects of human life, placing those aspects within a technical sphere, and destroying all possibilities for thinking or acting outside that sphere,"[33] then the situation appears hopeless: technology is so powerfully value-laden that humans have no ability to choose how they'll use it; our creative freedom to act how we want and construct what we want is abolished; technology appears to inevitably lead to totalitarianism and dystopia. This sounds like technological determinism at its worst! Not only does the system-level view of technology apparently assume that "the effect of technology on society [is] unidirectional and does not consider social factors,"[34] it seems to believe "that technology develops as the sole result of an internal dynamic, and then, unmediated by any other influence, molds society to fit its patterns."[35] Supposed pessimists like Ellul cannot appreciate or understand modern technology as a "new cultural constellation," and they try "to reject it nostalgically in demanding a return to some prior, seemingly more harmonious and idyllic relation assumed to be possible between nature and culture."[36] Here the social construction of technology cuts both ways: because all relations between nature and culture are socially constructed, all human experiences of life are technologically or artificially mediated; and because all social constructions are contingent and dependent on humans for their existence and stability, technologies are also contingent on our choices and dependent on us. On the one hand, there's no escape from technological domination; on the other hand, there's nothing to fear from technology anyway, because we're in charge of it.

Technological determinism, as we have seen, is a major flashpoint in STS, and some have argued that the discipline itself is defined by rejecting determinism. After all, the usual story which STS usually rejects is technologically determinist in a major way: "Technological determinism is imbued with the notion that technological progress equals social progress,"[37] a defining characteristic of modernist ideology. STS's debunking of the "equation of technological change with progress" is therefore consistent with its critique of the usual story and its own resistance towards determinism.[38] However, STS would find it difficult to deny that "technological change causes or determines social change," even though that is a common and very broad definition of technological determinism.[39] To oppose this would be to align with the usual story in accepting that technology is *not* value-laden. Therefore, "our guilty secret in STS is that really we are all technological determinists. If we were not, we would have no object of analysis...."[40] It might make more sense, then, to not simply label anyone who "treat[s] technology seriously"[41] as a determinist and thereby dismiss rather than assess the merits of their analysis.

Nevertheless, analyses like Ellul's have often been associated with the "strand of very pessimistic technological determinism," as if "technology does indeed follow an inexorable path" like progress in reverse.[42] Ellul was aware of this interpretation of his work, and he rejected it explicitly:

> I am neither by nature, nor doctrinally, a pessimist, nor have I pessimistic prejudices. I am concerned only with knowing whether things are so or not.... I only ask that the reader place [themself] on the factual level and address [themself] to these questions: "Are the facts analyzed here false?" "Is the analysis accurate?" "Are the conclusions unwarranted?" "Are there substantial gaps and

> omissions?" ... I do not seek to show, say, that [humanity] is determined [by technology], or that technique is bad, or anything else of the kind.[43]

As a sociologist, Ellul simply holds that individual actions or human freedom "are not discernible at the most general level of analysis, and that the individual's acts or ideas do not here and now exert any influence on social, political, or economic mechanisms."[44] At the sociological level of techno-logic, "We are dealing with collective mechanisms, with relationships among collective movements, and with modifications of political or economic structures."[45] He's trying to be as objective about our technological society as technoscientists try to be objective about the laws of nature, and at that level, human freedom is not a sociologically measurable factor.

Rather than a determinist who believes that the onward advance of technology in its current trajectory is inevitable and unstoppable, Ellul argues that techno-rationality is *autonomous*. He means this in the same sense as a self-driving car: technology moves in its own direction according to its own standards and programming.

> [T]echnology ultimately depends only on itself, it maps its own route, it is a prime and not a secondary factor, it must be regarded as an "organism" tending towards closure and self-determination: it is an end in itself.... The technological system, embodied, of course, in the technicians, admits no other law, no other rule, than the technological law and rule visualized in itself and of itself.[46]

Technology is autonomous because we think it is the only way to think and live. Technology doesn't simply determine us. We allow it to drive itself, which means we give it the freedom to take away ours.

Thinking and living as if technology were the only way to think and live is, of course, a socially constructed human idea, and it depends on human societies to actualize it. But when we let it have its own way, it masters us rather than us mastering it. When enframed by technological thinking, we only feel free to make decisions acceptable to techno-logical thinking. Nobody in a technological system is free to manage that system in any other way than how techno-logic demands that it be managed. Not even engineers or politicians have the freedom to control the technological system, because they are "spiritually taken over by the technological society; they are the most fervent adepts of that society. They have been profoundly technicized."[47] If they weren't, they wouldn't be given the authority or position to manage the system in the first place. On Ellul's analysis, therefore, modern human beings are mastered by their own autonomous technology, which is first and foremost an idea.

> **Technological Autonomy:** allowing technology to follow its own directions by believing that it provides the only way to think and live, rather than directing and controlling it in accordance with our own human purposes and values.

Freedom and Hope

Does technological autonomy mean that the technoscientific system actually achieves self-consciousness, intentionally directing society towards its own ends as if it literally had a mind of its own? No. Ellul clearly states that he is "not personifying in any way," but "simply using an accepted rhetorical shortcut"[48] in the same way that "liberal economists ... speak of the laws of the market."[49] It's odd, actually, that the usual story finds it so easy to speak of the economy as if it were a living, breathing organism, while it wants to treat everything else—including technological artifacts and the human systems that generate them—as a lifeless and inert collection of parts (see Chapter Eleven). There is much to be said about technology and economics (including the possibility that Ellul and others, like Langdon Winner, are economic rather than technological determinists),[50] but Ellul's main point is that technology *envelops* economics. Economics is a technological discipline, rather than technology being an economic discipline, which is why modern socialist and communist societies are just as dominated by techno-rationality as capitalist ones are. If we think the economic system is tough to control or resist, then the technological system would be even more so.

Recall that one of the main motivators behind social constructivism was to emphasize human freedom and control over science and technology. Ellul shares the same concern, in fact:

> [I]f each one of us ... abdicates [our] responsibilities with regard to values; if each of us limits [ourself] to leading a trivial existence in a technological civilization, with greater adaptation and increasing success as [our] sole objectives; if we do not even consider making a stand against these determinants, then everything *will* happen as I have described it, and the determinants *will* be transformed into inevitabilities.[51]

Humans will lose their freedom and control over science and technology if they surrender it to an autonomous technological system that will treat us like tools. The solution, however, is not to assert that the social construction of technology means that the technological system of modernity doesn't have momentum, autonomy, or embodied value; the solution is refusing to surrender our freedom to it. Technologies themselves, as objects, are not intrinsically bad for Ellul, but treating human beings like technologies is. Treating human beings like they are artifacts themselves means treating them like material mechanisms that can and should be controlled and determined by an external system of governance. In politics, this system of governance has a name: technocracy.

Technocracy is a political system where everyone is ruled by technical experts. All social questions are determined to be technical questions, and so technicians should answer them, not the voting population. In such a political system, the non-technical masses

> can no more be said to be agents of … choices than [we] can be said to have a choice when deciding whether four is quantitatively more than three. [Humans] are more appropriately understood to be apparatuses for registering the results to be obtained by different techniques. The choices are made for them by the requirements of the technical milieu in which they find themselves. Once one line of action is seen to entail maximum efficiency, the decision has been made.[52]

Technocratic rule by techno-logic is inherently anti-democratic. This is why Ellul is concerned when all our social problems are supposed to be solved technologically; we might succeed in making everyone happy thanks to our techniques of manipulation, but at the expense of forcibly harnessing everyone to a system that treats us all as material objects to be manipulated like slaves are. Adapting human beings to the technological system—instead of adapting the technological system to human beings—abolishes our freedom and is thereby dehumanizing, even if we're too drugged or brainwashed to feel anything but good about it.

Technocracy: a political system where the rulers are technical experts.

There is no technological solution to this problem, for the problem *is* technological solutionism. If there were a technological solution, the solution would perpetuate the problem rather than solve it. This is why it is so incredibly difficult to think of how we might extricate ourselves from the problem. The only solutions we're familiar with are technological ones, because the usual story's technological rationality already has a hold on our imaginations. But Ellul does not say that "no solution will be found; I only aver that in the present social situation there is not even a beginning of a solution, no breach in the system of technical necessity."[53] This means that "the technological society is a totalizing *potentiality*,"[54] one that we actualize depending on the actions we take, consciously or not. The solution to technological autonomy is to *stop believing in it*.

Put another way, the solution to Enframing is to replace the usual story with better informed perspectives on science and technology. This includes accepting that "machines … are relatively useful, but not all that important."[55] Otherwise, we'll never be in a position to start resisting the momentum embedded in technological systems. Technologies are important as objects of study within STS and related social analysis, but they're not as important as they usual story says they are. They're not the salvation of a free society, they're not what extracted humanity from the State of Nature, and they're not going to let humanity colonize the universe the way Francis Bacon wanted to. The critical perspectives of STS makes it "open to [us] to overcome [technological] necessity, and that … *act* is freedom."[56] In thinking through technological determinism and the possibility that technology is a system of thought in addition to being a collection of artifacts and synthetic items, STS opens up the possibility of acting freely with respect to technology rather than acquiescing to it. This is why, in this chapter, we've started working through the theoretical tensions in our culture's engagement with technology. STS "is a call for the sleeper to awake."[57]

Where does that leave us, then, with the suspicion we had at the start of this chapter, namely that technologies in our societies are advancing in a way that is out of control? First, we need to remember that technology isn't just a bunch of devices, but a way of thinking (itself a technique) which falsely proclaims itself as the only way to think, even though it is—at most—only a few hundred years old. Second, this way of thinking pervades the societies we live in, because our societies have been built *by* the technological system of thought and they now *embody* it. So it's not just that we've been *told* the usual story for a long time; we've literally been *living in it* as a social system for several generations already.

Third, the social system embodying techno-rationality presupposes a number of values that everyone is supposed to accept without question, especially mathematization and efficiency. Everything, including you, should be evaluated as a tool, and believing otherwise is supposed to be crazy. Fourth, that's why it feels like we are in a technologically determined situation, as if the system had a mind of its own that lies outside human control. The values of techno-thinking are embedded in all the systems of our modern world, and we rarely see how that came to be. We adapt to that situation before we even realize that's what we're doing, and we assume that the situation—our technological environment—is normal, natural, and inevitable. We think and we live as if it is in charge, and as if all we can do is acquiesce to it.

The usual story *wants* us to think that technology has a mind of its own, therefore, otherwise we might attempt to exert our own control over it, rather than letting it take its own course. We might actually think outside the enframing box of techno-logic and work towards living lives that *aren't* dominated by power-seeking control over every material particle in the universe. If we did that, however, it would create a 'problem' for the few people who already benefit from this constructed social system, and so they will use ideological weapons, including science itself (see Chapter Two), to ensure that we do not. Technology doesn't really have a mind of its own, but it does if most of us believe that it does. That's where its autonomy comes from.

Ellul thought we had only a very narrow window of time within which to reject the dominance of technological rationality. Perhaps he was right and that opportunity has passed. It may be unthinkable, now, to see how we as a society could re-engineer our systems so that they *don't* embody *la technique*. Even so, that doesn't make resistance futile. The usual story, even when it is literally built into our society's infrastructure, is not as invulnerable as it wants us to think. There are cracks and gaps in the machine; it could very well fail or collapse. Regardless of what will eventually happen to the system itself, individuals can still create moments of freedom within these gaps or cracks. We can always switch off our phones, for example, even though the technological system wants to make that difficult for us to do. And we can always organize ourselves, to work towards creating communities where the usual story does not dominate us. The usual story does not have to be the final word—not as long as there are human beings out there who can see through it.

Notes

1 Sergio Sismondo, *An Introduction to Science and Technology Studies*, 2nd ed. (Wiley-Blackwell, 2010), 9.

2 Anabel Quan-Haase, *Technology and Society: Social Networks, Power, and Inequality*, 2nd ed. (Oxford University Press, 2016), 60.

3 Sally Wyatt, "Technological Determinism Is Dead; Long Live Technological Determinism," in *The Handbook of Science and Technology Studies*, 3rd ed., ed. Edward J. Hackett et al. (MIT, 2008), 167.

4 Wyatt, "Technological Determinism," 168.

5 Wyatt, "Technological Determinism," 168.

6 US Energy Information Administration, "History of Gasoline," last updated 22 December 2023, https://www.eia.gov/energyexplained/gasoline/history-of-gasoline.php.

7 James Howard Kunstler, *The Geography of Nowhere: The Rise and Decline of America's Man-Made Landscape* (Simon & Schuster, 1993).

8 US Geological Survey, "Distance to Nearest Road in the Conterminous United States," Fact Sheet 2005-3011, January 2005, https://pubs.usgs.gov/fs/2005/3011/report.pdf.

9 Wyatt, "Technological Determinism," 168.

10 Wyatt, "Technological Determinism," 169.

11 Quan-Haase, *Technology and Society*, 51.

12 Quan-Haase, *Technology and Society*, 60.

13 Quan-Haase, *Technology and Society*, 51.

14 Jacques Ellul, *The Technological Society*, trans. John Wilkinson (Vintage, 1964), xxv.

15 Don Ihde, "Foreword," in *American Philosophy of Technology: The Empirical Turn*, ed. Hans Achterhuis (Indiana University Press, 2001), viii.

16 Jacques Ellul, "Ideas of Technology," trans. John Wilkinson, in *1984 and All of That*, ed. Fred H. Knelman (Wadsworth, 1971), 12; original emphasis.

17 Ellul, *Technological Society*, xxv.

18 Hans Achterhuis, "Introduction: American Philosophers of Technology," trans. Robert P. Crease, in *American Philosophy of Technology*, 3.

19 Jacques Ellul, "A Theological Reflection on Nuclear Developments: The Limits of Science, Technology and Power," in *Waging Peace: A Handbook for the Struggle to Abolish Nuclear Weapons*, ed. Jim Wallis (Harper and Row, 1982), 115–16.

20 Achterhuis, "Introduction," 8.

21 Sismondo, *Science and Technology Studies*, 140.

22 Robert L. Sexton et al., *Exploring Microeconomics*, 5th Canadian ed. (Nelson, 2016), 14–42.

23 Ellul, *Technological Society*, xxv; emphasis removed.

24 Ellul, "Ideas of Technology," 12.

25 Ellul, "Ideas of Technology," 18.

26 John D. Rockefeller (the richest person in modern history, with his personal wealth totalling between 2 and 3 percent of the entire American economy) is often credited with saying something like this, but there appears to be no reliable source for this attribution.

27 Ellul, "Ideas of Technology," 12.

28 Quan-Haase, *Technology and Society*, 127.

29 Martin Heidegger, *The Question concerning Technology and Other Essays*, trans. William Lovitt (Harper & Row, 1977).
30 Ellul, "Ideas of Technology," 12.
31 *Transcendent Man: The Life and Ideas of Ray Kurzweil*, directed by Barry Ptolemy, Ptolemaic Productions and Therapy Studios, 2009, https://transcendentman.com/product/transcendent-man-documentary/.
32 Ellul, "Ideas of Technology," 12–13; original emphasis.
33 Kevin Garrison, "Perpetuating the Technological Ideology: An Ellulian Critique of Feenberg's Democratized Rationalization," *Bulletin of Science, Technology, and Society* 30 (2010): 197.
34 Quan-Haase, *Technology and Society*, 60.
35 Langdon Winner, "Do Artifacts Have Politics?," *Daedalus* 109, no. 1 (1980): 122.
36 Achterhuis, "Introduction," 8.
37 Wyatt, "Technological Determinism," 168.
38 Wyatt, "Technological Determinism," 172.
39 Wyatt, "Technological Determinism," 168.
40 Wyatt, "Technological Determinism," 175.
41 Wyatt, "Technological Determinism," 176.
42 Wyatt, "Technological Determinism," 169.
43 Ellul, *Technological Society*, xxvii–xxviii.
44 Ellul, *Technological Society*, xxviii; emphasis removed.
45 Ellul, *Technological Society*, xxviii; emphasis removed.
46 Jacques Ellul, *The Technological System*, trans. Joachim Neugroschel (Continuum, 1980), 125–26.
47 Ellul, "Ideas of Technology," 15; see also p. 17.
48 Ellul, *Technological System*, 335.
49 Vincent Punzo, "Christian Hope in a Technological Age: Jacques Ellul on the Technical System and the Challenge of Christian Hope," *Proceedings of the American Catholic Philosophical Association* 70, no. 1 (1996): 23.
50 Wyatt, "Technological Determinism," 168.
51 Ellul, *Technological Society*, xxix.
52 Punzo, "Christian Hope," 23.
53 Ellul, *Technological Society*, xxxi.
54 Garrison, "Technological Ideology," 197.
55 Jacques Ellul, *Perspectives on Our Age: Jacques Ellul Speaks on His Life and Work*, trans. Joachim Neugroschel (Canadian Broadcasting Corporation, 1981), 109.
56 Ellul, *Technological Society*, xxxiii; original emphasis.
57 Ellul, *Technological Society*, xxxiii.

VII

Case Studies

Chapter Eighteen

Science and Religion

THIS BOOK IS AN INTRODUCTION TO THE SOCIAL STUDY OF SCIENCE AND technology, and eventually introductions come to an end. You've been introduced to the idea of a usual story about science and technology, a standardized narrative, familiar to most people in the globalized Western world. It claims technoscience to be the salvation of a free society and the deepest desire of the human spirit across time. You've been introduced to the contrary idea that prehistoric societies had virtually no interest in this narrative, but that its first seeds were planted in relatively recent agrarian civilizations that invented writing and mathematics for the management of land, weather, and slaves. You've seen distinctly Western forms of scientific rationality and technological power emerge in the slave economies of Greece and Rome, and their unification emerge in the Catholic countries of medieval Europe. You've been introduced to the likelihood that these developments were neither inevitable nor necessarily good, but socio-historically contingent and morally problematic. Nevertheless, modern Western technoscience found itself coalescing a few hundred years ago in an ideological soup of power-hungry, sexist, and racist progressivism and mechanical explanation. Therefore, the historical course of contemporary global technoscience was not "one of linear growth," but was "challenged as vacuous, or mean spirited, or blasphemous" by the likes of Romanticists and Luddites.[1] STS as an academic discipline arose out of technoscientific controversies in the twentieth century that included positivism, the history of scientific revolutions, and industrialized warfare. Finally, you've been introduced to controversies internal to STS itself: scientific social constructivism, technological value-ladenness, and technological autonomy.

With this background, you can start applying STS perspectives to any number of particular issues you might face. Much STS research is just this: case study analysis, or taking a careful look at a scientific issue (e.g., the Galileo affair) or technological change (e.g., the safety bicycle) and applying or revising STS theories as needed to better understand those examples. I propose that we do the same as we wrap up this book. In this last section, let's examine some controversial contemporary issues about science and technology in light of our newfound (albeit introductory) understanding of STS theory. Two hugely divisive issues immediately come to mind in my North American context: religion and climate change. Each of these should provide plenty of opportunity to practice our STS skills, helping us deal with massively important questions that continue to bedevil modern Western societies. For this chapter, then, we'll look at science and religion in STS.

Whose Science, Whose Religion?

It's no secret that science and religion are at each other's throats in America, if not elsewhere. It takes a split second internet search to find thousands of examples of religious-based skepticism towards science (one example: "Christians Against Dinosaurs")[2] and thousands of examples of science-based debunking of religious-based alternative science (another example: "Dinosaurs Against Christians Against Dinosaurs").[3] If ever there was a culturally rich motherlode for STS to mine, this would certainly be one of them.

But what science and what religion are at odds here? All of them? Is Mesopotamian astrology opposed to Vedic Hinduism? Is Renaissance alchemy set against Second Temple Judaism? What about Aristotelian metaphysics versus Greek polytheism? At least in the latter case, Aristotle (like Socrates before him) was accused of atheism and impiety, which feeds into the usual story of scientists as brave rational inquirers "persevering against religious interference and obscurantism."[4] On the other hand, we have seen in this book a complicated interplay between *mythos* and *logos*, such that Western and eventually modern technoscience *exist* because of the Catholicism of St. Francis and the Protestantism of Sir Francis Bacon, to say nothing of the theology of Sir Isaac Newton or the atheism of the logical positivists. There's a lot to untangle here.

> **Young Earth Creationism:** the belief that the Bible conveys scientific truths about the age of the Earth and the process by which life originated; therefore, God created the world and all life on it over the course of six 24 hour-long days roughly 6,000 years ago.

Even so, the current science and religion mayhem we might see around us is largely restricted to a particular subset of Christianity directed at a particular subset of modern Western technoscience. It isn't Buddhists who have a problem with dinosaurs, or even most Roman Catholics, for that matter. To understand the science and religion issue, therefore, we have to carefully delineate both the sciences and the religions at issue. The science and religion debate is not a conflict, for example, between quantum physics and Russian Orthodoxy. Rather, it is a dispute that fundamentalist and conservative evangelical Protestant Christianity in North America has with the evolutionary life and Earth sciences of the modern West (i.e., biology and geology). The science and religion in our case are modern and Western, even American, and the debate almost always boils down to evolutionary science versus so-called creationism. So what's that?

Young Earth creationists believe that the Bible should be interpreted literally and, as such, that it conveys divinely inspired and inerrant truths about the age of the Earth and the process by which life originated. On this view, God created the world and all life on it over the course of six 24 hour-long days roughly 6,000 years ago. Each biological species was created distinct from every other one, out of nothing, by God. Young Earth creationism, as a theory, dates back to the early twentieth century.[5] Old Earth creationists, meanwhile, interpret the Bible more figuratively, allowing that the 'days' of God's creating could refer to much longer epochs of time. However, they still hold that species do not evolve into other species, and as such there is no common biological descent from a single ancestral group. Again, God created each species as a distinct, unchanging lifeform category. Old Earth creationism wouldn't have been in conflict with biological science prior to Charles Darwin and Alfred Russel Wallace in the mid-nineteenth century, and with the advent of evolutionary biology in that period, many Christians abandoned Old Earth creationism. They accepted that God created life through an evolutionary process without abandoning their faith. In fact, "the leading Darwinian in [nineteenth-century] North America, who did more than any other scientist to get Americans to take Darwin seriously, was Asa Gray," a Harvard botanist who was also a strongly committed Congregationalist Christian.[6]

Old Earth Creationism: the belief that the Bible figuratively refers to long epochs of time as 'days,' allowing for consistency with modern geological theory while denying modern biology's theory of common descent from a single ancestral group.

Two Kinds of Literalism

When we see the science and religion debate in historic, geographic, and theological context, it becomes somewhat more manageable. While some in STS may want to investigate the conflict between Aristotle's science and Macedonian civic religion—and while we might learn valuable lessons from that study which could be applicable to contemporary situations—we can focus on the specifics of the contemporary situation so long as we avoid painting 'science' and 'religion' with overly broad brushes. Specifically American evangelical Christianity and specifically modern evolutionary biology and geology are at issue here, and some of those scientists think their science has religious implications, and some of those Christians think their Scriptures have scientific implications. In fact, each of them think their respective community provides facts which overcome the beliefs of the other community. The religious folk here think they have divinely revealed truths that obliterate the lies of modern science, while the scientific folk here think they have factual evidence that obliterates the quackery of fundamentalist religion. In both cases, these groups seem to be fighting over the same epistemological territory: who has privileged access to the literal truth about nature?

The biologist Lewis Wolpert, for example, considers himself "an atheist reductionist materialist" because he believes that science is "the best way to understand the world."[7] For Wolpert, science is 'materalist,' which in this context doesn't mean unmitigated consumerism, but rather the metaphysical conviction that everything that could possibly exist is made out of matter (see Chapter Eleven). That is also why he's a 'reductionist.' All possible experiences, phenomena, or facts about the world must ultimately be reduced to and explained in terms of their material components: mechanistic atomism, in other words. Finally, given this view of science, atheism follows. Unless there are gods made out of physical particles, there can be no divinities or anything else remotely immaterial in the universe. The *logos* of science (see Chapter Seven) literally describes the physical facts as they are, apart from which nothing else can be.

Scientific Absolutism: science is the best way to understand anything at all; everything that exists should only be explained in terms of material particles and forces.

On the other side, creationists are literalists about what they believe to be the infallible word of God found in the Bible. American evangelicals and fundamentalists have often been explicit and forthright about this: "a system of hermeneutics [textual interpretation] which is usually called literal" but also "normal or plain" was one of the defining features of proper theology, according to Charles Ryrie in 1965.[8] Thus, "the Bible is to the theologian what nature is to the man of science. It is his storehouse of facts; and his method of ascertaining what the Bible teaches, is the same as that which the natural philosopher adopts to ascertain what nature teaches...."[9] Interpreting the Bible literally to derive factual knowledge was seen as the same method as "that which is pursued in the science of physics" and in accordance with "the maxims [or rules of action] of Bacon and Newton," said Leonard Woods, Jr. in 1822.[10] Biblical literalism, therefore, sees itself as the application of modern scientific method to the activity of interpreting religious texts, and in this respect, both the fundamentalists and the 'atheist reductionist materialists' are using the same early modern approach to epistemology: absolutist literalism.

Biblical Literalism: the Christian Bible is the best way to understand anything at all; everything that exists should only be explained in terms of a literal reading of the biblical text.

The question remains whether this early modern approach to epistemology is potentially problematic (see Chapter Fourteen). Its insistence on literalism—be it materialist or supernaturalist—makes it incapable of allowing for other forms of explanation, and its insistence on absolutism makes it difficult to consider other perspectives. And yet it may be that we need multiple forms of reasonable explanation to do justice to human experience. The Romanticists asked, for example, if there was something more to life than just the repetitive and dead world of industrial drudgery (see Chapter Twelve). After all, poetry cannot be fully understood by reducing it to its material components of paper and ink. In this book we've also considered the comparative legitimacy of differing modes of discourse: *mythos* and *logos*, method and goal, or physics and metaphysics. With these distinctions in mind, we can recognize that the answer to a question like "why is the water boiling?" can be "because the water's temperature is +100°C" just as well as it can be "because I wanted to make a cup of tea." Neither answer is reducible to the other, but both appear to be reasonable and potentially true answers to the (ambiguous) question. Similarly, therefore, the science and religion debate might boil down (pun intended) to the issue of the legitimacy of different forms of explanation.

How vs. Why Questions

The 2001 winner of the Nobel Prize for Physics was Dr. Eric Cornell, along with Drs. Carl Wieman and Wolfgang Ketterle, granted for first synthesizing Bose-Einstein condensate (about which we need not concern ourselves here, thankfully). When Cornell was inducted into the American Academy of Arts and Sciences, he gave a speech on the relationship between science and religion. In a version of the speech published in *TIME* magazine, he said "my years of scientific research have made me a renowned expert on my topic: God. Just kidding."[11] This little joke made an important point: Cornell's Nobel Prize-winning work in physics had no scientific relevance to religious questions or answers. The inverse was also true, he claimed; religion had no scientific relevance to scientific questions or answers. Like my example of the boiling water above, the example Cornell used was also framed as a question: "Why is the sky blue?" As a physicist with expertise in optical phenomena, he could respond that the sky is blue because of Rayleigh scattering: the shorter wavelength of blue light means it predominates in the atmosphere while the other colors of the visible spectrum of light are directed away from the terrestrial observer. He notes, however, that one possible religious answer to the question would be "the sky is blue because blue is the color God wants it to be."[12] Cornell sees no conflict between these two answers because he is aware—unlike a reductionist or literalist—that the question of 'why' is ambiguous.

Ambiguity is when two meanings masquerade as one. In English, the word 'why' can elicit answers ranging from mechanical processes to personal choices to ultimate meaning. In Cornell's understanding, science does not provide answers about ultimate meaning or even personal choices, but rather mechanical processes. In English, however, there is another word that evokes mechanical processes less ambiguously: 'how.' Cornell could have said "how is the sky blue?" in which case the answer would unambiguously be Rayleigh scattering. Science, he says, is about "understanding nature and the reasons for things,"[13] or answering 'how (does it work)?' questions. On the other hand, if we restricted 'why' questions to just those about "values, ethics, morals or, for that matter, God," Cornell says that science cannot answer those kinds of questions.[14] And that's okay, he says.

In this way, Cornell is not a reductionist or a literalist, where only one form of explanation (science or religion) is possibly true. Rather, both could be true, but true about different 'things,' dimensions of being, modes of experience, or in different ways. This distinction mirrors the different Aristotelian causes (see Chapter Seven), where some explanations would be within the field of what we call technoscience (matter and efficacy) while others would be within the fields of what we call metaphysics or religion (essence and purpose). We could also utilize the *mythos-logos* distinction here (see Chapter Four). Religion functions in much the same way as myths and rituals, providing psychological satisfactions for non-material needs, while technoscience is supposed to provide a reliable picture of processes that can be used to effectively solve material problems. In this respect, there's no reason why one of the two should have to compete with or replace the other. Satisfactions of material needs do not have to satisfy non-material needs, and vice versa. On Cornell's understanding, science and religion could actually get along.

Cornell's model of the science–religion relationship isn't particularly new. It is substantively similar to the model proposed by numerous Western Christians over the past 1,500 years. In his 1615 letter to the Grand Duchess Christina of Lorraine, for example, Galileo (yes, that guy) argued that literalists

> want to extend, not to say abuse, [the] authority [of Scripture], so that for even purely physical conclusions which are not matters of faith one must totally abandon the senses and demonstrative arguments in favor of any scriptural passage whose apparent words may contain a different indication.... I think that in disputes about natural phenomena one must begin not with the authority of scriptural passages but with sensory experience and necessary demonstrations. For the Holy Scripture and nature derive equally from the Godhead, the former as the dictation of the Holy Spirit and the latter as the most obedient executrix of God's orders....[15]

Galileo's position was that religious scripture was no more an authority on astronomical matters than astronomy was an authority on religious matters. Rather, the authority on matters of physical science was sense experience (empirical induction) and logical argument (rational deduction; see Chapter Eight). In this respect, Galileo was relying on St. Augustine of Hippo, who thought that if Christians used their (often misunderstood) readings of the Bible to overrule the best science of the day, this only displayed their ignorance and made their faith disreputable:

> Usually, even a non-Christian knows something about the earth, the heavens, and the other elements of the world, about the motion and orbit of the stars and even their size and relative positions, about the predictable eclipses of the sun and moon, the cycles of the years and the seasons, about the kinds of animals, shrubs, stones, and so forth, and this knowledge he holds to as being certain from reason and experience. Now, it is a disgraceful and dangerous thing for an infidel to hear a Christian, presumably giving the meaning of Holy Scripture, talking nonsense on these topics; and we should take all means to prevent such an embarrassing situation, in which people show up vast ignorance in a Christian and laugh it to scorn.[16]

If St. Augustine, writing in the fifth century CE, had encountered the "Christians Against Dinosaurs" group, he would have sided with the dinosaurs.

The framework of Augustine and Galileo is known as a 'Two Books' theology, which is similar to Cornell's distinction between 'how' and 'why' questions. They thought that God spoke in the book(s) of the Bible, but also in another book: the 'book' of nature, which includes our natural capacities for sense experience and reasoning. Their religious scriptures, then, comprised the book of God's *words* (things God presumably *said*), whereas the natural world and our scientific understanding of it comprised the book of God's *works* (things God presumably *did*). The book of God's *word* would be (according to their religion) where one would find answers to 'why' questions about religious significance or meaning, while the book of God's *works* would be (also according to their religion) where one would find answers to 'how' questions about the way things operate. The way things operate—natural processes—is the way God made them (according to their religion), and the way to figure out how God made them operate (according to their religion) is to study those operations scientifically, not to dig around in the Bible. God didn't write the Bible to explain to people how natural processes work. Those operations are nevertheless how God achieved the goal of making those things work, which is why Cornell can say that "Rayleigh scattering [is] the method God has chosen to implement his colour scheme" for the sky.[17] Even some American evangelicals held a two-books theology in the second half of the nineteenth century: "Nature is as truly a revelation of God as the Bible; and we only interpret the Word of God by the Word of God when we interpret the Bible by science."[18] Another way to put the distinction comes from the sixteenth-century Cardinal Cesare Baronio (paraphrased by Galileo): the purpose of religion "is to teach us how one goes to heaven and not how heaven goes."[19] The purpose of science, therefore, is to teach us how the heavens (etc.) work, but not how to go to heaven.

Can Science and Religion Productively Interact?

STS research into science and religion has progressed to the point of becoming a sub-discipline in its own right, complete with its own specialist researchers and academic journals. We certainly cannot dive into all that in detail here, but generally the field of science and religion tests out various theoretical models of interaction between science and religion, trying to find the optimal one (if any). We can conclude this chapter by briefly surveying some of these proposed models.[20]

In fact, we've already encountered several of them. One model is *positivism* reborn (see Chapter Thirteen); it is a model of *conflict* between science and religion. Here the 'how' and the 'why' questions come into conflict because there can be only one form of explanation: scientific literalism, the materialist-reductionist description of empirical facts. How and why questions are not seen as fundamentally distinct. Richard Dawkins represents this model excellently when he says that "the question of whether there exists a supernatural creator, a God, is one of the most important that we have to answer. I think it is a scientific question. My answer is no."[21] This model of science and religion conflict doesn't hold up well under scrutiny, though, for reasons similar to the self-referential absurdity of positivism: on the one hand, science is supposed to be the only way to think properly, but on the other hand, scientific method is explicitly limited to natural phenomena. Therefore, supernatural possibilities cannot be scientific questions! Positivism is a model of science and religion conflict because it overstates what science can do, namely answering questions that it is not equipped to answer. Religious issues are metaphysical, and science does not examine metaphysics, so (in Cornell's words) "neither I nor any scientist ... has anything scientific to say about ... the God answer. Not to say that the God answer is unscientific, just that the methods of science don't speak to that answer."[22] It's impossible for the existence of a supernatural creator to be a scientific question, and Dawkins ought to know better.

The other conflict model of science and religion doesn't hold up well to scrutiny either. *Biblical literalism* is the idea that any statement in the Bible about the material world is absolutely true because it is divinely inspired, and any scientific statement which appears to contradict a divine statement about nature must therefore be false. So when Dawkins complains that "there are plenty of places where religion does not keep off the scientific turf,"[23] he's referring to fundamentalist Christians who think that the Bible *is* science. The turf they're fighting over is absolute truth, and both positivism and literalism want to be the only things that occupy that ground. Even if the Bible is infallible because it is divinely inspired (and that is a very big 'if'), it is not at all clear that the purpose of divine scripture is to convey scientific truths about material processes to human readers. It is also not at all clear that it should be interpreted literally—as if it were a scientific textbook—when religion is offering a "description of who God was, who we are and what our relationship is supposed to be with God."[24] In these crucially central respects, religious scriptures provide a mythological and even ritualistic function because they seek to provide a form of non-material satisfaction regarding ultimate meaning and significance. Of course, the usual story has tended to see mythic

> **Conflict Models:** seeing science and religion as interacting antagonistically because both are seen as trying to do the same thing: describing the facts of the universe literally to arrive at absolute truth.

expression as illegitimate and false, so in that context it makes sense that some religious believers would want their faith to occupy the holy ground of literally true *logos*. However, it's not clear that anything can adequately occupy the ground of absolute truth (it failed for positivism). Religious literalism is a conflict model of science and religion because it is trying to take the place of positivism in the usual story, and it doesn't realize that the usual story is deeply flawed in a wide variety of ways. Trying to get religion to fulfill the pipe dreams of the usual story is very likely a dead end.

A third model of science and religion interaction tries to avoid conflict by dispensing with contact altogether. Proposed by the Harvard paleontologist Stephen Jay Gould, *non-overlapping magisteria* (NOMA) is a model of interaction that sees science and religion as each having their own proper authority and domain (that's what he means by "magisteria"). Each has a different area of expertise with different aims in mind, much like the distinction between 'how' and 'why' questions we saw earlier. When both science and religion look at the same event (say, perhaps, a flood), the results of their investigations will differ (e.g., the flood was caused by climate change, and the flood was divine retribution against sin). Though scientific and religious explanations differ, the two explanations are compatible: one is a causal account describing material processes, the other is a purposive account about meaning and significance. In principle, science and religion cannot come into conflict in the NOMA model because 'how' and 'why' questions are completely disambiguated.

Gould's NOMA Model: science and religion each have their own proper authority and domain (i.e., a "magisterium") with correspondingly different areas of expertise and goals; neither magisterium can overlap, so conflict between science and religion is improper.

Of course, both positivists and literalists fail to "embody the principles [of NOMA] precisely,"[25] but that isn't a problem for the NOMA model: it's just a problem for positivists and literalists. A more significant problem for the NOMA model is that it "sets up an artificial wall between the two worldviews that doesn't exist in my life," according to Francis Collins, director of the human genome project and an American Protestant Christian.[26] If a scientist can be a religious believer at the same time, why can't there be any possible connection between these two very important aspects of their lives? Lived human experience may integrate the two in far more complex ways than the NOMA model allows. Perhaps answers to how and why questions should be better integrated with each other, rather than separated.

Intelligent Design as a 'God of the Gaps' Model: religion can provide explanatory hypotheses that fill gaps in scientific explanatory frameworks.

One way in which science and religion might come into fruitful contact or even dialogue with each other is a model known as the *God of the gaps*. This model proposes that when science struggles to find an adequate hypothesis for explaining a natural phenomenon (say, the extreme precision of the gravitational and other universal constants required for the possibility of life), religion can offer God as an explanatory hypothesis that fills that gap in scientific knowledge. Sometimes this is referred to as 'intelligent design' (ID) theory, where certain features of the physical universe are supposed to be so difficult for science to explain that those features constitute inductive evidence for the existence of a divine creator. The problem with this model, however, is that as science increases its understanding of the universe, the role of God as an explanatory hypothesis correspondingly shrinks. In this respect, the God of the gaps reduces to a conflict model like the ones found wanting above. According to the God of the gaps model, religion first replaces science, but later science replaces religion, as the latter discovers better explanations for how things work than the former, and the gaps close.

There is another understanding of 'intelligent design' that doesn't succumb to this problem, however. According to Cornell,

> [a]s a theological idea, intelligent design is exciting. Listen: If nature is the way it is because God wants it to be that way, then, by looking at nature, one can learn what it is that God wants! The microscope and the telescope are no longer merely scientific instruments; they are windows into the mind of God.[27]

Natural Theology as an Alternate Model of Intelligent Design: scientific explanations of natural phenomena can have religious implications, giving indications of the mindset of God the way a work of art gives an indication of the mindset of the artist.

Scientific knowledge, according to this model, is neither evidence for nor against the existence of God. However, if there is a divine creator, then science can give us some indications of what that divine creator is like. This is called *natural theology* (see Chapter Eight). Collins finds that "studying the natural world is an opportunity to observe the majesty, the elegance, the intricacy of God's creation."[28] Somewhat ironically, Dawkins also learned something of the mind of God through natural theology:

> If God wanted to create life and create humans, it would be slightly odd that he should choose the extraordinarily roundabout way of waiting for 10 billion years before life got started and then waiting another 4 billion years until you got human beings capable of worshipping and sinning and all the other things religious people are interested in.[29]

From what science has discovered about the natural world, religion can derive the why-level implications that God is majestic, elegant, interested in intricacies, and also extraordinarily patient, slightly odd by our standards, and quite possibly not as anthropocentric as many religious people seem to be. Dawkins thinks this is 'slightly odd,' but that's just a hang-up he seems to have. Natural theology means that science can modify what religion thinks about God (etc.), and that's not necessarily a bad thing. Natural theologians embrace what they discover about God from studying nature.

Finally, however, the study of religion can modify our understanding of science, just as much as the study of science can modify religious understanding. There isn't really a name for this as a model, though. We have already seen in this book how science has metaphysical commitments, embodies various paradigms or worldviews, and even can serve (or tries to serve) religious functions (see Chapter Two). Attending to these unusual dimensions of science is one of the basic tasks of STS, and the skill set of religious studies can be particularly useful in this regard. (One of the most famous STS scholars, Bruno Latour—see Chapter Fifteen—did his Ph.D. in philosophical theology!) Scholars of religion have developed sophisticated understandings of different forms of human thought and behavior that go beyond the literal and the mathematical. Awareness of these dimensions allows science to be interpreted in light of religious or philosophical categories. For example, the theologian John Haught argues that "religion is in a very deep way supportive of the entire scientific enterprise."[30] Many scientists speak in broadly religious terms about "something incredibly grand and incomprehensible and beyond our present understanding" that science might reveal, "the most wonderful range of future possibilities, which I cannot even dream about, nor can you, nor can anybody else."[31] Even when Dawkins refers to "the spirit of science," he's using an explicitly religious concept.[32] Wittgenstein might remind him that there is an entire mythology deposited in our science,[33] and that is not the usual story!

Overall, it should be clear that the relationship between science and religion is not a scientific question. It cannot be resolved by the application of scientific methods. Indeed, though we have consulted scientists in this chapter (Cornell, Dawkins, and Collins), none of them were operating within their areas of scientific expertise or performing experiments to find their answers. They were doing STS. The relationship of science and religion is *metascientific*, and that lies within the expertise of the liberal arts, humanities, social sciences, and even the scholarly study of religion. The religious dimension of science is one of the areas where STS can be effectively applied to provide insight into difficult issues of pressing concern. Another such area is anthropogenic climate catastrophe, to which we now turn.

Notes

1 Steven L. Goldman, "Science, Technology, and God: In Search of Believers," *Bridges* 2, no. 1–2 (1990): 51.

2 "Christians Against Dinosaurs–CAD," Facebook, accessed 30 August 2022, https://www.facebook.com/ChristiansAgainstDinosaurs/.

3 "Dinosaurs Against Christians Against Dinosaurs (DACAD)," Facebook, accessed 30 August 2022, https://www.facebook.com/groups/1534830836769348/.

4 Goldman, "Science, Technology, and God," 47.

5 Mark A. Noll, *The Scandal of the Evangelical Mind* (Eerdmans, 1994), 189–90.

6 Noll, *Evangelical Mind*, 179.

7 Lewis Wolpert, *Six Impossible Things before Breakfast: The Evolutionary Origins of Belief* (W.W. Norton & Company, 2007), x.

8 Charles Ryrie, *Dispensationalism Today* (Moody Press, 1965), 44–46; quoted in Noll, *Evangelical Mind*, 118.

9 Charles Hodge, *Systematic Theology*, 3 vols. (Eerdmans, 1952; originally published in 1872–73), 1:10–11; quoted in Noll, *Evangelical Mind*, 98.

10 Leonard Woods, Jr., *Letters to Unitarians* (Andover, 1822), 18–21; quoted in Herbert Hovencamp, *Science and Religion in America, 1800–1860* (University of Philadelphia Press, 1978), 61.

11 Eric Cornell, "What Was God Thinking? Science Can't Tell," *TIME* (Canadian ed.), 14 November 2005, 72.

12 Cornell, "What Was God Thinking?," 72.

13 Cornell, "What Was God Thinking?," 72.

14 Cornell, "What Was God Thinking?," 72.

15 Galileo Galilei, "Letter to the Grand Duchess Christina," in *The Galileo Affair: A Document History*, ed. and trans. Maurice A. Finocchiaro (University of California Press, 1989), 90, 93.

16 Augustine, *The Literal Meaning of Genesis*, 2 vols., trans. John Hammond Taylor (Newman, 1982), 1:42–43; quoted in Noll, *Evangelical Mind*, 202.

17 Cornell, "What Was God Thinking?," 72.

18 Charles Hodge, "The Bible in Science," *New York Observer*, 26 March 1863, 98–99; quoted in Noll, *Evangelical Mind*, 183.

19 Galileo, "Letter," 96.

20 For fuller accounts of these various models, see Ian G. Barbour, *Religion in an Age of Science* (Harper, 1990) and John F. Haught, *Science and Religion: From Conflict to Conversation* (Paulist Press, 1995).

21 Richard Dawkins, interviewed by David Van Biema, "God vs. Science," *TIME* (Canadian ed.), 13 November 2006, 35.

22 Cornell, "What Was God Thinking?," 72.

23 Dawkins, "God vs. Science," 36.

24 Francis Collins, interviewed by David Van Biema, "God vs. Science," 38.

25 Collins, "God vs. Science," 39.

26 Collins, "God vs. Science," 35.

27 Cornell, "What Was God Thinking?," 72.

28 Collins, "God vs. Science," 35–36.

29 Dawkins, "God vs. Science," 36.

30 Haught, *Science and Religion*, 21.

31 Dawkins, "God vs. Science," 37, 39.

32 Dawkins, "God vs. Science," 36.

33 Ludwig Wittgenstein, *Remarks on Frazer's Golden Bough*, ed. Rush Rhees, trans. A.C. Miles (Brynmill, 1979), 10e: "A whole mythology is deposited in our language."

Chapter Nineteen

Climate Change

IT WOULD BE OVERLY BOLD TO HOPE THAT A SINGLE SHORT CHAPTER could survey all the ways that STS can improve our understanding of the controversies surrounding climate change, let alone attempt to resolve those controversies. That is why this chapter will not try to do any of those things. Instead, what we'll do here is *apply* what we've learned to the issue of climate change, to see how an STS perspective can illuminate what might otherwise be impenetrable and irresolvable confusions about it.

If it's real, climate change is the greatest existential threat civilization has ever faced. We have to be precise about our terms, though. Civilization is a type of human society characterized by densely populated urban centers (*civitas* is the Latin word for "city") and dependence on agriculture to feed all those people. However, the climactic instability and warming predicted by climate science will, if allowed to occur, make agriculture and thus civilization impossible (even though humanity could survive in much smaller numbers using much different forms of culture and subsistence).[1] Therefore, if you want to perpetuate the status quo of the current arrangement of human culture we broadly call civilization, then you will not want climate change to happen the way it is predicted to happen.

There are a number of ways for climate change to *not* happen. One way is for climate science to be *wrong* in its predictions. That way, climate change won't happen because it wasn't happening in the first place and wasn't going to happen either. Another way is to *prevent* climate change from happening, by mitigating its causes: that is, by reducing the emissions and (eventually) atmospheric concentrations of greenhouse gases (GHGs).

Another way is *adapting* to the symptoms of climate change (like building higher sea walls to counteract rising sea levels), or even *geoengineering* the planet so that the GHGs don't trap as much heat (like injecting aerosols into the atmosphere that reflect sunlight back out into space). So whether or not you think climate science is wrong, or GHGs should be reduced, or that technological solutions will eventually save civilization, climate change is clearly entangled with critical social questions about science and technology. The stakes could not be higher, nor STS any more relevant.

There is no question that many people are skeptical about the science behind climate change (i.e., the role of GHGs in the atmosphere) and the role of technology in causing it (i.e., the industrial combustion of fossil fuels), while being optimistic about technological solutions to potential problems that might arise from climate change (e.g., carbon sequestration and storage, nuclear power, etc.). But who are these people, and why do they have this particular mix of beliefs about climate science and technology? According to a recent study, only 18 percent of the population of India are *not* sure that climate change is a serious threat or that it's caused by human activity.[2] By comparison, 46 percent of the US population falls into the same category of doubt.[3] This is a significant difference. Why might Americans be so much more skeptical about climate science than Indians? Or, put more broadly, how is skepticism or inaction about climate change connected to human social systems?

Internalist STS

The STS scholars Naomi Oreskes and Erik M. Conway argue that "the people of Western civilization knew what was happening" to the climate, but that they did not "[act] upon what they knew."[4] Even though 54 percent of the American population and 82 percent of the Indian population *are* sure that climate change is a serious threat caused by human activity, we still need to ask why these (and other) majorities have not lead to a reduction in GHG emissions or atmospheric concentrations.

Oreskes and Conway suggest that at least some of the reasons why knowledge (like climate science) doesn't necessarily lead to effective climate action are internal to Western technoscience itself. For starters, science is a conservative body, "plac[ing] the burden of proof on novel claims, including those about climate."[5] It is difficult for new science to become accepted by the scientific community because, as Thomas Kuhn argued, normal science *defends* existing paradigms rather than trying to overturn them (see Chapter Fourteen). Thus, while climate science was slowly becoming understood, greenhouse gasses kept being emitted.

Moreover, Oreskes and Conway claim that Kuhnian conservativism is amplified by Western science's roots in medieval religious institutions (see Chapter Eight):

> Just as religious orders of prior centuries had demonstrated moral rigor through extreme practices of asceticism in dress, lodging, behavior, and food—in essence, practices of physical self-denial—so, too, did natural scientists of the

> twentieth century attempt to demonstrate their intellectual rigor through intellectual self-denial. This practice led scientists to demand an excessively stringent standard for accepting claims of any kind, even those involving imminent threats.[6]

These "excessively stringent standards" included a definition of statistical significance where anything less than 95 percent certain was considered to be 'unproven.' Another was the belief that it is better to mistakenly *fail to believe* in something that is *true* (a type II error) than it is to mistakenly *believe* in something that is *false* (a type I error). The conservativism of normal science was codified in these epistemological standards, both making it harder for climate science to become generally accepted, and (once it was generally accepted), making the Intergovernmental Panel on Climate Change (IPCC) habitually underestimate the extent and severity of GHG emissions. These exceptionally high—and monastic—disciplinary standards created the impression that we have more time to act than we actually have, because it appeared that there was no 'certainty' about the climate science (yet). If Oreskes and Conway are right, the medieval and Kuhnian characteristics of modern Western technoscience actually make it *easier* for nearly half of all Americans to dismiss or deny the science of climate change! It doesn't help explain, however, why so few Indians are similarly skeptical.

Externalist STS

Of course, nothing is *certain* in science; that's why scientific beliefs are supposed to be held tenuously (see Chapter Two). For climate science, the question is rather *how much* certainty we need before we can *decide* that it's *worth* the expense and inconvenience of reducing our emissions or investing in technological solutions. This is not a scientific question. It might be an economic one, a political one, or even an ethical one. We might wish to apply a precautionary principle here, where we adjust the level of certainty required for action in accordance with how catastrophic the outcomes of inaction are likely to be. Taking big risks isn't so bad if the consequences of failure aren't so bad, but if climate change is going to be as bad as the scientists expect, it might not be wise to gamble on the possibility that they're wrong. There are, however, plenty of social interests that want us to take the chance that climate science is wrong because if it isn't, those social interests will have to be curtailed—if civilization in some recognizable form is going to continue into the twenty-second century. But which social interests?

> **Precautionary Principle:** "where there are threats of serious or irreversible damage, lack of full scientific certainty shall not be used as a reason for postponing cost-effective measures to prevent environmental degradation."—Principle 15 of the 1992 Rio Declaration of the United Nations

Oreskes and Conway point out what might be obvious to some: "power [does] not reside in the hands of those who [understand] the climate system, but rather in political, economic, and social institutions that [have] a strong interest in maintaining the use of fossil fuels."[7] Cutting back enough on GHG emissions will invariably mean that industries of oil and coal (etc.) are going to take an economic hit. Technological solutions at the scale necessary to reconfigure planetary systems will also be immensely expensive. It would be far more convenient if climate science were incorrect, because then there'd be no need to spend all that money. It's a lot cheaper to spend some of the immense amount of money made by the carbon combustion complex (i.e., fossil fuel producers, service industries, transportation, and finance) to actively discredit climate science. Making climate science look even more uncertain than it actually is would cost these industries less than reducing emissions or paying for climate adaptations. Many so-called think tanks are hired to do this discrediting, and some of them are the same ones that were hired by tobacco companies to discredit the link between smoking and cancer.[8]

There are, therefore, powerful and self-serving social interests external to climate science that want to see climate science fail, or to at least paralyze public opinion about it. That paralysis is generally what think tank 'counter-science' is supposed to do: make people doubt that climate scientists know what they're talking about, because some set of 'alternative facts' are possibly true. Without getting into detail about these 'alternative facts,' it is clear that climate science does not occur within a political or economic vacuum, unaffected or unimpeded by the social world surrounding it. Climate science exists in a complex social world which includes immensely powerful social forces arrayed directly against it.

STS encourages us to see science as politically and economically embedded, rather than neutral, but this is not how the usual story sees science. The natural sciences still tend to see their work "in isolation from social systems,"[9] which could be another reason why the IPCC has underestimated GHG emissions: it did not "anticipate China's economic growth, or resistance by the United States and other nations to curbing greenhouse gasses."[10] These are social factors that have direct bearing on climate science, but social factors are not normally within the investigative range of natural science. They are within the investigative range of STS, and can be tough to notice without the lens STS provides. As we've seen above, some of these social forces can be amplified by active opposition to climate science, funded by massively wealthy industries with vested interests in climate science being false or doubtful. If climate science is in the business of predicting global GHG emissions, it will have to start doing STS: incorporating social factors into its analyses.

Another external social force arrayed against climate action can be the government itself. Certain political parties or their electorates can share the same vested interests against climate science as the fossil fuel industry (not least because they are often economically dependent on those very same fossil fuel industries). Even before the first Trump presidency—which saw the removal of references to climate change in press releases and government websites[11]—four American states formally or informally restricted the language or teaching of climate change.[12] Climate scientists were also muzzled by the Canadian government under Prime Minister Stephen Harper.[13] If governments prevent climate change from being mentioned by scientists, let alone taught in schools, it's unsurprising that large segments of the population won't be 'sure' that it's real or concerning.

Inversely, government funding for climate change research is—when available—almost exclusively directed towards science itself, rather than the myriad of ways that science might be communicated to the broader society. Indeed, the dissemination of climate research to non-scientists is not a typically scientific endeavor either. Rather, such communication is usually the purview of literature, music, film, sculpture, or journalism—all various forms of artistic production. The comparative lack of government funding for artistic endeavors allows climate science to be largely hidden behind specialized jargon and journals, rather than incentivizing the communication of science using skills typically found outside science itself. Furthermore, this funding structure directs money away from those disciplines which focus on the social factors that need to be incorporated into the natural sciences. STS is among the disciplines underfunded compared to science, technology, engineering, and medicine (STEM disciplines). Here, government may unintentionally restrict action on the basis of climate science because it is focused too much on technoscience itself, and not the broader scope of how social beings think and communicate.

In terms of externalist STS, then, climate science as a form of knowledge does not necessarily translate into effective action, in part because a number of external factors interfere with it. Huge amounts of money are available to pay think tank employees to sow doubts about climate science in the minds of many non-scientists. Ill-meaning governments might actually prevent climate science from being mentioned at all. And even well-meaning funding agencies may focus funding on STEM disciplines rather than the artistic techniques that could be used to communicate climate science more effectively or (gasp!) do STS. Usual-story thinking about science does not help focus attention on these external social factors. It is rather an ideological framework that stands in the way of turning knowledge about climate change into action on climate change. STS, by contrast, forces us to think critically about ideologies hidden within our sciences and our technology.

Progressive Ideology

Climate change is a threat to the usual story about science and technology. At root, this is because climate change does not map onto the Doctrine of Progress. According to the usual story of Progress (see Chapter Ten), knowledge is supposed to keep on growing and getting closer to the absolute truth, while technology is supposed to capitalize on this knowledge by giving us ever increasing power over the material world. Combining these two factors, everything is supposed to get inevitably better forever. The Doctrine of Progress mapped onto the Industrial Revolution almost seamlessly: the factory system of production seemed to be the newest stage in the inevitable advance of scientific discovery and the rational organization of labor (see Chapters Eleven and Twelve). Industrial production of scientifically advanced goods like chemicals and machines is supposedly why, after thousands of years of superstition and poverty, 'we' have more money, more technology, and more intelligence than ever before. So goes the usual story.

However, anthropogenic climate catastrophe is also positively correlated with industrialization. As industrialization ramps up, so do GHG emissions. A graph of the increase of progressive 'good' things—scientific knowledge, technological power, etc.—is more or less the same exponential curve as atmospheric GHG concentrations and the resulting global average temperatures. According to what we know about climate change, however, more of *that* isn't better; higher atmospheric concentrations of GHGs, higher global average temperatures, and higher sea levels are *bad* things. Progress, however, is the deep-seated belief that more of the same (i.e., more Western technoscience) is always *better*. Otherwise, we'd be 'going backwards,' right?

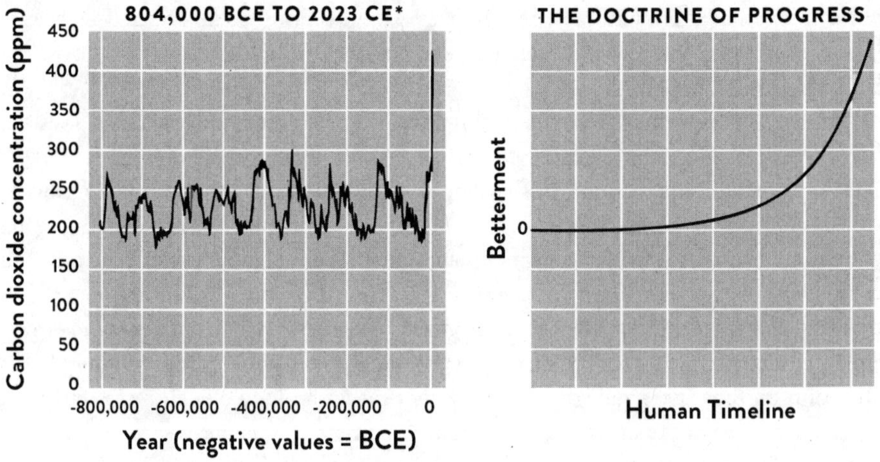

Two "hockey stick" graphs: CO_2 concentrations and progress.

Fossil fuel funded think tanks clearly believe that fighting climate change is going backwards: "The fuels that produce CO_2 have freed us from a life of backbreaking labour," says a TV advertisement from the Competitive Enterprise Institute, funded in part by Exxon-Mobil.[14] If we were to cut back significantly on our GHG emissions, we would (supposedly) end up with "less people, less affluence, less technology: We call that death, poverty, and ignorance"—at least according to Fred Smith, the head of that think tank.[15] To cut back on fossil fuel use would be to go against Progress, and that sounds like heresy. Climate science does not fit very well with a political ideology that says the best kind of human society to date is the one that emits over 34 billion tonnes of CO_2 per year. If you want to save industrial civilization, you can't cut back on the very fuels that make industrial civilization possible (and make Exxon-Mobil a lot of money), right?

So on the one hand, climate science seems to require that industrialization slow down rather than speed up: we need to generate *less* energy the usual way, not more; we need to cut back, reduce, or limit all kinds of things, rather than develop or expand them indefinitely. This is where we see many political conservatives pushing back against climate science; they're just as committed to the ideology of Progress (especially economic and technological progress) as 'progressives' are.

On the other hand, political progressives (usually liberals) are also disappointed in climate science. It appears to have *failed* to bring about progress—in terms of replacing ignorance with knowledge, crisis with salvation, dirty fuels with magically clean and powerful superfuels, etc. For example, Oreskes and Conway fear that Western civilization is entering a new Dark Age (they don't seem concerned with the possibility that there never was a Dark Age in the first place; see Chapter Eight).[16] Even though scientific knowledge about the climate certainly has developed and improved over time, climate change is still happening. Progress has not turned out the way it was supposed to, and so climate change threatens the usual story even from the left-of-center side of the political spectrum. Science and technology do not appear poised to save us from disaster. It's not even clear that we're progressing any more (did we ever?). Progressives may argue, therefore, that we need more technoscientific progress, not less, in order to fight climate change. They just disagree with conservatives about what that technoscientific progress entails.

Positivism and Baconianism

On the one hand, if progressive ideology makes you resist significant limitations on the fossil fuel economy, then it will stand in the way of GHG emission reductions. On the other hand, if progressive ideology makes you naively think that the human situation will inevitably improve as scientific knowledge and technological power increase, then it will leave you surprised and disappointed when knowledge doesn't lead to power, and when power (generated by burning fossil fuels) turns out to be the cause of the problem, instead of the solution. Thus, even though Oreskes and Conway consider "themselves children of the Enlightenment,"[17] they recognize that the usual stories of positivism and Baconianism (which are part of the so-called Enlightenment) inhibit effective climate action. Oddly, they don't seem to have any problem with the positivist notion that science is "absolutely, positively true" (see Chapter Thirteen), but rather with the Baconian notion "that this knowledge would empower its holder" (see Chapter Nine).[18] They know that it doesn't.

As we saw earlier in this chapter, there are both internal and external reasons for why knowledge of climate science doesn't necessarily lead to meaningful climate change policies. The Baconian usual story about technoscience fails to have an explanation for this, because on that account, knowledge *is* power. If knowledge is power, then it's impossible for climate science to *not* turn into climate action! And yet that's what we get: inaction. Climate science is politically weak, not powerful. Baconian ideology thus prevents people from seeing science (and technology) in a social way, a way that would help them identify and overcome those internal and external obstacles to transforming knowledge into power. Simply having scientists provide (absolutely true?) information to society is not going to do what the Baconian usual story thinks it will.

Positivism will also prevent you from seeing science (and technology) in a social way, because 'absolutely, positively true' facts are not things which are supposed to be subject to internal or external limitations or obstacles. Positivistic truth is just there; it is what it is. Positivism is thus uninterested in and inattentive to social complexities that might infect such truth and make it less than absolute. Positivists can get rather upset at fellow citizens who deny that scientists know what they're talking about, but positivism doesn't have the resources for resolving the problem of denial. All it can do is continue to insist that science is absolutely true (evidentially warranted, statistically probable, etc.) and label naysayers as irrational, superstitious, or ignorant. Positivism and Baconianism are mystified by anyone not agreeing with science, and have no way to explain why simple facts don't automatically transform into rational human action. Positivism and Baconianism are ideological impediments to understanding how to get anything done about climate science, because they can only conceive of facts in isolation from human sociality. They aren't socially constructivist enough!

Market Fundamentalism

Baconian and positivist ideologies are often found within contemporary technoscience, and as such constitute internal social factors which prevent technoscience from seeing itself as it really is. That is, positivism and Baconianism inhibit critical self-awareness of science about science, which is important if we're to understand why climate science has such trouble turning into climate action. Another ideological impediment to climate action is more external but often just as difficult to see, because it is so pervasive in Western and especially American culture: market fundamentalism. This is the political and economic view that free markets are *both* the best way to satisfy human material needs *and* the only way to maintain personal political freedom.[19] Classical economic liberalism championed the first part of market fundamentalism—free markets for material prosperity—whereas *neo*-liberalism added the second part: free markets as the only way to avoid totalitarian dictatorship.

Neoliberalism: the socio-economic theory which holds that free markets are the best way of generating wealth for the most people, and also the best way to ensure political freedom for everyone.

Neoliberal economic theory is typically found towards the right-of-center side of the political spectrum, which is why (ironically) it is so venomously opposed to political liberalism towards the other side of the political spectrum. It is also closely connected to climate change skepticism. Here's why: climate change does not appear solvable using free markets. The Earth's atmosphere is a limited 'sink,' meaning we can only dump so much gas into it before it becomes overloaded. However, the atmosphere is not a market. You might have to pay to dump garbage in a landfill, but nobody has to pay to put GHGs into the air. Of course, some jurisdictions are trying to turn the atmosphere into a market (e.g., with carbon taxes or cap-and-trade systems), but the market didn't demand these measures: activists, scientists, and politicians did. Even if carbon taxes (etc.) work (they might not), they're regulations imposed by the government onto the market. As such, the market is no longer completely free and we're on the path to a socialist dystopia. If so, climate change science wants to take away our freedom.

That's how the market fundamentalist story goes, at least. First, climate change is a problem that calls for intervention in the market by strong, centralized socio-economic systems. Second, strong and centralized socio-economic systems are evil. Therefore, climate change cannot be real (enter the think tanks, see above). Given the failure of technological liberalism (see Chapter Sixteen), it should not be surprising that climate science has politics too: it may have implications that fit with some political systems better than others. After all, climate change was *caused* by a unique and particular set of socio-economic frameworks, so why wouldn't its solutions require an inversely unique and particular set of socio-economic frameworks? The neoliberals are not wrong about their first point: climate change *is* a problem that calls for market intervention on a global scale, because the free market did not (and could not) respond adequately to the problem of climate change. The real issue is with their second point: if we get market intervention, we (supposedly) get the end of liberty.

Friedrich Hayek, one of the main proponents of neoliberalism, claimed that

> [i]t was men's submission to the impersonal forces of the market that in the past has made possible the growth of a civilization which without this could not have been developed; it is by thus submitting [to the market] that we are every day helping to build something that is greater than any one of us can fully comprehend.[20]

In his estimation, human civilization is a result of humans submitting to the laws of the market, rather than making the market submit to human laws. But because the free market has given us climate change and no solution for it, it looks as if civilization will have to end either way: either we stop climate change by putting limits on the market and destroy civilization thereby, or we don't put limits on the market and let climate change heat the planet to the point where civilization cooks itself to death.

We might dispute Hayek's theory of civilization (I sure would), but we don't need to because it turns out that government intervention *can* occur within neoliberalism without undermining civilization. According to Hayek,

> the price system becomes ... ineffective when the damage caused to others by certain uses of property cannot be effectively charged to the owner of that

> property. [Therefore] some method other than competition may have to be found to supply the services in question.... [C]ertain harmful effects of deforestation, of some methods of farming, or of the smoke and noise of factories [cannot] be confined to the owner of the property in question or to those who are willing to submit to the damage for an agreed compensation. In such instances we must find some substitute for the regulation by the price mechanism. But the fact that we have to resort to the substitution of direct regulation by authority where the conditions for the proper working of competition cannot be created does not prove that we should suppress competition where it can be made to function.[21]

If neoliberals like Hayek can accept market regulations that prevent the harmful effects of the noise and smoke of factories, they should be able to accept market regulations that prevent the harmful climactic effects of factory emissions (etc.). Even more so, if they can accept that we can create conditions where market competition can "be made to function," then they can accept that we can create a regulated market for GHG emissions using taxes or carbon credits or whatever. Doing so will not create a totalitarian dictatorship where our political liberty is erased. In actual fact, therefore, market fundamentalism is only an obstacle to effective climate action because the fundamentalists aren't paying careful attention to what their own neoliberal ideology actually says. Then again, fundamentalist ideologues aren't typically concerned with rational consistency when they've got other axes to grind—in this case, maintaining the fossil fuel status quo.

Techno-Rationality

Let us consider one last way in which the usual story injects itself into climate change: *we can fix it, right?* If climate change is a problem, the usual story thinks technology should be able to solve it, just like it thinks it has solved every other problem humanity has ever faced. Technological solutionism is an ideology shared by people on the left side of the political spectrum just as much as on the right. The main difference is that right-wingers usually hold out hope that there won't be a need to solve climate change technologically because climate change isn't happening (or isn't that bad), whereas left-wingers wonder why we haven't already technologically solved this problem with our obviously awesome technics and techniques. On both sides of the spectrum, technological progress is supposed to be able to meet the challenge of climate change and allow us to continue living our lives in more-or-less the same way as always—except better, of course.

Technological solutions to climate change are technological largely because they are *not* behavioral. A technological solution ideally should not require anyone to act or live differently. So if there's a technological solution to GHGs emissions coming from cars, it should not require me to drive my car less or start riding a bicycle instead; it will rather allow me to continue driving a car, only my new car will be electric or hydrogen powered instead of gasoline or diesel powered. In some ways, then, technological solutions can address the cause of climate change: GHG emissions. If we can suck up the carbon emitted by factory smokestacks or automotive exhaust pipes and pump it underground (aka "carbon capture and storage"), then we might reduce atmospheric GHG concentrations even while burning fossil fuels. Or if I can power my car with electricity generated by wind turbines or solar panels, I can keep driving without adding to atmospheric GHG concentrations—assuming the math works out (calculating the net reduction in GHGs with these various technological workarounds is notoriously difficult).

Other technological solutions address the causal link between GHGs and negative consequences, rather than the GHGs themselves. These methods are known as 'geoengineering,' because they propose changing the nature of the atmosphere or the ocean's chemistry to make the planet more tolerable *in spite of GHG emissions*. Solar radiation management, for example, includes the injection of aerosols (usually sulfur dioxide) into the upper atmosphere to create white clouds (of sulfuric acid) that reflect sunlight away from the atmosphere and into space, rather than trapping its heat inside the greenhouse that the GHGs have made. That way, the Earth won't warm as much, even though the atmospheric concentration of GHGs hasn't been reduced. Another example isn't related to warming, but rather the carbonation of the oceans (around 30 percent of anthropogenic CO_2 emissions are absorbed by the oceans each year). To prevent ocean acidification (the CO_2 turns into carbonic acid when it dissolves in seawater), we could design unmanned drone ships to seed the seas with iron so that phytoplankton rather than water sequesters the carbon. The CO_2 would still be emitted, though, unless emissions are reduced by other methods.

A third form of technological solutionism doesn't even try to change the nature of the planet. It merely adapts human society to the greenhouse that the planet is changing into. One example is building higher seawalls to protect coastal areas from rising sea levels as the polar icecaps melt. In this case, GHG emissions would continue to rise *and* their role in melting polar ice would continue, but coastal cities and farmland wouldn't be as vulnerable to flooding. Other examples might include increasing access to air conditioning, filtering indoor air to remove smoke from burning forests, covering roofs in vegetation (aka "green roofs") to reduce urban heat island effects, and genetically modifying agricultural crops to be more drought resistant. Here the symptoms of climate change, rather than the causes, are mitigated by the technologies.

So why wouldn't we do any or all of these things? That is one of the questions that bothers Oreskes and Conway: "Western civilization [has] the technological know-how and capability to effect an orderly transition to renewable energy, yet the available technologies [are] not [being] implemented in time."[22] The answer to their question involves all the internal, external, and ideological reasons we've already surveyed, but the question itself hides its own ideological presupposition. We already assume that Western technoscience has—and should have—the knowledge and the power to rebuild the physical world so that globalized Western culture does not have to reconsider how it lives. But does it, and should it?

Asking this question reminds us that Western technology is itself a way of thinking and living (see Chapter Seventeen), and not simply the high point of human history to date or the only way to think and live. Oreskes and Conway recognize that geoengineering techno-solutions likely have "widespread support from wealthy nations anxious to preserve some semblance of order,"[23] but they don't acknowledge that all technological solutions to climate change do this: each one tries to save modern civilization from having to significantly change, especially behaviorally. *Any* technological 'solution' is going to be embedded in a non-neutral social framework for whom that technology is a solution, but it may not be a solution for others (see Chapter Twelve). So is viewing this planet as an object for (some) humans to manipulate the price to pay for Western civilization continuing on into the future? Techno-logical rationality thinks yes, it's easier to imagine changing nature than changing modern Western culture. Oreskes and Conway don't seem to have a problem with this, which is why they're more Baconian than they let on.

So what? Maybe as long as these technological solutions work well enough, 'we' will be okay (because 'we' all want to be in Western civilization, supposedly). Some people might object that our industrial technologies (and maybe even Western techno-logic itself) are what got us into this mess in the first place, and so it would be a mistake to try to use the same tools and ideas to solve the problems they've caused. Technological solutionism embraces this criticism, however:

> Each new act of salvation will result in new unintended consequences, positive and negative, which will in turn require new acts of salvation ... for as long as humans inhabit [the Earth].... The solution to the unintended consequences of modernity is, and always has been, more modernity....[24]

Humans, in this picture of the world, are defined as modern (and Western), so living outside that cultural framework is assumed to be impossible nowadays. If we accept the usual story that *everybody* has *always* done what 'we' are currently doing, we can embrace the likelihood that technological solutions to climate change (like geoengineering) will cause unforeseen problems more optimistically. That's because the problems of the future will be *better* than the problems of the past:

> These new problems will largely be better than the old ones, in the way that obesity is a better problem than hunger, and living in a hotter world is a better problem than living in one without electricity.[25]

If you say so, dude. But anyway (so the usual story goes), that's just how it is. Human life is just like that, so get over it. We can't solve all the problems, but progress is better than going backwards. Modern Western culture is as good as it gets—and it always gets better. Just keep hanging on, hold on to your faith in technoscience, and don't ask *who* technological rationality is better *for*.

The fact is that rapid industrial development only happened because Western agrarian societies thought it was a good idea to dig up millions of years of solar power that had been compressed and stored in geological formations. With that amount of easily accessible energy at their fingertips, it's no wonder that they (in particular) started to think that there were no limits to human growth or power. Techno-optimism is an ideology fueled by fossil fuels. Renewable energy sources like sun, wind, and muscle don't have anywhere near this energy concentration, and so it seems unlikely that the geologically brief energy intensity of the past century's carbon burn could be sustained by renewables. Globally (but inequitably distributed), we have a high-energy superculture, and it's not clear that it can be fueled by low-energy sources. There may be no sustainable technological solution to fossil fuels when we're trying to maintain the culture that fossil fuels made possible in the first place.

Technological solutions to climate change aren't simply a solution to an objective problem; they're a solution *for* an idea of what human culture should be, an idea that most of us don't even realize we have. That's what an ideology is: a way of seeing the world that operates so much in the background that we don't know that it's our operating system. Technological solutions to climate change are trying to save the industrialized culture that caused climate change, the culture that will change the planet before it changes itself, that says doing so is human nature even when it's not, that wants power (for space travel or machine learning or whatever) to magically come from solar panels like it used to come gushing out of the ground. Technological solutions are there to save the usual story, because otherwise climate change is going to require that we learn new stories about ourselves and our place on this Earth.

So when the usual story of human nature, scientific knowledge, and technological power tells us that "[g]iven the emergency conditions [of impending climate disaster], the world [has] no choice but to take the risk" of climate geoengineering,[26] a critical STS perspective will remind us why we supposedly have no choice. We have no choice because our ideology gives us none. It created the emergency conditions, and now it's selling itself as the only solution that will perpetuate its own status quo. In Ellul's terminology (see Chapter Seventeen), our technological ideology has the autonomy and we don't, because we don't know how to think—let alone live—differently. It tells us what's necessary, as if it were a force of nature instead of an idea in our heads. In Hacking's terminology (see Chapter Fifteen), it's conditionally inevitable that we have to fix climate change in every possible way *except* living differently. We don't have to think or live the way we do, but *if* we live that way, *then* our only options are these technologies that will save 'us' (i.e., the dominant system). Climate change *and* its technological solutions are socially constructed, but we too easily forget this because the usual story denies that it's socially constructed. The usual story of Progress is undermined by the science of progressively worse GHG emissions, and technological rationality is supposed to hide that fact from us. Technological solutions to climate change are equally there to save the usual story from embarrassing defeat. That in itself is a barrier to effective action, because 'effective action' is ultimately about saving an ideology and a very peculiar way of life: a modern and Western life for some people, not 'the planet,' or 'civilization,' or even 'everybody.' That's how enframed we are: it's nearly impossible to imagine anything outside this ideological framework we don't realize we have.

***

STS is weird. It isn't 'helpful' for solving problems, if by that you mean something that will assist us in *not* thinking critically about our usual routines but instead help us maintain them. Climate change is an existential threat to our usual routines, and there are all kinds of ways that STS can show how our usual routines make it tough for climactic techno-science to make sense. Our routines hide from us the conservativism and monasticism of modern science, the vested financial and political power opposed to climate science, the funding disparity between the arts and sciences, and the progressive, positivist, Baconian, neoliberal, and technological ideologies that inhibit doing much if anything about climate change. An STS perspective can reveal these things to us, climate change being just one example among many.

If you want to make sense of the world of science and technology outside the lock the usual story has on our minds, STS will help. You're ready to use it. If you don't want to, well then never mind.

Notes

1 John Gowdy, "Our Hunter-Gatherer Future: Climate Change, Agriculture, and Uncivilization," *Futures* 115 (2020): art. 102488.

2 Anthony Leiserowitz et al., "Global Warming's Four Indias, 2022: An Audience Segmentation Analysis," Yale Program on Climate Change Communication, 4 May 2023, https://climatecommunication.yale.edu/publications/global-warmings-four-indias-2022-an-audience-segmentation-analysis/.

3 Anthony Leiserowitz et al., "Global Warming's Six Americas, 2022," Yale Program on Climate Change Communication, 14 March 2023, https://climatecommunication.yale.edu/publications/global-warmings-six-americas-december-2022/.

4 Naomi Oreskes and Erik M. Conway, "The Collapse of Western Civilization: A View from the Future," *Daedalus* 142, no. 1 (2013): 41.

5 Oreskes and Conway, "Collapse of Western Civilization," 44.

6 Oreskes and Conway, "Collapse of Western Civilization," 44.

7 Oreskes and Conway, "Collapse of Western Civilization," 49.

8 Naomi Oreskes and Erik M. Conway, *Merchants of Doubt: How a Handful of Scientists Obscured the Truth on Issues from Tobacco Smoke to Global Warming* (Bloomsbury, 2010).

9 Oreskes and Conway, "Collapse of Western Civilization," 54 note 1.

10 Glenn Scherer, "How the IPCC Underestimated Climate Change," *Scientific American*, 6 December 2012, www.scientificamerican.com/article/how-the-ipcc-underestimated-climate-change/.

11 Julia Belluz and Umair Irfan, "The Disturbing New Language of Science under Trump, Explained," *Vox*, updated 30 January 2018, www.vox.com/2017/12/20/16793010/cdc-word-ban-trump-censorship-language.

12 Tanya Lewis, "Florida Isn't the Only State to 'Ban' Climate Change," *Live Science*, 9 March 2015, www.livescience.com/50085-states-outlaw-climate-change.html.

13 "It's Official—the Harper Government Muzzled Scientists. Some Say It's Still Happening," *CBC News*, last updated 23 March 2018, www.cbc.ca/news/health/second-opinion-scientists-muzzled-1.4588913.

14 Quoted in Joel Achenbach, "The Tempest," *The Washington Post*, 28 May 2006, W08.
15 Fred Smith, quoted in Achenbach, "The Tempest," W08.
16 Oreskes and Conway, "Collapse of Western Civilization," 40.
17 Oreskes and Conway, "Collapse of Western Civilization," 43.
18 Oreskes and Conway, "Collapse of Western Civilization," 49.
19 Oreskes and Conway, "Collapse of Western Civilization," 49–53.
20 Friedrich A. Hayek, *The Road to Serfdom* (University of Chicago Press, 1944), 204.
21 Hayek, *Road to Serfdom*, 38–39.
22 Oreskes and Conway, "Collapse of Western Civilization," 48–49.
23 Oreskes and Conway, "Collapse of Western Civilization," 47.
24 Michael Shellenberger and Ted Nordhaus, "Evolve," in *Love Your Monsters: Postenvironmentalism and the Anthropocene*, ed. Michael Shellenberger and Ted Nordhaus (Breakthrough Institute, 2011), 9, 13.
25 Shellenberger and Nordhaus, "Evolve," 14–15.
26 Oreskes and Conway, "Collapse of Western Civilization," 47.

Concluding Untechnoscientific Postscript

SO WHAT WAS I TRYING TO DO IN THIS BOOK? I WAS TRYING TO GET YOU to start doing STS. Does that mean I wanted you to reach the conclusion that science and scientists are not to be trusted, or that all modern technologies are bad? No. Sometimes it's easy to reach despairing conclusions like these, though, when exposed to STS for the first time. If modern Western technoscience is socially constructed, that might sound like science isn't telling the truth any more than an influencer in your social media feed trying to get you to buy something. If government scientists nonconsensually tested nutritional theories on intentionally malnourished Indigenous children,[1] that might suggest that government scientists shouldn't be trusted when they want you to get vaccinated. If artificial intelligence companies nonconsensually use other people's copyrighted artwork and writing to train their for-profit machine learning models, that might be just one example among many of how modern technology seems to be developed without moral consideration of its social implications. On top of all that, it's also difficult that STS is an ongoing work-in-progress, especially if we want it to give us finalized declarations of truth. Why bother doing STS when nobody really knows what the answers are to all these questions yet?

The truth is, however, that nobody will ever get the final truth. That is the human condition. The trick is to do a good job of being human when you cannot get all the answers. That is, I suggest, how we should understand technoscience and STS. They're both bodies of argument. Arguments are sets of reasons assembled in support of a conclusion, and some reasons support conclusions better than others. There are ways to determine which reasons and arguments are more successful; in general, we call this logic. Science, technology, and STS all use arguments, reason, and logic to advance their conclusions, as I have done in this book and am still trying to do (right now!). Some of these attempts work better than others, and all of them should be scrutinized to see which ones do in fact work better than others. No set of arguments will ever be perfect, but we can hope that with sustained criticism over time, we'll get closer to something like what has been called 'the truth.'

Technoscience specifically is a set of arguments about the natural or physical world. Technoscience is also a socially constructed system with particular (usually Western and modern) interests in the natural or physical world. Arguments are complicated things, because they always exist in a broader context, and it's probably impossible for us to get outside that context to see if the arguments technoscience makes about the world actually line up with how the universe really is. But at the same time, technoscience is not alone in the universe, as if all we ever bumped into were the arguments it makes. One can never be completely sure, but it's highly probable that there is something out there beyond our social constructions of everything. Let's call that 'the real world.' We can never really know what that is—that would involve absolute knowledge that stepped outside of contexts altogether—but sometimes the real world pushes back. Sometimes our social constructions about the world don't mesh very well with whatever is behind them. We bump into the real world, or better yet, it bumps into us, even if we cannot know what it is objectively, in itself. If we're being honest about this, we should feel the need to adjust our constructions so that they can fit better with what might be the real world out there.

I think technoscience, at its best, is a set of arguments about how nature pushes back when technoscience pokes at it with experiments. When these arguments do justice to the way the real world appears to be responding to experimental interference with it, I think we should take that seriously. I don't think technoscientists are just making that up.[2] If technoscience has a solid body of evidence suggesting that infections are better treated with antibiotics than with apple cider vinegar, then I think we should go with antibiotics if we want to get rid of infections. If not, then have at it with the vinegar.[3]

But why do we have infections, and why do we want to get rid of them? Why do we have antibiotics or antibiotic resistance, and why do some people want to use apple cider vinegar for various ailments? This might not be a set of questions that technoscience asks or has a way of answering. We might need to find a different set of arguments, with different epistemological methods than experimentation, to answer questions like those. And in the process, we might also end up asking why technoscience is busy poking and prodding the real world anyway.[4] We might, in other words, have to start asking STS questions about the value, purpose, and social function of modern Western technoscience—even if we justifiably take antibiotics to kill a bacterial infection.

When we ask honest but difficult questions about what values technoscience assumes, serves, and embeds in our everyday lives, we're doing STS. We do the same when we inquire into the possibility of using technoscience in ways consistent with—or contrary to—those sets of values assumed, served, and embedded by technoscience. For example, can we use artificial intelligence without massively disrupting the careers of people working in the creative industries? Maybe, maybe not. Either way, human bias and values inevitably affect technoscience, and it's crucial that we know how to deal well with this fact of the human condition, rather than wish it away by lying to ourselves that technoscience (or something else) gives you or anyone else access to the absolute truth or the absolute good. We need to learn how to identify better and worse arguments about technoscience in society, and that's what STS is all about.

When I started writing this book, COVID-19 didn't exist. Now that the pandemic is over, nobody even wants to think about it (understandably). But in the middle of it, there sure were a lot of arguments about technoscience and social values! People who didn't want to wear masks, socially distance themselves, or get vaccinated were accused of caring more about their personal convenience and economic wellbeing than the lives of their elderly neighbors. One of my close friends from high school lost his father to COVID-19, but he thinks his dad died from vaccine shedding coming off the nurses in the hospital. These are arguments that need to be assessed critically, and it's not easy, but nobody said STS was easy.

That difficulty doesn't mean we can't discover good answers, though, ones that do justice to the pushback of the real world against our theories about it, and ones that helpfully prioritize our moral duties in a multilayered and complicated moment in human history. Being able to do this is a human excellence, and it doesn't imply that medical technoscientists are wrong about COVID-19 killing my friend's unvaccinated father. But it might also make us think about why vaccines were so important for keeping people working indoors (where the virus is most likely to infect us) in an economy that funnels most of the money it generates upwards into the hands of a few, while those who do most of the work can't afford groceries or rent.

Human values inevitably affect technoscience, therefore, but this can be good or bad. Bad values can infect technoscience, like the racism and imperialism that drove Western technoscience during the ages of colonization (one of which we're still in). But good values can push technoscience in better directions, like eradicating polio or making insulin affordable for people with diabetes. Still further, however, human values can push us to reconsider how we do technoscience itself, or whether we should do it at all. Just because we can poke and prod the natural world doesn't always mean that we should. Figuring out *if we should* is something we need to get better at doing.

The usual story does *not* help us to figure that out, which brings me back to what I was trying to do in this book. Doing STS, I think, requires what (in Chapter Two) Richard Lewontin called a reasonable skepticism about the sweeping claims of technoscience. Those sweeping claims are the usual story, and I have argued that *they're* wrong, even if (for example) technoscience *isn't* wrong about viruses or bacterial infections. Maybe you think *I'm* wrong, in which case it's your turn to develop what are hopefully better arguments than mine. Either way, this requires careful, critical thinking about science, technology, history, philosophy, society, and ethics, which is exactly what this book encourages you to do.

Notes

1 Ian Mosby, "Administering Colonial Science: Nutrition Research and Human Biomedical Experimentation in Aboriginal Communities and Residential Schools, 1942–1952," *Histoire sociale/Social History* 46, no. 91 (2013): 145–72.

2 Nathan Kowalsky, "Science and Transcendence: Westphal, Derrida, and Responsibility," *Zygon* 47, no. 1 (2012): 132–34.

3 Darshna Yagnik et al., "Antimicrobial Activity of Apple Cider Vinegar against *Escherichia coli*, *Staphylococcus aureus* and *Candida albicans*; Downregulating Cytokine and Microbial Protein Expression," *Scientific Reports* 8, no. 1732 (2018), https://doi.org/10.1038/s41598-017-18618-x.

4 Kowalsky, "Science and Transcendence," 135–36.

Bibliography

Achterhuis, Hans, ed. *American Philosophy of Technology: The Empirical Turn.* Indiana University Press, 2001.

Arendt, Hannah. "The 'Vita Activa' and the Modern Age." In *Philosophy of Technology—The Technological Condition: An Anthology*, edited by Robert C. Scharff and Val Dusek, 352–68. Blackwell, 2003.

Barbour, Ian G. *Religion in an Age of Science*. Harper, 1990.

Bauchspies, Wenda K., Jennifer Croissant, and Sal Restivo. *Science, Technology, and Society: A Sociological Approach*. Blackwell, 2006.

Baumgardt, Carola, ed. *Johannes Kepler: Life and Letters*. Victor Gollancz, 1952.

Bavli, Itali, and David S. Jones. "Race Correction and the X-Ray Machine: The Controversy over Increased Radiation Doses for Black Americans in 1968." *New England Journal of Medicine* 387, no. 10 (2022): 947–52.

Becker, Carl. "Progress." In *Encyclopaedia of the Social Sciences*, vol. 12, 495–99. Macmillan, 1934.

Bekar, Cliff, and Richard G. Lipsey. "Science, Institutions, and the Industrial Revolution." *The Journal of European Economic History* 33, no. 3 (2004): 709–53.

Berger, Peter L., and Thomas Luckman. *The Social Construction of Reality: A Treatise in the Sociology of Knowledge*. Doubleday, 1966.

Bernard, L.L. "The Significance of Comte." *Social Forces* 21, no. 1 (October 1942–May 1943): 8–14.

Best, Joel. "Historical Development and Defining Issues of Constructionist Inquiry." In *Handbook of Constructionist Research*, edited by James A. Holstein and Jaber F. Gubrium, 41–64. Guilford, 2008.

Bijker, Wiebe E., Thomas P. Hughes, and Trevor J. Pinch, eds. *The Social Construction of Technological Systems: New Directions in the Sociology and History of Technology*. MIT Press, 1987.

Binfield, Kevin, ed. *Writings of the Luddites*. Johns Hopkins University Press, 2004.

Boas, George, trans. *The Mind's Road to God*. Liberal Arts Press, 1953.

Bowles, Samuel. "Cultivation of Cereals by the First Farmers Was Not More Productive Than Foraging." *Proceedings of the National Academy of Sciences* 108, no. 12 (2011): 4760–65.

Braun, Lundy, Anne Fausto-Sterling, Duana Fullwiley, et al. "Racial Categories in Medical Practice: How Useful Are They?" *PLOS Medicine* 4, no. 9 (2007): 1423–28.

Bronk, Richard. *Progress and the Invisible Hand: The Philosophy and Economics of Human Advance*. Little Brown, 1998.

Buchan, Bruce, and Linda Anderson Burnett. "Knowing Savagery: Australia and the Anatomy of Race." *History of the Human Sciences* 32, no. 4 (2019): 115–34.

Caiazza, John C. *The Crisis of Progress: Science, Society, and Values*. Transaction, 2016.

Capon, Robert Farrar. *The Third Peacock: The Goodness of God and the Badness of the World*. Image/Doubleday, 1972.

Carnap, Rudolf, Hans Hahn, and Otto Neurath. "The Scientific Conception of the World: The Vienna Circle." In *Philosophy of Technology—The Technological Condition: An Anthology*, edited by Robert C. Scharff and Val Dusek, 86–95. Blackwell, 2003.

Carson, Rachel. *Silent Spring*. Houghton Mifflin, 1962.

Carter, Dee. "Unholy Alliances: Religion, Science, and Environment." *Zygon* 36, no. 2 (2001): 357–72.

Cauvin, Jacques. *The Birth of the Gods and the Origins of Agriculture*. Translated by Trevor Watkins. Cambridge University Press, 2000.

Chiang, Ted. "Tower of Babylon." In *Stories of Your Life and Others*, 1–28. Vintage Books, 2002.

Clarke, Arthur C. "Clarke's Third Law on UFO's." *Science* 159, no. 2812 (1968): 255.

Cornell, Eric "What Was God Thinking? Science Can't Tell." *TIME* (Canadian edition), 14 November 2005.

Crossan, John Dominic, with Richard G. Watts. *Who Is Jesus? Answers to Your Questions about the Historical Jesus*. HarperCollins, 1996.

Daston, Loraine. "Can Liberal Education Save the Sciences?" *The Point*, 25 May 2016. https://thepointmag.com/2016/examined-life/can-liberal-education-save-the-sciences.

Dawkins, Richard. "The Moon Is Not a Calabash." *Times Higher Education Supplement*, no. 1143 (1994): 17.

Diamond, Jared. *Guns, Germs and Steel: A Short History of Everybody for the Last 13,000 Years*. Random House, 1997.

Dreyer, Edward L. *Zheng He: China and the Oceans in the Early Ming Dynasty, 1405–1433*. Pearson, 2007.

Eco, Umberto. *The Name of the Rose*. Translated by William Weaver. Harcourt Brace Jovanovich, 1983.

Ede, Andrew, and Lesley B. Cormack. *A History of Science in Society: From Philosophy to Utility*. 4th ed. University of Toronto Press, 2022.

Ellul, Jacques.. *The Technological Society*. Translated by John Wilkinson. Vintage, 1964.

—. "Ideas of Technology." Translated by John Wilkinson. In *1984 and All of That*, edited by Fred H. Knelman, 11–20. Wadsworth, 1971.

—. *The Technological System*. Translated by Joachim Neugroschel. Continuum, 1980.

—. *Perspectives on Our Age: Jacques Ellul Speaks on His Life and Work*. Translated by Joachim Neugroschel. Canadian Broadcasting Corporation, 1981.

—. "A Theological Reflection on Nuclear Developments: The Limits of Science, Technology and Power." In *Waging Peace: A Handbook for the Struggle to Abolish Nuclear Weapons*, edited by Jim Wallis, 114–21. Harper and Row, 1982.

Faria, Miguel A., Jr. "America, Guns, and Freedom: Part II—An International Perspective." *Surgical Neurology International* 3, no. 135 (2012): n.p.

Feenberg, Andrew. *Questioning Technology*. Routledge, 1999.

Feigl, Herbert. "The Scientific Outlook: Naturalism and Humanism." *American Quarterly* 1, no. 2 (1949): 8–18.

Ferré, Frederick. "Technological Faith and Christian Doubt." *Faith and Philosophy* 8, no. 2 (1991): 214–24.

Feyerabend, Paul. "Theses on Anarchism." In *For and Against Method: Including Lakatos's Lectures on Scientific Method and the Lakatos-Feyerabend Correspondence*, edited by Matteo Motterlini, 113–18. University of Chicago Press, 1999.

Fichman, Martin. *Science, Technology, and Society: A Historical Perspective*. Kendall Hunt, 1993.

Galileo Galilei. "Letter to the Grand Duchess Christina." In *The Galileo Affair: A Document History*, edited and translated by Maurice A. Finocchiaro, 87–118. University of California Press, 1989.

Garrison, Kevin. "Perpetuating the Technological Ideology: An Ellulian Critique of Feenberg's Democratized Rationalization." *Bulletin of Science, Technology, and Society* 30 (2010): 195–204.

Gillispie, Charles Coulston. *The Edge of Objectivity: An Essay in the History of Scientific Ideas*. Princeton University Press, 1960.

Goldman, Steven L. "Science, Technology, and God: In Search of Believers." *Bridges* 2, no. 1–2 (1990): 43–53.

Goldstone, Jack A. "Efflorescences and Economic Growth in World History: Rethinking the 'Rise of the West' and the Industrial Revolution." *Journal of World History* 13, no. 2 (2002): 323–89.

Gowdy, John. "Our Hunter-Gatherer Future: Climate Change, Agriculture, and Uncivilization." *Futures* 115 (2020): art. 102488.

—. *Ultrasocial: The Evolution of Human Nature and the Quest for a Sustainable Future.* Cambridge University Press, 2021.

Grant, George. *Technology and Justice.* University of Notre Dame Press, 1986.

Gray, John. "Can Religion Tell Us More Than Science?" *BBC News*, 16 September 2011. http://www.bbc.com/news/magazine-14944470.

Guthrie, W.K.C. *The Greek Philosophers: From Thales to Aristotle.* Harper & Row, 1950.

Hacking, Ian. *The Social Construction of What?* Harvard University Press, 1999.

Haught, John F. *Science and Religion: From Conflict to Conversation.* Paulist Press, 1995.

Hawking, Stephen. *The Universe in a Nutshell.* Bantam, 2001.

Hayek, Friedrich A. *The Road to Serfdom.* University of Chicago Press, 1944.

Hovencamp, Herbert. *Science and Religion in America, 1800–1860.* University of Philadelphia Press, 1978.

Irving-Stonebraker, Sarah. "From Eden to Savagery and Civilization: British Colonialism and Humanity in the Development of Natural History, c. 1600–1840." *History of the Human Sciences* 32, no. 4 (2019): 63–79.

Jonas, Hans. "Toward a Philosophy of Technology." *The Hastings Center Report* 9, no. 1 (1979): 34–43.

Kimmerer, Robin Wall. *Braiding Sweetgrass: Indigenous Wisdom, Scientific Knowledge, and the Teachings of Plants.* Milkweed Editions, 2013.

Kluger, Jeffrey. "The Cathedral of Science: The Elusive Higgs Boson Is at Last Found and the Universe Gets a Little Less Mysterious." *TIME*, 23 July 2012.

Kover, Tihamer R. "The Domestic Order and Its Feral Threat: The Intellectual Heritage of the Neolithic Landscape." In *Nature, Space and the Sacred: Transdisciplinary Perspectives*, edited by Sigurd Bergmann, P.M. Scott, M. Jansdotter Samuelsson, and H. Bedford-Strohm, 235–45. Ashgate, 2009.

Kowalsky, Nathan. "Science and Transcendence: Westphal, Derrida, and Responsibility." *Zygon* 47, no. 1 (2012): 118–39.

Krauss, Lawrence. *The Greatest Story Ever Told ... So Far: Why Are We Here?* Atria, 2017.

Kuhn, Thomas S. *The Structure of Scientific Revolutions.* 2nd ed. University of Chicago Press, 1970.

Kunstler, James Howard. *The Geography of Nowhere: The Rise and Decline of America's Man-Made Landscape.* Simon & Schuster, 1993.

Lakatos, Imre. "Falsification and the Methodology of Scientific Research Programmes." In *Criticism and the Growth of Knowledge*, edited by Imre Lakatos and Alan Musgrave, 91–196. Cambridge University Press, 1970.

Lasch, Christopher. *The True and Only Heaven: Progress and Its Critics*. W.W. Norton, 1991.

Latour, Bruno. *We Have Never Been Modern*. Translated by Catherine Porter. Harvard University Press, 1993.

—. "A Collective of Humans and Nonhumans: Following Daedalus' Labyrinth." In *Pandora's Hope: Essays on the Reality of Science Studies*, 174–215. Harvard University Press, 1999.

Lee, Richard B. *We Have Never Been Modern*. Translated by Catherine Porter. Harvard University Press, 1993.Latour, Bruno. "A Collective of Humans and Nonhumans: Following Daedalus' Labyrinth." In *Pandora's Hope: Essays on the Reality of Science Studies*, 174–215. Harvard University Press, 1999.

—., and Richard Daly, eds. *The Cambridge Encyclopedia of Hunters and Gatherers*. Cambridge University Press, 1999.

Leeson, Peter T., and Jacob W. Russ. "Witch Trials." *The Economic Journal* 128, no. 6 (2018): 2066–105.

Leiserowitz, Anthony, Edward Maibach, Seth Rosenthal, et al. "Global Warming's Six Americas, 2022." Yale Program on Climate Change Communication, 14 March 2023. https://climatecommunication.yale.edu/publications/global-warmings-six-americas-december-2022/.

Leiserowitz, Anthony, Jagadish Thaker, Matthew Goldberg, et al. "Global Warming's Four Indias, 2022: An Audience Segmentation Analysis." Yale Program on Climate Change Communication, 4 May 2023. https://climatecommunication.yale.edu/publications/global-warmings-four-indias-2022-an-audience-segmentation-analysis/.

Lewontin, R.C. *Biology as Ideology: The Doctrine of DNA*. Anansi, 1991.

Lindberg, David C. *The Beginnings of Western Science: The European Scientific Tradition in Philosophical, Religious, and Institutional Context, 600 B.C. to A.D. 1450*. University of Chicago Press, 1992.

Linton, David. "Luddism Reconsidered." *Et Cetera* 42, no. 1 (1985): 32–36.

Lund, David H. *Making Sense of It All: An Introduction to Philosophical Inquiry*. 2nd ed. Prentice Hall, 2003.

Lyotard, Jean-François. *The Postmodern Condition: A Report on Knowledge*. Translated by Geoff Bennington and Brian Massumi. University of Minnesota Press, 1984.

Matthews, Steven. *Theology and Science in the Thought of Francis Bacon*. Ashgate, 2008.

McGinn, Robert E. *Science, Technology, and Society*. Prentice Hall, 1991.

McMillan, Alan D., and Eldon Yellowhorn. *First Peoples in Canada*. 3rd ed. Douglas & McIntyre, 2004.

Merchant, Brian. *Blood in the Machine: The Origins of the Rebellion against Big Tech*. Little, Brown, and Company, 2023.

Merchant, Carolyn. *The Death of Nature: Women, Ecology, and the Scientific Revolution*. 40th anniversary ed. HarperOne, 2020.

—. "Mining the Earth's Womb." In *Machina Ex Dea: Feminist Perspectives on Technology*, edited by Joan Rothschild, 99–117. Pergamon, 1983.

Midgley, Mary. *Science as Salvation: A Modern Myth and Its Meaning*. Routledge, 1992.

Mosby, Ian. "Administering Colonial Science: Nutrition Research and Human Biomedical Experimentation in Aboriginal Communities and Residential Schools, 1942–1952." *Histoire sociale/Social History* 46, no. 91 (2013): 145–72.

Mumford, Lewis. *Technics and Civilization*. Harcourt Brace, 1934.

—. "The First Megamachine." *Diogenes* 14, no. 55 (1966): 1–15.

Murphy, Brian. "If Time Existed … Theoretical Physics at the UofA Would Be 50." *Folio*, 24 September 2010.

Nadler, Steven. "Doctrines of Explanation in Late Scholasticism and in the Mechanical Philosophy." In *The Cambridge History of Seventeenth-Century Philosophy*, edited by Daniel Garber and Michael Ayers, 513–52. Cambridge University Press, 1998.

Nagel, Thomas. *The View from Nowhere*. Oxford University Press, 1986.

Noll, Mark A. *The Scandal of the Evangelical Mind*. Eerdmans, 1994.

Novella, Steven. *The Skeptics' Guide to the Universe: How to Know What's Really Real in a World Increasingly Full of Fake*. Grand Central, 2018.

Oreskes, Naomi, and Erik M. Conway. "The Collapse of Western Civilization: A View from the Future." *Daedalus* 142, no. 1 (2013): 40–58.

—. *Merchants of Doubt: How a Handful of Scientists Obscured the Truth on Issues from Tobacco Smoke to Global Warming*. Bloomsbury, 2010.

Ortega y Gasset, José. *Meditations on Hunting*. Translated by Howard B. Wescott. Wilderness Adventures Press, 1995.

Pickering, Andrew. *Constructing Quarks: A Sociological History of Particle Physics*. University of Chicago Press, 1984.

Postman, Neil. *Amusing Ourselves to Death: Public Discourse in the Age of Show Business*. Penguin, 1985.

Punzo, Vincent C. "Jacques Ellul on the Technical System and the Challenge of Christian Hope." *Proceedings of the American Catholic Philosophical Association* 70 (1996): 17–31.

Qianlong Emperor. "Qianlong's Rejection of Macartney's Demands: Two Edicts." In *The Search for Modern China: A Documentary Collection*, edited by Pei-kai Cheng and Michael Lestz with Jonathan D. Spence, 103–09. W.W. Norton, 1999.

Quan-Haase, Anabel. *Technology and Society: Social Networks, Power, and Inequality*. 2nd ed. Oxford University Press, 2016.

Raedts, Peter. "Wat is middeleeuws?" In *Cultuurgeschiedenis van de middeleeuwen: Beeldvorming en perspectieven*, edited by Rob Meens and Carine van Rhijn, 13–29. Wbooks/Open Universiteit, 2015.

Rajagopalan, Ramya M., Alondra Nelson, and Joan H. Fujimura. "Race and Science in the Twenty-First Century." In *The Handbook of Science and Technology Studies*,

4th ed., edited by Ulrike Felt, Rayvon Fouché, Clark A. Miller, and Laurel Smith-Doerr, 359–88. MIT Press, 2016.

Restivo, Sal, and Jennifer Croissant. "Social Constructionism in Science and Technology Studies." In *Handbook of Constructionist Research*, edited by James A. Holstein and Jaber F. Gubrium, 213–29. Guilford, 2008.

Sagan, Carl. *Demon-Haunted World: Science as a Candle in the Dark*. Random House, 1995.

Saleeby, Caleb Williams. *Parenthood and Race Culture: An Outline of Eugenics*. Moffat Yard, 1916.

Scharff, Robert C., and Val Dusek, eds. *Philosophy of Technology: The Technological Condition—An Anthology*. Blackwell, 2003.

Schnaiberg, Allan. *The Environment: From Surplus to Scarcity*. Oxford University Press, 1980.

Scott, James C. *Against the Grain: A Deep History of the Earliest States*. Yale University Press, 2017.

Sebastiani, Silvia. "A 'Monster with Human Visage': The Orangutan, Savagery, and the Borders of Humanity in the Global Enlightenment." *History of the Human Sciences* 32, no. 4 (2019): 80–99.

Segal, Robert A. *Myth: A Very Short Introduction*. Oxford University Press, 2004.

Serpell, James. *In the Company of Animals: A Study of Human-Animal Relationships*. Basil Blackwell, 1986.

Sexton, Robert L., Peter N. Fortura, and Colin C. Kovacs. *Exploring Microeconomics*. 5th Canadian ed. Nelson, 2016.

Shellenberger, Michael, and Ted Nordhaus. "Evolve." In *Love Your Monsters: Postenvironmentalism and the Anthropocene*, edited by Michael Shellenberger and Ted Nordhaus, 8–16. Breakthrough Institute, 2011.

Shepard, Paul. *Nature and Madness*. University of Georgia Press, 1982.

Simantoni-Bournia, Eva, and Lina Mendoni, eds. *Archaeological Atlas of the Aegean: From Prehistoric Times to Late Antiquity*. Ministry of the Aegean–University of Athens, 1999.

Sioui, Georges E. *An Amerindian Autohistory*. Translated by Sheila Fischman. McGill-Queen's University Press, 1992.

Sismondo, Sergio. *An Introduction to Science and Technology Studies*. 2nd ed. Wiley-Blackwell, 2010.

Skakoon, Elizabeth. "Nature and Human Identity." *Environmental Ethics* 30, no. 1 (2008): 37–79.

Smith, Mick. "The State of Nature: The Political Philosophy of Primitivism and the Culture of Contamination." *Environmental Values* 11 (2002): 407–25.

Steeves, Paulette F.C. *The Indigenous Paleolithic of the Western Hemisphere*. University of Nebraska Press, 2021.

Stenmark, Mikael. *Scientism: Science, Ethics and Religion*. Ashgate, 2001.

Subramaniam, Banu, Laura Foster, Sandra Harding, Deboleena Roy, and Kim Tallbear. "Feminism, Postcolonialism, Technoscience." In *The Handbook of*

Science and Technology Studies, 4th ed., edited by Ulrike Felt, Rayvon Fouché, Clark A. Miller, and Laurel Smith-Doerr, 417–44. MIT Press, 2016.

Szostak, Rick. *Restoring Human Progress*. Cranmore, 2012.

Taibbi, Matt. "The Facebook Menace." *Rolling Stone*, 19 April–3 May 2018.

Van Biema, David. "God vs. Science." *TIME* (Canadian edition), 13 November 2006.

Wajcman, Judy. *Feminism Confronts Technology*. Pennsylvania State University Press, 1991.

Wheelwright, Philip, ed. *The Presocratics*. Prentice Hall, 1997.

White, Lynn, Jr. "The Historical Roots of Our Ecologic Crisis." *Science* 155, no. 3767 (1967): 1203–07.

Wilson, Rob A., Matthew J. Barker, and Ingo Brigandt. "When Traditional Essentialism Fails: Biological Natural Kinds." *Philosophical Topics* 35 (2007): 189–215.

Wilson, Shawn. *Research Is Ceremony: Indigenous Research Methods*. Fernwood, 2008.

Winner, Langdon. "Do Artifacts Have Politics?" *Daedalus* 109, no. 1 (1980): 121–36.

Wittgenstein, Ludwig. *Remarks on Frazer's* Golden Bough. Edited by Rush Rhees, translated by A.C. Miles. Brynmill, 1979.

Wolpert, Lewis. *Six Impossible Things Before Breakfast: The Evolutionary Origins of Belief*. W.W. Norton & Company, 2007.

Wright, Ronald. *A Short History of Progress*. Anansi, 2004.

Wyatt, Sally. "Technological Determinism Is Dead; Long Live Technological Determinism." In *The Handbook of Science and Technology Studies*, 3rd ed., edited by Edward J. Hackett, Olga Amsterdamska, Michael Lynch, and Judy Wajcman, 165–80. MIT, 2008.

Yagnik, Darshna, Vlad Serafin, and Ajit J. Shah. "Antimicrobial Activity of Apple Cider Vinegar against Escherichia coli, Staphylococcus aureus and Candida albicans; Downregulating Cytokine and Microbial Protein Expression." *Scientific Reports* 8, no. 1732 (2018). https://doi.org/10.1038/s41598-017-18618-x.

Credits

p. 30. Daston, Lorraine. From "Can Liberal Education Save the Sciences?" *The Point* magazine, 25 May 2016. Reprinted by permission of The Point.

p. 77 image. My cousin helping a cow give birth on his ranch. Photo by Kirsty Kurpjuweit. Reproduced by permission of the photographer.

p. 88 image. Stele of Minnakht, chief of the scribes, during the reign of Ay (c. 1321 BCE). Photo by Clio20. Wikimedia Commons, https://commons.wikimedia.org/wiki/File:Minnakht_01.JPG. CC BY-SA 3.0 (https://creativecommons.org/licenses/by-sa/3.0/deed.en).

p. 89 image. Cuneiform tablet: record of a lawsuit (c. 20th–19th century BCE). Photo by the Metropolitan Museum of Art. Wikimedia Commons, https://commons.wikimedia.org/wiki/File:Cuneiform_tablet-_record_of_a_lawsuit_MET_DP162268.jpg. CC0 1.0 (https://creativecommons.org/publicdomain/zero/1.0/deed.en).

p. 94 image. Ancient Greece Hoplite warrior on ceramic plate. Photo by Gary Lee Todd. Wikimedia Commons, https://commons.wikimedia.org/wiki/File:Ancient_Greece_Hoplite_Warrior_on_Ceramic_Plate_(28119732964).jpg. CC0 1.0 (https://creativecommons.org/publicdomain/zero/1.0/deed.en).

p. 103 image. Head of Aristotle, Vienna, Museum of Art History, Collection of Classical Antiquities (c. 320 BCE). Photo by Sergey Sosnovskiy, Wikimedia Commons, https://commons.wikimedia.org/wiki/File:Head_of_Aristotle.jpg. CC BY-SA 2.0 (https://creativecommons.org/licenses/by-sa/2.0/deed.en).

p. 105 image. Oxybeles: ancient missile weapon launching bolts and being a larger type of crossbow, used by the Greeks starting in 375 BCE. Illustration by Arz, Wikimedia Commons, https://commons.wikimedia.org/wiki/File:Oxebeles.jpg. CC BY-SA 3.0 (https://creativecommons.org/licenses/by-sa/3.0/deed.en).

p. 112 image. Lorica Hamata. Photo by Greatbeagle, Wikimedia Commons, https://commons.wikimedia.org/wiki/File:Lorica_Hamata.jpg. CC BY-SA 3.0 (https://creativecommons.org/licenses/by/3.0/deed.en).

p. 117 image. St. Benedict of Nursia. Photo by Gerd A.T. Müller, Wikimedia Commons, https://commons.wikimedia.org/wiki/File:Benedikt_von_Nursia_20020817.jpg. CC BY-SA 3.0 (https://creativecommons.org/licenses/by-sa/3.0/deed.en).

p. 119 image. Architectural details in Alhambra, Granada. Photo by Michal Osmenda, Wikimedia Commons, https://commons.wikimedia.org/wiki/File:Architectural_details_in_Alhambra,_Granada_(6930669668).jpg. CC BY 2.0 (https://creativecommons.org/licenses/by/2.0/deed.en).

p. 120 image. Lithograph of Avicenna (18th century). Lithograph by Bauer, printed by J. Ratch, Wikimedia Commons, https://commons.wikimedia.org/wiki/File:Avicenna_lithograph_-_cropped.png.

p. 121 image. *Saint Thomas Aquinas* by Carlo Crivelli (1476). Photo by the National Gallery. Wikimedia Commons, https://commons.wikimedia.org/wiki/File:St-thomas-aquinas.jpg.

p. 129 image. *The Extasy of St. Francis* (detail) by Bartolomé Esteban Murillo (17th century). Photo by Jl FilpoC, Wikimedia Commons, https://commons.wikimedia.org/wiki/File:San_Francisco_de_As%C3%ADs,_fragmento_de_la_pintura_El_%C3%89xtasis_de_San_Francisco_de_As%C3%ADs_(Murillo).jpg. CC BY-SA 4.0 (https://creativecommons.org/licenses/by-sa/4.0/deed.en).

p. 143 image. Francis Bacon, Viscount St Albans. Stipple engraving by J. Posselwhite after J. Houbraken, 1738, Wellcome Collection, https://wellcomecollection.org/works/cz5v5pdd.

p. 145 image. Portrait of René Descartes. Engraving by William Holl, 1807–87, Smithsonian Museum and Archives, https://library.si.edu/image-gallery/73402.

p. 149 image. *Adam and Eve* by Rembrandt van Rijn (1638), Cleveland Museum of Art, https://www.clevelandart.org/art/1973.37.

p. 151 image. An ornithopter by Leonardo da Vinci (15th century), Wikimedia Commons, https://commons.wikimedia.org/wiki/File:Ornithopter_Leonardo_da_Vinci.png.

p. 170 image. Inuk in a kayak (c. 1929). Photo by Edward S. Curtis, Wikimedia Commons, https://commons.wikimedia.org/wiki/File:Inuit_man_by_Curtis_-_Noatak_AK.jpg.

p. 180 image. *Portrait of Isaac Newton* by Godfrey Kneller (1689), Wikimedia Commons, https://commons.wikimedia.org/wiki/File:Portrait_of_Sir_Isaac_Newton,_1689_(brightened).jpg.

p. 197 image. Zheng He's ship compared to Columbus's. Photo by Lars Plougmann, Flickr, https://www.flickr.com/photos/criminalintent/361639903. CC BY-SA 2.0 (https://creativecommons.org/licenses/by-sa/2.0/deed.en).

p. 200 image. *Portrait of Jean-Jacques Rousseau* by Maurice Quentin de La Tour (18th century). Photo by Paris Musées, Wikimedia Commons, https://commons.wikimedia.org/wiki/File:Maurice_Quentin_de_La_Tour_-_Portrait_de_Jean-Jacques_Rousseau_(1712-1778),_%C3%A9crivain_et_philosophe_-_P210_-_Mus%C3%A9e_Carnavalet.jpg.

p. 210 image. Daguerreotype of Comte taken in 1849. Anonymous photographer, Wikimedia Commons, https://commons.wikimedia.org/wiki/File:Auguste_Comte_daguerreotype.jpeg.

p. 220 image. *David Hume* by Allan Ramsay (1766). Wikimedia Commons, https://commons.wikimedia.org/wiki/File:David_Hume_Ramsay.jpg.

p. 224 image. Karl Popper in 1990. Photo by Lucinda Douglas-Menzies, Wikimedia Commons, https://commons.wikimedia.org/wiki/File:Karl_Popper2.jpg.

p. 226 image. Portrait of the philosopher of science Thomas Samuel Kuhn. Illustration by Davi.trip, Wikimedia Commons, https://commons.wikimedia.org/wiki/File:Thomas-kuhn-portrait.png. CC BY-SA 4.0 (https://creativecommons.org/licenses/by-sa/4.0/deed.en).

p. 280 image. French sociologist and technology critic Jacques Ellul in his house in Pessac, France. Photo taken by Jan van Boeckel as part of the filming of the documentary *The Betrayal by Technology* by Rerun Productions, Wikimedia Commons, https://commons.wikimedia.org/wiki/File:Jacques_Ellul_(cropped).jpg. CC BY-SA 4.0 (https://creativecommons.org/licenses/by-sa/4.0/deed.en).

p. 286 image. Raymond Kurzweil, an American academic and author. Photo by null0, Wikimedia Commons, https://commons.wikimedia.org/wiki/File:Raymond_Kurzweil,_Stanford_2006_(square_crop).jpg. CC BY-SA 2.0 (https://creativecommons.org/licenses/by-sa/2.0/deed.en).

Index

About the Publisher

THE WORD "BROADVIEW" EXPRESSES A GOOD DEAL OF THE PHILOSOPHY behind our company. Our focus is very much on the humanities and social sciences—especially literature, writing, and philosophy—but within these fields we are open to a broad range of academic approaches and political viewpoints. We strive in particular to produce high-quality, pedagogically useful books for higher education classrooms—anthologies, editions, sourcebooks, surveys of particular academic fields and sub-fields, and also course texts for subjects such as composition, business communication, and critical thinking. We welcome the perspectives of authors from marginalized and underrepresented groups, and we have a strong commitment to the environment. We publish English-language works and translations from many parts of the world, and our books are available world-wide; we also publish a select list of titles with a specifically Canadian emphasis.

broadview press

This book is made of paper from well-managed FSC® - certified forests, recycled materials, and other controlled sources.

FSC www.fsc.org MIX Paper | Supporting responsible forestry FSC® C103567

PCF

BIO GAS ENERGY

PERMANENT